Technology and Power in the Early American Cotton Industry: James Montgomery, the Second Edition of His "Cotton Manufacture" (1840), and the 'Justitia' Controversy about Relative Power Costs

David J. Jeremy

American Philosophical Society
Independence Square   Philadelphia
1990

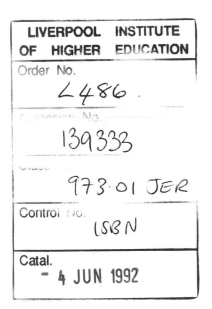
*Publication of this volume has been subsidized in part by the
Jayne Fund of the American Philosophical Society.*

Cover illustration: Nineteenth-century engraving of Graniteville Cotton Factory.
From: *The Textile Industry in Antebellum South Carolina* by Ernest M. Lander, Jr.

Used with permission of the Louisiana State University Press.

Library of Congress Catalog Card No.: 87-73044
International Standard Book No.: 0-87169-189-2
US ISSN: 0065-9738 .

*Technology and Power in the
Early American Cotton Industry:
James Montgomery, the Second Edition
of His "Cotton Manufacture" (1840),
and the 'Justitia' Controversy about
Relative Power Costs*

Frontispiece: York Factories, Saco, Maine from original version of Montgomery's *Cotton Manufacture.*

CONTENTS

PART I
JAMES MONTGOMERY AND HIS
COTTON MANUFACTURE

v

PART II
THE JUSTITIA CONTROVERSY

List of Illustrations

List of Tables

ACKNOWLEDGMENTS

Twenty years ago the Kress Library of Business and Economics at the Harvard Graduate School of Business Administration acquired the interleaved and annotated copy of the *Cotton Manufacture of the United States* (1840) by James Montgomery, a Glasgow cotton mill manager who had emigrated to New England. The book's original appearance was a major event. For the first time a precise and reliable comparison, at both technical and commercial levels, was made between cotton production in Britain, the first industrial nation, and cotton production in the USA, its youthful, and adventuresome, rival. Montgomery's data and assessments were subjected to close scrutiny, especially in New England where his estimates of power costs favoured water power rather than the new technology of steam power. Controversy erupted in the Boston newspapers and in response Montgomery prepared an interleaved version of his book in readiness for issuing a second edition. For reasons long since forgotten that edition was never published, although Montgomery made his amendments over a period of two or three years. Until the Kress Library obtained the volume its existence was unknown to scholars.

Kenneth E. Carpenter, then Curator of the Kress Library, who purchased Montgomery's interleaved volume, generously invited me to edit it. I agreed, though I had to postpone work until I had completed another study, *Transatlantic Industrial Revolution*. The work which follows was completed while I was engaged on a very different project, as editor of the *Dictionary of Business Biography* in the Business History Unit at the London School of Economics.

I am most grateful to Ken Carpenter for the opportunity to edit Montgomery's work, for providing a photocopy of the manuscript, for discussing editorial methods and, most of all, for persistent encouragement. He also arranged for one of his assistants, Geraldine Pratt, to search the Boston newspapers for the letters in the "Justitia" debate over power and production costs: my thanks to her for a thorough piece of research. Mary Chatfield, Librarian of the Baker Library, kindly organised a grant-in-aid to support my expenses. In the Business History Unit the Director, Leslie Hannah, welcomed the completion of this study; two early members of the *DBB* team, Greta Edwards and Jill Gosling, came to my rescue in a hectic week of typing; and my colleague Jonathan Liebenau helped me to understand the essential changes in the German translation of Montgomery's *Cotton Manufacture*. For help in selecting illustrations I am obliged to Clare M. Sheridan, Librarian of the Museum of American Textile History (formerly the Merrimack Valley Textile Museum) and to John Hume of the Scottish Development Department, Historic Buildings and Museums section. For undertaking the publication of this, as well as of my first scholarly edition of the work of a British textile manufacturer in the USA *(Henry Wansey and His American Journal, 1794)*, I am indebted to the American Philosophical Society, where Carole N. LeFaivre has been a responsive editor. As in the past, Theresa my wife has buttressed my endeavours, not least by typing and checking most of the manuscript, while my daughters Rebecca and Joanna helped me complete the tedious task of checking. My thanks to all—who are of course absolved from any responsibility for what remains.

ABBREVIATIONS

CM1   James Montgomery, *A Practical Detail of the Cotton Manufacture of the United States of America; and the State of the Cotton Manufacture of that Country Contrasted and Compared with that of Great Britain; with Comparative Estimates of the Cost of Manufacturing in Both Countries* (Glasgow: John Niven, Jr., 1840). Pp. xi + 219.

CM2   Interleaved edition of *CM1* with annotations by the author; copy in the Kress Library of Business and Economics, Harvard Business School.

CSMA1   (James Montgomery), *The Carding and Spinning Master's Assistant: or the Theory and Practice of Cotton Spinning* (Glasgow: John Niven, Jr., 1832). Pp. viii + 282. Preface signed "J.M."

CSMA2   Second edition of *CSMA1*; same title, same publisher; 1833. Pp. x + (13–)332. Preface signed "James Montgomery."

CSMA3   Third "enlarged and improved" edition of *CSMA1*, published as *The Theory and Practice of Cotton Spinning; or the Carding and Spinning Master's Assistant* (Glasgow: John Niven, Jr., 1836). pp. xiv + (15–)348. Montgomery's name appears on the title page of this edition.

DAB   *Dictionary of American Biography*

DNB   *Dictionary of National Biography*

GB   Great Britain

JM   James Montgomery

NA   National Archives, Washington, D.C.

PP   *Parliamentary Papers*

RG   Record Group

RIHS   Rhode Island Historical Society

TIR   David J. Jeremy, *Transatlantic Industrial Revolution: the*
            Diffusion of Textile Technologies between Britain and
            America, 1790–1830s (Cambridge, Massachusetts:
            MIT Press; Oxford: Basil Blackwell, 1981).

PART I
JAMES MONTGOMERY AND HIS
COTTON MANUFACTURE

I. INTRODUCTION

I. James Montgomery (1794–1880): The Circumstances of His Career

(a) Birth and Scottish background

James Montgomery was born in 1794 at Blantyre Mills, then a newly colonised mill village about eight miles up the Clyde valley from Glasgow, and southeast of the city.[1] The mills were grafted on to an older community of about five hundred people scattered through a flat, wooded but cultivated section of the valley where, fifty years earlier, the local mineral springs made it a summer resort for Glasgow families.[2] The first mill at Blantyre, one of the early Arkwright-type mills in Scotland, was built on the banks of the Clyde by David Dale, the Glasgow merchant and banker—and eventually the leading cotton capitalist in Scotland—and his partner James Monteith. Containing 4,096 spindles, it began working in 1787. Its workforce of 368 men, women, and children was housed in a cluster of dwellings nearby. Blantyre's population

---

[1] For his birthplace, see obituaries in *Boston Daily Advertiser* and *Boston Evening Transcript*, 3 January 1881.

[2] Sir John Sinclair, *The Statistical Account of Scotland. Drawn up from the Communications of the Ministers of the Different Parishes* (21 vols., Edinburgh: William Creech, 1791–1799) II, 213–222.

1

Fig. 1. Blantyre Mills: The mills in which Montgomery worked. Courtesy of Scottish Development Department, Historic Buildings and Museums section.

increased with additions to the mills: after 1791 when a second mill, planned for 15,000 mule spindles, was built; and a decade later after a dye works, famous for its Turkey-red yarn dyeing, was set up; and yet again after 1813 when a weaving shed, partly steam-powered, was added.[3] By 1816 the Blantyre Cotton Works, now owned by H. Monteith, Bogle & Co., was the second largest employer among forty-one cotton manufacturing firms in Scotland, with 255 males and 505 females, most crowded into tenement accommodation.[4]

Montgomery's arrival in this thriving mill colony was recorded in the parish register in a single unpunctuated sentence: "James Son lawfull to William Montgomery Taylor Cotton-mill and Helen Reid his spouse was Born the Second and Baptized the Seventh day of December 1794 years by Mr. Stevenson."[5] Either at the parish school or under the school master employed by the cotton mills, he would have learned to read and write.[6] At the same time, he grew up in the first generation of children to be inducted into the machine-dominated discipline of the cotton factory, and immersed in its distinctive sights, sounds, and smells, and its frequently harsh conditions. In this environment the young Montgomery gained an intuitive understanding of cotton fiber behaviour and the characteristics of the various processing machines. His significant achievement lay in distilling (not without the aid of the community of industrial artisans of course) this instinctive knowledge into objective, frequently mathematical, rules and in publishing them. The combined spinning, weaving, and finishing plant at Blantyre, assuming that Montgomery spent his early career there, gave him the chance of a comprehensive practical education in cotton manufacturing which, forty years later, brought him to eminence among the mill managers in the Glasgow cotton industry.

Over those forty years cotton manufacturing in Glasgow grew and approached its peak, before declining during the railway age

---

[3] Ibid. and The Ministers of the Respective Parishes, *The New Statistical Account of Scotland* (15 vols., Edinburgh and London: W. Blackwood & Sons, 1845) VI, 322–324.

[4] GB, *PP (Commons)* 1816 (397) III, "Report from the Committee on Children Employed in Manufactories," 240–241.

[5] Blantyre parish register, this date, General Register Office of Scotland, Edinburgh.

[6] Sinclair, *Statistical Account* II, 216, 219. By 1816 less than two percent of the 10–18 year old males at Blantyre could not read: GB, *PP (Commons)* 1816 (397) III, 240–241. Blantyre's most famous son, David Livingstone, the African missionary–explorer, started his working life in the mills there, nearly two decades after Montgomery was born. See Tim Jeal, *Livingstone* (London: Book Club Associates, 1973).

in the face of foreign and domestic competition and displacement as a regional specialisation by the coal, iron, shipbuilding, and engineering industries.[7] Even at its height in the early 1830s, the Glasgow cotton district, compared to Lancashire, was a secondary one. It was much smaller. The Factory Commissioners' Returns of 1835 showed 159 cotton mills in Scotland, compared to 779 in Lancashire and Cheshire, figures which reliably indicated relative, though not absolute, size.[8] Smaller in size, the Glasgow district had greater difficulty in developing those regional external economies on which Lancashire's expansion partly rested. This was especially true of machine building. In 1824 and again in 1833, Scottish manufacturers complained that they had to go to Lancashire for machinery of the latest design and the best quality.[9] The complaints seem justified in part because less than a fifth of Britain's patented cotton manufacturing inventions of the period 1790–1830 came from Scotland, compared to two-thirds from Lancashire and Cheshire.[10]

Besides being much smaller and much less innovative than the Lancashire cotton region, the Glasgow cotton district moved towards two distinctive specialisations in the 1820s and 1830s. A minority of factory spinning firms, like the Houldsworths, spun fine yarns for the numerous branches of hand weaving still untouched by the powerloom, chiefly the fancy and figured work of Glasgow and Paisley muslins, gauzes, and shawls.[11] The majority of factory spinners made low count yarns for coarse goods manufacture, a trend which increased with the spread of the powerloom.[12] By 1833 there were, it was reported, 15,000 powerlooms in Scotland (compared to 85,000 in England), a good proportion

---

[7] For older introductions to the history of the Scottish cotton industry, see W. H. Marwick, "The Cotton Industry and the Industrial Revolution in Scotland," *Scottish Historical Review* XXI (1923–1924) 207–218; G. M. Mitchell, "The English and Scottish Cotton Industries. A Study in Interrelations," ibid. XXII (1924–1925), 101–114. Good for the eighteenth century scene is John Butt, "The Scottish Cotton Industry during the Industrial Revolution, 1780–1840," in L. M. Cullen and T. C. Smout, *Comparative Aspects of Scottish and Irish Economic and Social History, 1600–1900* (Edinburgh: John Donald Publishers, 1977) 116–128.

[8] David T. Jenkins, "The Validity of the Factory Returns, 1833–1850," *Textile History* IV (1973) 37.

[9] GB, *PP (Commons)* 1824 (51) V, "Six Reports from the Select Committee on Artizans and Machinery," 378. GB, *PP (Commons)* 1833 (690) VI, "Report from the Select Committee on the Present State of Manufactures, Commerce, and Shipping in the United Kingdom," 310, 314, 325.

[10] *TIR*, chapter 3.

[11] GB, *PP (Commons)* 1833 (690) VI, 73–74, 334, 668–672, 698.

[12] Ibid., 311, 321.

of which were probably in combined spinning and weaving firms: of 64 Scottish firms engaged in manufacturing cotton yarn, thread, or goods in 1834, 13 were integrated spinning and weaving firms and seven powerloom firms only.[13]

Although chiefly engaged in coarse spinning, the Glasgow cotton manufacturers made their weft or filling yarn on mules, which required skilled adult male operatives; they made their warp yarns, which were harder twisted, on throstle frames, tended by unskilled teenagers or women. Houldsworth in 1833 reckoned that there were about 800 spinners in Glasgow and "Generally speaking, there are about five hands that are not spinners to one spinner," which suggests that he was referring to mule spinners.[14] Each of these, if Lancashire experience applied to Scotland, ran 700–840 spindles, making an aggregate mule spindleage for Glasgow in the region of around 640,000 spindles, an estimate perhaps over-generous because 200–300 Glasgow spinners were then reportedly unemployed.[15] Coarse spinners and goods manufacturers, as one of their number disclosed in 1833, resisted the northern New England practice of spinning entirely on throstles. To counter rising labour costs, Glasgow manufacturers employed light, part-powered mules, operated by women; or else, and this was a new move in 1833, they installed mules equipped with Richard Roberts's headstock, which were self-actors and required less actual operative skill (excluding maintenance tasks). The manufacturer who described these Glasgow practices made "a common printer's cloth," which probably explains his attachment to mule spinning for it produced a softer, more uniform yarn, essential for high definition of print designs, than did throstle spinning.[16]

When in the early 1830s the British cotton industry faced difficulties associated with declining cotton yarn and goods prices and falling manufacturers' costs and profit margins, the Glasgow manufacturers complained of two special problems.[17] First, they

---

[13] Edward Baines, Jr., *History of the Cotton Manufacture in Great Britain* (1835, reprinted London: Frank Cass, 1966) 235–237. GB, *PP (Commons)* 1834 (167) XX, "Reports from the Commissioners: Factories Inquiry, Part II," Section A.1, this editor's summary of returns.

[14] GB, *PP (Commons)* 1833 (690) VI, 312.

[15] Ibid., 324, 675.

[16] Ibid., 321, 323, 324, 332 (William Graham's evidence).

[17] For the economic pressures affecting the British cotton industry in the 1830s, see Robert C. O. Matthews, *A Study in Trade-Cycle History: Economic Fluctuations in Great Britain, 1833–1842* (Cambridge: University Press, 1954) 127–151.

said, their labour costs were artificially higher than in Lancashire. For example, the wage for a certain number of hanks of No. 16s yarn was 3s 6d at Stalybridge, Cheshire, near Manchester, but 4s 11d in Glasgow.[18] This was attributed to the influence of a strong union organisation in Glasgow which enforced the closed shop and paid spinners unemployment money or emigrant passage money to inflate mill wages.[19] Second, competition from the new American cotton industry particularly hurt Scottish manufacturers because, like the Americans, they were newcomers to industrial cotton manufacturing and were starting with coarse goods manufacture, with its large volume demand but low unit profit margins. Given his high wage bill and the duties on imported cotton, one Glasgow manufacturer predicted that he could not continue "in low goods": a cloth sample of American cotton goods sent to him from the Mexican market had been costed in his factory and it transpired that he could not make an equivalent fabric for less than $1/8$d under the price per yard at which the American goods were sold in Mexico, an unacceptably narrow margin.[20] Price for price, American goods offered a more durable quality and for this reason were gaining in markets, seemingly "everywhere"—in South America primarily but also Turkey, India, and the East and West Indies.[21] When Montgomery emigrated in 1836, this difficult period of the early 1830s in Scotland had been followed by two years of industrial expansion deriving from the recovery of demand, rising profit margins and the influx of capital for new mills. In the light of these broader features of his Scottish perspective, Montgomery's emigration to New England in the mid-1830s represented a move from an infant industry to one of its thriving rivals.[22]

(b) Montgomery's Scottish mill management: theory and practice

Montgomery, in both his publications and his private contacts with visitors,[23] opposed the secretiveness which permeated Brit-

[18] GB, *PP (Commons)* 1833 (690) VI, 322–324.

[19] Ibid., 311–313, 322–324.

[20] Ibid., 324, 330.

[21] Ibid., 55, 120, 321, 325, 331.

[22] Matthews, *Study*, 134–136. In the early 1830s capital requirements made it hard for an individual to set up in the cotton industry as owner rather than manager, and this may have influenced Montgomery. See GB, *PP (Commons)* 1833 (690) VI, 316.

[23] As did Daniel Treadwell, for whom see *DAB* and below.

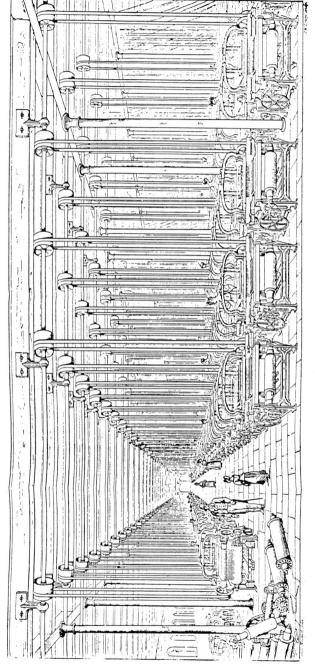

Andrew Ure, *The Philosophy of Manufactures: or, an Exposition of the Scinetific, Moral, and Commercial Economy of the Factory System of Great Britain* (London: Charles Knight, 1835), p. 1.

POWER LOOM FACTORY OF THOMAS ROBINSON ESQ.[R]
STOCKPORT.

Fig. 2.   View of interior of British cotton powerloom factory, Thomas Robinson's factory at Stockport, showing the overhead shafting (generally gear driven from the steam engine or waterwheel) from which power was taken to individual looms by belting. From Andrew Ure, *The Philosophy of Manufactures* (London: Charles Knight, 1835), p. 1.

ain's leading manufacturing districts. Glasgow manufacturers seem to have been less protective than those in Lancashire, and several general technical books were published there before 1832.[24] Since the Glasgow district lagged behind Manchester, it presumably had less to conceal. On the other hand when John Smith, an ex-Glasgow cotton machinery maker, returned on a visit to the city in 1826 after ten years' absence in Andover, Massachusetts, he found "there was no way I could make out to get into the works, unless I found some person or friend to introduce me."[25]

The publication of Montgomery's first book, *The Carding and Spinning Master's Assistant: or the Theory and Practice of Cotton Spinning*, in 1832 helped to shatter the effectiveness of much of this kind of secretiveness and in retrospect was a momentous event. Abraham Rees's monumental *Cyclopaedia: or Universal Dictionary of Arts, Sciences and Literature* (45 vols., London: Longman et al., 1802–1820) had drawings and descriptions of the new industrial textile technology, and lesser works in this genre followed.[26] But in the technical handbook tradition there were only warp tables, which were not in any case exclusive to industrial technology.[27]

In his Preface, Montgomery described how, almost by chance, he came to write his expository volume. It originated in his practice of taking notes about technical matters and then discussing them further with managers and mechanics—unusual but surely not that rare. Friends, seeing his notes, urged him to publish them: in the trade the need for such a technical handbook had long been felt and, presumably, it would save both the chore of compiling notes and the time and money otherwise lost in unnecessary learning by doing. Montgomery defended publication of what after all were trade secrets in terms of breaking the monopoly on knowledge (sometimes imperfect) held by experienced manufacturers, and of helping the industry's managers both in training assistants and in assessing new innovations.[28]

He saw his work as "theory" because it would "assist the master, manager, or artisan, in acquiring a correct and systematical knowl-

---

[24] GB, *PP (Commons)* 1824 (51) V, 384. Glasgow's technical publications included Robert Brunton's *A Compendium of Mechanics* (1824) and William Grier's *The Mechanic's Calculator* (1832), both quoted in the Justitia controversy.

[25] "Journal of John Smith (1796–1886), from 28 November 1826 to 9 March 1827," *Essex Institute Historical Collections* CVI (1970), 101.

[26] *TIR*, chapter 3.

[27] Ibid.

[28] Paragraph derived from *CSMA* 1, Preface.

edge of the *real* principles of the business."[29] That is, he aimed to impart a reliable, comprehensive, and deductive understanding of the practical and technical aspects of cotton manufacturing. His view was reliable because he had taken care to consult various articles on his subject and to solicit the opinions of "some experienced mechanics," as well as having his own experience going back thirty years.[30] It was comprehensive because it presented a complete treatment of manufacturing, arranged in the sequence of processing stages: hitherto this knowledge had been fragmented, part known to carders, part to spinners, part to mechanics, and so on. And it offered a deductive understanding of cotton manufacturing because it set out the basic rules by which overseers and managers operated spinning equipment and mills. After each machine description—in which he gave some of its history, its topology, and its function, much like the Rees's *Cyclopaedia* entries—Montgomery laid out rules, illustrated with worked examples, for calculating speeds, drafts and settings for component cylinders, rollers, spindles, and other moving parts, and ways of adjusting them to suit various cotton qualities or to produce various end or yard finenesses: the key information that mill overseers and managers generally kept in jealously guarded notebooks or else locked behind a profound reticence. After all, such knowledge was normally acquired only through long hard years of experience, usually stretching from childhood to middle age.

In *The Carding and Spinning Master's Assistant* Montgomery paid special attention to a number of recent innovations. Noticeably most of them came from the United States: Asa Arnold's fly frame differential, George Danforth's tube roving frame, Charles Danforth's cap spindle, and the American dead spindle. Surprisingly, the latest Manchester improvements, like Johann Georg Bodmer's preparatory innovations or Richard Roberts's self-acting mule found no place in Montgomery's book.[31] Indeed he claimed, two years after Roberts perfected his mule, that "self acting mules . . . seem so unimportant, as to be seldom spoke of, and therefore it is unnecessary to take further notice of them in this place."[32] In part this reflected Glasgow's backwardness in relation to Manchester. His American interest, however, demonstrated a keen

[29] Ibid., i.
[30] Ibid., iii.
[31] For these, see information and sources quoted in *TIR*, chapter 3.
[32] *CSMA* 1, 170

awareness of the technical progress being made by Glasgow's
major foreign competitor.

Besides the technical sections in his first book, Montgomery's
chapter on "Management and Government of Spinning Factories"
offered valuable advice to aspiring spinning factory managers.
After stressing the importance of "early and long" practice in the
business, Montgomery emphasized the equally necessary require-
ment of all-around experience in every department of the factory.[33] Only then could the manager appraise properly the performance of subordinate overseers and perform the calculations
in machine speeds and drafts: necessary not only for varying combinations of cottons or altering yarn sizes, to meet changing market prices, but also for maintaining an approximate balance in the
production line. Because of the diversity of opinion on this last
matter, Montgomery drew up a table showing an average balance
of equipment over a range of yarn sizes for mule spindles, from
which the balance in carding engines and drawing and roving
frames could then be calculated.

Montgomery finished his chapter on mill management with a
section on the treatment of workers. A child of the mills, he came
down on the side of a humane paternalism, influenced perhaps
by the example of Robert Owen at New Lanark.[34] "*Unnecessary
severity*" should be avoided, without diminishing "proper authority."[35] Furthermore, "I do not hesitate to assert, that a Spinning Factory can never be managed more profitably, and more
to the satisfaction of the proprietors, than when there exists a
good feeling and a good understanding between the manager and
workers."[36] He advised that the manager be "firm and decisive in
all his measures, but not overbearing and tyrannical; not too distant and haughty, but affable and easy of access, yet not too familiar."[37] Few orders in few words, pleasantly delivered and supported by strict justice and impartiality, was Montgomery's dictum.

He then briefly commented on three issues of contention between operatives and managers: fines, which he suggested should
be paid to a charitable institution rather than the mill proprietor;

---

[33] Ibid., 210.
[34] For the debate over factories, see John L. and Barbara Hammond, *Lord Shaftesbury*
(Harmondsworth: Pelican Books, 1939), 19–60 and J. T. Ward, *The Factory Movement,
1830–1855* (London: Macmillan & Co., 1962).
[35] *CSMA* 1, 219.
[36] Ibid.
[37] Ibid., 220.

yarn sizes, over which he advised adherence to a clearly defined scale of piece rates; and the manager's treatment of operatives, which, he reiterated, ought to be free of overbearing conduct and degrading language towards the workers. Lastly he acknowledged that there would be differences in managerial style between carding and spinning masters. The latter's operatives, mule spinners on piece rates, needed less close supervision than the cardroom girls and women who were on time rates. The whole book Montgomery rounded off with chapters on the history of cotton spinning (which "if not useful, may, at least, be found interesting"), on the different varieties of cotton, and on Britain's cotton trade.[38]

The *Carding and Spinning Master's Assistant* evidently met a widespread need, for a year later the publisher, John Niven, Jr., brought out a second edition.[39] In this Montgomery took the opportunity to re-evaluate Roberts's self-acting mule, but the book was still generally limited by its Scottish perspective. When in April 1836 Montgomery had the chance to visit Lancashire, he seized the opportunity to bring out a third edition, enlarged as well as revised. Now under its old sub-title, *The Theory and Practice of Cotton Spinning*, the handbook contained a number of comparisons, mostly at machine level, of Scottish and English manufacturing practices.[40] One significant addition to the third edition was an estimate of manufacturing costs (fixed and overhead, the latter for a fortnight's work) in a Glasgow spinning and weaving mill.[41] This Montgomery later used as the basis for a comparison between British and American manufacturing costs. His citation of American statistics,[42] as originally published by the Friends of Domestic Industry in their *Report on the Production and Manufacture of Cotton*, showed that he was well aware of the manufacturing strength of the country to which, at that date, he soon intended to emigrate.

The other title Montgomery published before he emigrated was *The Cotton Spinner's Manual: or a Compendium of the Principles of Cotton Spinning* (1835), again with his Glasgow publisher John Niven, Jr. Intended for the pocket of mill overseer or manager, it was a handy compilation of the rules, calculations, and worked

---

[38] Ibid., vi.

[39] *CSMA* 2, Preface. *CSMA* 2 is a very rare book and I am grateful to Mrs. Ruth R. Rogers, Curator of the Kress Library, for noting differences between *CSMA* 1 and *CSMA* 2 for me.

[40] *CSMA* 3, 76–86 (carding machines), 173–176 (throstle spindles), 196–208 (self-acting mules), for example.

[41] Ibid., 248–256.

[42] Ibid., 310–311.

examples found in his larger, first book. As he explained in the Preface, it was written in response to criticisms that his first work, because of its price, size, and literary approach, was more appropriate for proprietors and managers than for departmental overseers. It was not quite the first of its kind for only the year before *The Cotton Spinner's Companion: Containing the Methods of Calculating the Different Machines Used in a Cotton Spinning Factory,* written by George Galbraith, was published in Glasgow by W. & W. Miller.

Neither this work by Galbraith nor the assistance Montgomery received in preparing his books can detract from his pioneering achievement in reducing empirically discovered technical knowledge to a reliable and systematised corpus of operating rules compiled for the purpose of publication. The appearance of his work, especially his first volume, marked a major step in eroding industrial secrecy and in diffusing new technology.

In 1835, James Montgomery was working for MacLeroy, Hamilton & Co. of Calton in Glasgow, presumably (in view of his publications and subsequent employment in New England) as a mill manager.[43] Glimpses of mill administration in Montgomery's firm were relayed by one of the proprietors in a report to the Factory Commissioners in 1834.[44] By the standards of the day, the firm demonstrated enlightened treatment of its operatives. Every window had a ventilating pane controlled by the operatives themselves. All machinery was fenced off. Room temperatures varied between 40–70 degrees for spinning; 50–60 for weaving; and 80 for dressing (a department with only ten operatives). Although the factory was lit by gas, no night work occurred. But the working week, now regulated by an effective Factory Act, was typically long: a twelve hour day, six days a week. Only once in the previous twelve months had the twelve hour day been exceeded, due to steam engine failure, and then by just fifteen minutes. Everyone received three-quarters of an hour for breakfast

---

[43] For his mill address, see James Montgomery to Daniel Treadwell, 6 July 1835 and 6 January 1836, Treadwell Papers, Harvard University Archives, Cambridge, Massachusetts.

[44] GB, *PP (Commons)* 1834 (167) XX, A1, No. 103. The identification of W Hamilton's cotton spinning and powerloom weaving factory at Calton, Glasgow, with MacLeroy, Hamilton & Co. (Montgomery's firm), seems confirmed by the size of its steam engine (50 h.p.) and the single date of mill construction (1825), particulars not reported by any other Glasgow firm in 1834. In 1825 the only 50 h.p. steam engine reported as driving a combined spinning and weaving mill in Glasgow belonged to MacLeroy, Hamilton & Co., according to James Cleland. See *The Franklin Journal* I (1826), 107.

and the same time for lunch (a statutory regulation). Discipline was maintained by threat of dismissal. Corporal punishment was forbidden. The youngest employee was nine years old but the mill conformed to the Factory Act in not employing children under the age of nine; in limiting hours of work to forty-eight hours a week (or nine a day fixed between 5.30 a.m. and 8.30 p.m.) for those under twelve (later thirteen) and to sixty-nine hours a week (or twelve a day) for those under eighteen; and in permitting children under thirteen to attend school for two hours a day.[45] This of course was the official face of the factory. Exactly how manager and overseers ran their production line and treated their workers within these boundaries remains unknown. All that can be said is that the managerial style that Montgomery advocated in 1832 was consonant with the publicly reported administration of the mills in which two years later he was a manager.

(c) Montgomery's emigration to the USA: an episode in the transatlantic transfer of technology

While the details of his transatlantic passage remain hazy, because of the loss of American passenger lists,[46] James Montgomery's recruitment to New England is well documented. In essence, his migration was accomplished through a network established by an American visitor to Britain, a typical early nineteenth-century method of recruiting British skill.

The American visitor was Daniel Treadwell, a well-known inventor and entrepreneur of Boston who went to England in 1835.[47] He had several reasons for going. First, his appointment

---

[45] The first effective Factory Act was 3 & 4 Wm. IV, c. 103 (1833).

[46] Montgomery recorded that he arrived in the USA in June 1836 (*Boston Courier*, 27 March 1841). The Boston passenger lists for this and the previous five months are missing (and for checking these I am grateful to my good friends George and Grace Fielding of Alexandria, Virginia). The New York lists include a James Montgomery who arrived on 2 June 1836 on board the brig *Czar* from Greenock. He gave his age as 40 and brought two children with him, James aged 12 and Agnes aged 13. However, he described himself as a farmer. Other Montgomerys reached New York that month: Richard a weaver aged 23, Andrew a farmer aged 22, and Elizabeth aged 50, arrived from Liverpool on the ship *Caroll of Carollton* on 2 June; Anna aged 24 and Elizabeth aged 17 came from Liverpool on the ship *Nimrod* on 19 June. See US National Archives, New York Passenger Lists, film M-237, reel 30, nos. 426, 427 and 514 (my thanks to Mary Fleming for a loan of this film).

Because of the discrepancies in age and trade, we cannot be sure that this was the Glasgow cotton manufacturer and his family: there was certainly no legal reason in 1836 why he should have concealed his true occupation.

[47] For Treadwell, see *DAB*.

in 1834 as Rumford Professor and Lecturer on the Application of Science to the Useful Arts at Harvard required that he support his annual allocation of forty lectures with "direct experiments" and for these he needed apparatus and machine models, some of which were obtainable only in England.[48] Secondly, Treadwell faced a number of difficulties relating to the patenting of his hemp spinning machine. In 1831 he had simultaneously taken out American and British patents for his invention, but the English patent, arranged through William Newton, the London patent attorney, and granted in the name of Joshua Bates, an American partner in the mercantile firm of Barings, had been inaccurately drawn up. The British specification erroneously claimed that the machine could spin other fibers besides hemp and flax, and that it twisted them although its action was confined to drafting. Treadwell feared that these inflated claims would lead to the voidance of his British patent.[49] Apparently he also felt he should take a more personal interest in the British situation because seven years of developmental work on his hemp spinning machines had culminated in four more American patents, granted in 1834, which urgently needed protection in England.[50] Already one of his workmen, Samuel Couillard, had picked up enough ideas in Treadwell's machine shop to patent a rival machine.[51] Treadwell appears to have taken Couillard to court over this but a version of Couillard's device, for combing wool, was patented in England in sum-

---

[48] Treadwell to Josiah Quincy (president of Harvard University), 1 January 1835, Treadwell papers.

[49] His American patent was dated 11 October 1831; the specification is in the National Archives, RG 241, Specifications XIII, 291–309. Only one drawing survives: ibid., Drawings, No. 6,794. An identical drawing and the complete patent, but with parts renumbered, is GB Pat. No. 6,185 (27 October 1831) granted to Joshua Bates.

For Treadwell's misgivings about the arrangements for the British patent for his invention, see Treadwell to an un-named London correspondent, 7 June 1834 and a draft of this letter in Treadwell Papers. Treadwell's experience shows that British patent agents could modify American patent specifications, if not drawings, in order to popularise the invention and so increase its patent returns.

Newton is identified in GB, *PP (Commons)* 1829 (332) III, "Report from the Select Committee on the Law Relative to Patents for Inventions," 66; Bates appears in GB, *PP (Commons)* 1833 (690) VI, 45.

[50] Dated 3 February (spinning flax and hemp), 5 February (cordage), 18 August (two: cordage; hatcheling flax and hemp), 1834. These, and his 1831 patent covered Treadwell's "Gypsey," and its accessories, amounting to a greatly improved system for making rope which gained world-wide acceptance. See *DAB*, s.v. Daniel Treadwell.

For the commencement and progress of Treadwell's development work, see deposition of Sargent M. Davis, machinist of Boston, 19 May 1831, Treadwell Papers.

[51] Ibid.

mer 1833 under Joshua Bates's name.[52] Unless Treadwell soon secured his American inventions in Britain, the fruits of his costly experimental work would be lost through a rival's piracy or an agent's ignorance or connivance.

When news of Treadwell's intended visit to Britain spread through the Boston mercantile–manufacturing community, he received a number of commissions to recruit skilled workers or to purchase equipment. At least five assignments were in his pocket when he sailed from New York on 2 March 1835.[53] Charles Russell Lowell, treasurer of the Lycoming Coal Co. of Pennsylvania, wanted workers who knew how to smelt iron with coke (the Darby process, well known in England since the 1750s).[54] H. Gray & Co. requested "a minute account" of methods for making iron sheets and hoops and an assessment of chances of obtaining machinery or workers.[55] Josiah Loring asked Treadwell to purchase globes and compasses in London.[56] Amos and Abbott Lawrence, brothers with a Boston selling house who were on the point of extending their Lowell investments with the formation of the Boott Manufacturing Co., wanted particulars of English machinery for making cotton twist for export, and also commercial information about manufacturing costs and yarn sizes suited for markets in northern Europe or China.[57]

Greater significance for the recruitment of James Montgomery lay in the commissions given to Treadwell by Samuel Batchelder (1784–1879), agent of the York Manufacturing Co., Saco, Maine.[58] Batchelder, a self made and highly expert cotton manufacturer who had previously been agent (manager) of the Hamilton Manufacturing Co. at Lowell, had several concerns. Firstly, he wanted

---

[52] Couillard's American patent was dated 30 March 1833; the patent is in the National Archives, RG 241, Specifications XVI, 1–10 and Drawings No. 7,501. The British patent, with an identical drawing and stated to be from a foreigner, was GB Pat. No. 6,459 (13 August 1833), granted to Joshua Bates.
      Couillard took out another American patent on 7 July 1835, again for a combing device.
[53] Treadwell to "Adda," 1 March 1835, Treadwell Papers.
[54] Charles Russell Lowell to Treadwell, 28 February 1835, ibid.
[55] H. Gray & Co. to Treadwell, 21 February 1835, ibid.
[56] Josiah Loring to Treadwell, 24 February 1835, ibid.
[57] A. & A. Lawrence to Treadwell, 19 February 1835, ibid.; *Annual Statistics of Lowell Manufactures, January 1, 1837* (Lowell, 1837); Hannah Josephson, *The Golden Threads: New England's Mill Girls and Magnates* (New York: Duell, Sloan and Pearce, 1949), 155–159.
[58] *DAB*; William R. Bagnall, "Samuel Batchelder," *Contributions of the Old Residents' Historical Association, Lowell, Massachusetts* III (1885), 187–211.

Fig. 3.   Samuel Batchelder, portrait from *The History of New Ipswich* (Boston: Gould & Lincoln, 1852). Courtesy of the Museum of American Textile History.

a clear picture of the comparative performance of British and American cotton manufacturing in terms of techniques and costs. To gain this he sent Treadwell a list of specific questions, and in one of these referred to Montgomery's *Carding and Spinning Master's Assistant*, asking whether Montgomery's method of counting cards was correct. Montgomery's first book, it should be noted, would have been known to American readers by 1836 at the latest, for that year the biographer of Samuel Slater hailed it as a pioneering achievement.[59] At another point Batchelder asked, "How far does the English practice differ from ours in any of these particulars?"[60]

Secondly, Batchelder sent Treadwell a box of spinning frame components—a cast steel front roll, an iron back roll, couplings, a weight, a tube bearing, spindles, flyers, bobbins, and washers—and wanted him to obtain samples and costs of comparable components from British machine makers.[61] Batchelder, then equipping a second mill, suspected that it might be cheaper to import key parts from England and tested that market to the extent of authorising Treadwell to spend up to £200 on machine parts.[62]

Whether Treadwell received payment for any of these assignments is unknown. He did, however, come to an agreement with Batchelder, apparently on a visit to Batchelder's factory in Maine, whereby he would try to promote the patenting of several American devices in England and in return have a share in any profits. Batchelder sent Treadwell £250 but cautioned him not to take action unless the American devices were significantly better than anything in use in Britain, unless they stood a good chance of being accepted there, and unless a dependable person could be found to install them in a factory. The devices in question were Batchelder's latch-type stop motion, applicable to the drawing frame, and a dead spindle, improved by Batchelder, neither of which was covered by an American patent.[63] Batchelder realised that the dead spindle had been patented in England by Robert Montgomery (a cotton spinner of Johnstone, just west of Paisley in Scotland; any relationship with James Montgomery, unknown) but "something might be done with the patentee . . . which would be for the interest of both parties."[64] Since Robert Montgomery's patent was for "the old Waltham spindle," Batchelder guessed that his British rivals had greater difficulty in obtaining unpatented American inventions than some imagined.[65] Whether

---

[59] George S. White, *Memoir of Samuel Slater* (Philadelphia: the author, 1836), 305.

[60] Samuel Batchelder to Treadwell, 13 February 1835, Treadwell Papers.

[61] Same to same, 16 February 1835 (letter no. 2: which starts, "The box . . ."), ibid.

[62] Pliny Cutler to Treadwell, 20 February 1835, ibid.; George S. Gibb, *The Saco-Lowell Shops. Textile Machinery Building in New England, 1813–1949* (Cambridge, Massachusetts: Harvard University Press, 1950), 110–111.

[63] Batchelder to Treadwell, 16 February 1835 (letter no. 1: which starts, "On further consideration . . ."), Treadwell Papers.

[64] Ibid. and GB Pat. No. 6,261 (26 April 1832), patent from a foreigner but granted to Robert Montgomery.

[65] Batchelder to Treadwell, 16 February 1835 (1), Treadwell Papers.

Treadwell took models or drawings, or both, of these devices was
not clear. Certainly Batchelder sent him a pair of temples, of the
self-acting rotary type, in case he could promote them in England
also.[66]

  Treadwell reached London on 25 March 1835, spent a month
there, then three weeks in Wales; in late May he was in Manchester; about this time he went to Staffordshire and in late June
paid a rapid visit to Glasgow. He returned to London on 9 July
and was then thinking about returning to Staffordshire before
sailing for Boston in early August.[67] Clearly he knew his way
around. He had been in England before and he had a number
of letters of introduction. One, which he needed in his professional capacity, was addressed to Michael Faraday.[68]

  On the industrial front his forays into Britain's manufacturing
districts met a mixed reception. For the Lycoming Coal Co. he
hired Edward Thomas of Merthyr Tydfil, a miner and iron furnace manager.[69] For H. Gray & Co. he reported on Staffordshire
methods of iron rolling.[70] But secretiveness and caution prevailed
among both merchants and manufacturers at Manchester, so
much so that the former even moved goods in and out of their
warehouses by back entrances and at night. Letters of introduction
gained Treadwell an entrance to cotton mills but his method of
espionage was limited to observation of mill rooms. Consequently
there were "many particulars which I could not find out, as I was
obliged to be exceedingly guarded in my inquiries, as any question
that should have led them to think that I knew anything about
the manufacture practically, would have spoiled all."[71] These protective attitudes forced Treadwell to abandon any hopes of promoting American patents in Manchester, though he observed that

[66] Same to same, 16 February 1835 (2), ibid.
[67] Treadwell to Batchelder, 1 June 1835; George Thomson to Treadwell, 7 July 1835;
Treadwell to H. Gray & Co., 9 July 1835, all ibid.
[68] Petty (?) Vaughan (Fenchurch Street, London) to Dr. Faraday, FRS, 29 March 1835,
ibid. Treadwell went to London in 1819–1820 to patent his printing press (GB Pat. No.
4,433, 25 January 1820).
[69] Agreement between Edward Thomas of Merthyr Tydfil of the first part and Gerard
Ralston and Daniel Treadwell, American citizens in London, on behalf of the Lycoming
Coal Co., May 1835; also Treadwell to Thomas, 9 May 1835; Treadwell's rough notes
on iron making; and Treadwell to (Thomas?), 6 May 1835, regarding the recruitment
of workers for the Lycoming Co. and information on furnaces, all Treadwell Papers.
The secret of smelting iron ore with anthracite coal was the technology sought by the
Americans. See Darwin H. Stapleton, *The Transfer of Early Industrial Technologies to America*
(Philadelphia: American Philosophical Society, 1987), 169–210.
[70] Treadwell to H. Gray & Co., 9 July 1835, ibid.
[71] Treadwell to Batchelder, 1 June 1835, ibid.

drawing frames lacked stop motions and that powerlooms, running at 120 picks per minute, still incorporated hand fixed temples.[72]

While he did not leave Manchester completely empty handed, Treadwell found a much more open stance among the Glasgow manufacturers. With one of them, Henry Houldsworth & Sons, Treadwell entered into an agreement to secure two patents, one for England and one for Scotland, for Batchelder's drawing frame stop motion. The two Americans shared the expense and profit equally with the Houldsworths and then held their share jointly.[73] Treadwell's confidence in the Houldsworths may have been related to their close connections with other New England manufacturers and the visit of one of the Houldsworths to the USA in the early 1820s.[74]

In pursuing answers to Batchelder's questions, Treadwell naturally sought out James Montgomery to settle the point about measuring sizes of card clothing. Montgomery, committed by his first book to an unreserved sharing of technical information, must have welcomed the chance of a discussion about technical change in New England factories, his keen rivals. What Montgomery learned is unclear; Treadwell certainly collected much of the information that Batchelder needed, through meeting Montgomery and one of his colleagues on the commercial side of the firm, George Thomson. Two letters record some of the information they gave. Montgomery provided details of the methods and equipment, including assorted technical specifications and costs, of "our general process of cotton spinning," which treated, in sequence, the stages of scutching, carding, drawing, and spinning.[75] Thomson gave Treadwell commercial information about output, wages, and manufacturing and power costs in the

---

[72] Ibid.

[73] Treadwell to Gerard Ralston, 2 August 1835, ibid. The patent, GB Pat. No. 6,951 (9 December 1835), was taken out in the name of John Houldsworth, cotton spinner of Glasgow. It did not incorporate the spring bolt action patented in America by Lewis Cutting of Lowell in 1834, for which see *PD2* chap. 2, note 14. For further comment on stop motions, see below.

[74] Henry Houldsworth, Jr., visited the USA in 1822–1823 and, on behalf of the proprietors of the Taunton Manufacturing Company of Taunton, Massachusetts, patented in Britain the roving frame differential gear, unbeknown to its inventor, Asa Arnold of Rhode Island. See GB, *PP (Commons)* 1824 (51) V, 384 and Theodore Z. Penn, "The Introduction of Calico Cylinder Printing in America: a Case Study in the Transmission of Technology," in *Technological Innovation and the Decorative Arts*, ed. Ian M. G. Quimby and Polly Anne Earl (Charlottesville, Va.: University of Virginia Press, 1974), 247–250.

[75] Montgomery to Treadwell, 6 July 1835, Treadwell Papers.

MacLeroy, Hamilton & Co. factory, which then ran 11,000 spindles and 200 powerlooms, all driven by a 50 horsepower steam engine, which Treadwell inspected.[76]

Treadwell left Scotland and Montgomery heard no more from him until 30 December 1835, when he received Treadwell's letter from Boston inviting him to America to become superintendent of the York Manufacturing Co. in Maine, an invitation originating with Samuel Batchelder.[77] What assets did Montgomery offer to the New England manufacturer? And why did Montgomery accept the invitation? Obviously, one asset Montgomery possessed was his long and tested experience as a cotton mill manager in a system of manufacturing similar in many respects to that developed in northern New England. If he was managing 11,000 spindles and 200 powerlooms engaged in producing coarse goods, then his skills matched those of the agents of some of the smaller vertically integrated firms at Lowell, like the Appleton, Suffolk, and Tremont companies, and certainly the York mills.[78] Batchelder in 1836 ran two mills and planned to add a third one to the York Manufacturing Co.'s capacity. Since the mills at Saco had on the order of 3,000 to 6,000 spindles each, Montgomery's experience suited the American company's needs with respect to quality of product, scale of operation, and organisational structure.[79] While his most recent experience of power sources related to steam and not water, he cannot have been wholly ignorant of the traditional water wheel technology which had driven the Blantyre mills of his youth. And though he probably did not have such a high proportion of females in his Glasgow workforce as he would meet in New England, Montgomery's advocacy and presumed practice of moderate, fair but firm labour policies would have commended his managerial style to New England proprietors. Yankee farm-girl operatives and their families would never tolerate the kind of brutal treatment reported in some British cotton mills.[80] In addition, Montgomery's publications—the most recent

[76] George Thomson to Treadwell, 7 July 1835, ibid. The address Thomson gave, 47 Ingram Street, Glasgow, was the trading address of MacLeroy, Hamilton & Co. given in the *Post Office Annual Directory* for Glasgow, 1829–1835.

[77] Montgomery to Treadwell, 6 January 1836, Treadwell Papers.

[78] For Lowell, see *Annual Statistics of Lowell Manufactures January 1, 1836* (Lowell, 1836).

[79] For the York mills, see Gibb, *Saco-Lowell Shops*, 109–112, 123, 133.

[80] For American mill discipline, see Carl Gersuny, "'A Devil in Petticoats' and Just Cause: Patterns of Punishment in Two New England Textile Factories," *Business History Review* L (1976) 133–152; and Thomas Dublin, *Women at Work: The Transformation of Work and Community in Lowell, Massachusetts, 1826–1860* (New York: Columbia University Press,

one of which, *The Cotton Spinner's Manual,* the author had given to Treadwell[81]—proved him to be one of the most highly informed and articulate of Scottish cotton mill managers with a keen interest in the latest innovations and committed to the free, international exchange of technical information.

Montgomery's reasons for taking a position in the United States are unknown and the circumstances of his decision are only partly discernible. His letter of acceptance disclosed that he was already searching for a superintendent's post and that he had been offered one in Britain "with a fixed salary and a share in the profits."[82] He took the job in New England with surprisingly little knowledge of the details of his new situation. Having accepted Treadwell's offer in one paragraph, Montgomery went on to ask:

> Is it a populous district, where the works of the York Manufacturing Coy. are situated: and is there any difficulty in obtaining work people already acquainted with cotton mill work? What particular branch of the business is carried on at these works: is it mule or throstle spinning or both? What are the range of Nos. generally spun? How many hours pr day does the workers labour in these works? Is Mr Batchelder practically acquainted with the internal management of cotton Factories: or is he a shareholder? What may be the extent of the mill already in operation: and the one which you say is now nearly ready? Are there any cotton Factories in the neighbourhood of those belonging to this company in the state of Main?[83]

Clearly Treadwell had not discussed Batchelder or his mills when he met Montgomery in Scotland. Nor is there mention of salary in the surviving letters. Presumably it was an improvement on the salary Montgomery was receiving from MacLeroy, Hamilton & Co.

Whatever his motives, Montgomery did not emigrate in total ignorance and insecurity. He consulted his friends before making his decision,[84] and a constant movement of spinners and mechanics from Scotland to the USA, some of whom returned, provided

---

1979), 58–85. For the British experience, see Sidney Pollard, *The Genesis of Modern Management. A Study of the Industrial Revolution in Great Britain* (Cambridge, Massachusetts: Harvard University Press, 1965), 186.

[81] Montgomery to Treadwell, 6 July 1835, Treadwell papers.

[82] Same to same, 6 January 1836, ibid.

[83] Ibid.

[84] Ibid.

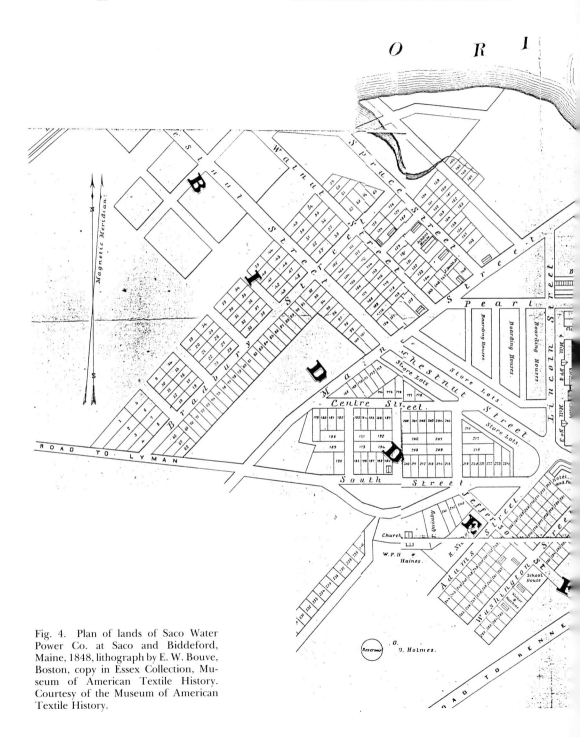

Fig. 4. Plan of lands of Saco Water Power Co. at Saco and Biddeford, Maine, 1848, lithograph by E. W. Bouve, Boston, copy in Essex Collection, Museum of American Textile History. Courtesy of the Museum of American Textile History.

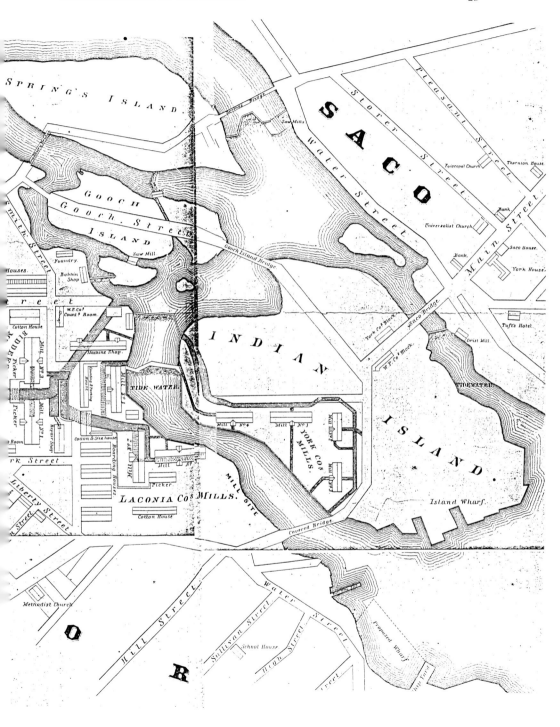

a reservoir of intelligence about American conditions.[85] Further-more, he planned to take his family with him and possibly other relatives:

> I have a Brother who is very anxious to accompany me to America. He has been for a number of years managing cotton mills. And I can with confidence assert regarding his qualifications as a practical man, that he is not exceled by any in Glasgow. He just now occupies a very good situation, as manager to an establishment where both spinning and weaving are carried on simultaneously. Do you think he might find a situation in America?[86]

Whether his brother followed him to America or not, James Montgomery headed a sizeable household living on Stour Street, Saco, Maine, at the time of the 1840 census of population: besides himself, two males, one aged 20–30 and the other 15–20; and six females, one 40–50, presumably his wife, two 15–20, one 10–15, one 5–10 and one under 5 years of age.[87] Lending further security to his move to New England was the support of Daniel Treadwell, influential engineer–inventor and Harvard professor. Certainly Montgomery looked to him to ease his migration: "I will come to America, relying intirely upon Mr. Daniel Treadwell, whom, I will ever be proud to call a friend, and I hope you shall never have cause to regrate having interested yourself in my behalf."[88] If, as it seems, Samuel Batchelder came to his defence in the newspaper controversy that erupted over his third book, Montgomery was not to be disappointed by his mentors.

Whether they realised the magnitude of their success or not, Batchelder and Treadwell had secured the services of an out-standing British cotton manufacturer. Montgomery's emigration, as an industrial manager in his early forties and accompanied by family, was distinctly untypical of early nineteenth century British textile migration to America and suggested rather unique circumstances which have been probed as far as the evidence will allow.[89]

---

[85] GB, *PP (Commons)* 1833 (690) VI, 311–312, 317–318, 322–324. For examples of emigrants' letters, see Charlotte Erickson, *Invisible Immigrants: The Adaptation of English and Scottish Immigrants in Nineteenth Century America* (Coral Gables, Florida: University of Miami Press, 1972).
[86] Montgomery to Treadwell, 6 January 1836, Treadwell Papers.
[87] Sixth Census of the United States, 1840, National Archives microfilm, M-704, roll 154, sheet 178 (York County, Maine).
[88] Montgomery to Treadwell, 6 January 1836, Treadwell Papers.
[89] *TIR*, chapter 8.

(d) Montgomery's American career

Apart from his last book and the controversy it generated, few particulars are known about Montgomery's career in the United States. He stayed at Saco, Maine, for seven years.[90] By the end of this period the York Manufacturing Co. had four mills in operation and Batchelder had organised a separate Water Power Co. and machine shop, indicating his hopes of building Saco into a manufacturing centre modelled on Lowell.[91] Montgomery left in 1843 to take up another post in New York state, apparently.[92]

Then in 1850, possibly sharing current optimism about the prospects of a slave-based cotton industry in the Southern states, he joined the sporadic migration of skilled workers from New England to the South, to become superintendent and treasurer of the newly built Graniteville Manufacturing Co.'s factory, situated on a tributary of the Savannah River in the lower part of South Carolina.[93] At this date, a mill superintendent was reckoned to be earning $1,200–1,500 a year in the South.[94] The company had been organised by William Gregg (1800–1867) the well-known promoter of manufacturing in the South. When Gregg's mill started in summer 1849, it ran nearly 10,000 spindles and 300 powerlooms, driven by water turbines, which produced low count (No. 14) yarns and coarse shirtings, sheetings, and drillings.[95] Operated by rural white families, the mill required just the managerial skills which Montgomery had successfully practised in Glasgow and then in New England.

Montgomery stayed at Graniteville until 23 April 1863.[96] Though it was in the midst of the Civil War (Grant was at Vicksburg and the tide was just beginning to turn against the South), the company recorded handsome tributes to Montgomery on his departure:

[90] Obituary, *Boston Daily Advertiser*, 3 January 1881.
[91] Gibb, *Saco-Lowell Shops*, 123.
[92] Graniteville Manufacturing Company records, information communicated to the editor by William C. Lott, company vice-president, 23 January 1974.
[93] Ibid.
[94] "Manufactures in South Carolina," *De Bow's Review* VIII (1850) 29. I am grateful to Mrs. William C. Morris of Greenville County Library for this reference.
[95] For Gregg and Graniteville, see Broadus C. Mitchell, *William Gregg, Factory Master of the Old South* (Chapel Hill, North Carolina: University of North Carolina, 1928); Ernest M. Lander, Jr., *The Textile Industry in Antebellum South Carolina* (Baton Rouge, Louisiana: Louisiana State University Press, 1969); *DAB*.
[96] Graniteville Manufacturing Company, Directors' Minutes, 23 April 1863, kindly transmitted to me by William C. Lott.

WILLIAM GREGG.

Fig. 5. William Gregg, portrait from Broadus Mitchell, *William Gregg, Factory Master of the Old South* (Chapel Hill, NC: University of North Carolina Press, 1928), frontispiece. Courtesy of the Museum of American Textile History.

1st That the Graniteville Company cannot allow the withdrawal of James Montgomery as superintendent of their Factory to pass unnoticed.

2d That his faithfulness and integrity have won for him a high place in our esteem and that the regret with which they part with him is only allayed by the belief of their late superintendent that this act entirely of his own accord is for his own advantage.

3 That the President [William Gregg] be directed to tender Mr Montgomery on behalf of the stockholders of the Graniteville

Fig. 6.   Graniteville Manufacturing Company mills.

>   Company such a present as shall testify to the cordial sentiments
>   entertained towards him.
>
> 4   That the Secretary of the Company be directed to furnish Mr
>   Montgomery with a copy of these resolutions.[97]

By this time Montgomery was approaching the age of seventy
but, according to his obituarist, he did not give up "his business"
until 1874 when he would have been eighty.[98] In 1867 he was
recorded as superintendent of the Batesville Manufacturing Company in Greenville County, South Carolina, in the Piedmont well
above the humidity of the lower region in which Graniteville was
located.[99] The Batesville company, one of the most successful of
the state's antebellum operations, had been founded in 1833 by
William Bates, one of the group of Rhode Island mechanics who
first brought the new cotton manufacturing technology to the state
more than a decade earlier.[100] When James Montgomery managed
it, the firm ran 1,260 spindles and 36 looms, all tended by 50
operatives: clearly a less onerous position than at Graniteville
where the physical plant was eight times larger.[101]

Almost no traces of Montgomery's personality remain beyond
his characteristics as a professional cotton mill manager. At his
death he was remembered as "decided in his religious convictions"
but "always charitable and broad in his views respecting other
than his own Presbyterian convictions."[102] His wife predeceased
him by four years and when he died on 27 December 1880, he
was living at Pliny in Greenville County.[103]

In looking over the broad sweep of James Montgomery's career,
one feature becomes evident. His migrations closely paralleled the
movement of the manufacturing frontier from Britain to New
England, and thence to the American South. To this westward
movement, chiefly of knowledge and skill, Montgomery himself
substantially contributed both through his publications and
through his practice as a cotton factory manager.

---

[97] Ibid.
[98] *Boston Daily Advertiser*, 3 January 1881. The obituary was signed "W. M."—William (?)
    Montgomery(?).
[99] August Kohn, *The Cotton Mills of South Carolina* (Columbia, S.C.: South Carolina De-
    partment of Agriculture, Commerce and Immigration, 1907) 19. I am indebted to
    Professor Tom E. Terrill, University of South Carolina, for this reference.
[100] Lander (see above, note 95) 19–21, 77, 107.
[101] *Boston Daily Advertiser*, 3 January 1881.
[102] Ibid.
[103] Ibid.

II. Montgomery's *Cotton Manufacture:* Background, Importance, Editions, Editorial Methods

(a) Background

Montgomery's *Cotton Manufacture* is one of the most important technical classics of the early industrial period. To understand its original and continuing importance, the book has to be placed in its contemporary setting, considered on its intrinsic merits, and then set in a larger perspective. Montgomery's study emerged partly from a background of mounting British anxiety, clearly justified, about American competition in overseas textile markets, and partly from Montgomery's own interests and circumstances.

Prior to the publication of the *Cotton Manufacture,* the British mercantile and manufacturing community heard little more than sparsely informed tocsins of alarm about American competition; certainly no inside views from the northern New England giants were published before 1840. Public complaints about American rivals were first aired by British manufacturers before a Parliamentary Select Committee in 1833, at a time when manufacturers were trying to explain falling profit margins in the cotton industry. Witnesses from manufacturing and trade, from the USA as well as Britain, affirmed the spread of American competition and attributed it to America's comparative advantages, cheaper raw cotton and power, and also to artificial costs borne by British manufacturers, like import duties on flour and cotton or variations in British wage levels.[1] And in 1833 James Montgomery was one of the first British writers to announce to the British textile community that their American rivals were now running 1.2 million spindles and 33,500 looms (presumably mostly powerlooms) in nearly 800 cotton mills: an estimate published in the United States in the *Report on the Production and Manufacture of Cotton* (1832)

---

[1] GB, *PP (Commons)* 1833 (690) VI. Witnesses commenting on the American competition were Kirkman Finlay, Joshua Bates, Timothy Wiggin, William Graham, and James Kempton. For the economic and political background, see Matthews, *Study,* and Lucy Brown, *The Board of Trade and the Free-Trade Movement, 1830–1842* (Oxford: Clarendon Press, 1958). Details of the import duty on cotton are to be found in GB, *PP (Commons)* 1840 (299) V. "Report of the Select Committee on Import Duties," 299, 309.

prepared for the Friends of Domestic Industry by one of their leaders, the Boston capitalist Patrick Tracy Jackson. Montgomery reckoned that Britain was then operating 16.6 million cotton spindles.[2]

The duty on cotton, amounting to under a halfpenny a pound, came under a fresh attack in 1836. The Glasgow manufacturers, who mostly operated at the lower end of the market and therefore both paid more in duty (based on weight not value) and suffered more from foreign competition than their Lancashire counterparts, lent their support to the publication of Alexander Graham's *The Impolicy of the Tax on Cotton Wool* (Glasgow: Associated Cotton Spinners, 1836). Graham printed excerpts from the Parliamentary reports and then reinforced his free trade arguments with more than forty affidavits sworn by partners or agents of British mercantile houses, all testifying to the nature and extent of American penetration of foreign markets, mostly in South America and the Far East.[3] He then analysed the respective advantages of Britain and her rivals. British manufacturers had access to more skilled operatives, lower interest rates, and, above all, "the long advance of all other nations . . . in the cotton manufacture"[4]—technological leadership. American manufacturers enjoyed lower freight and insurance rates, fewer middlemen, cheaper cotton, cheaper flour (for dressing and bleaching), abundant waterpower, the absence of trade unions, a lower cost of living, and longer working hours.[5]

But where did the net advantage lie when Britain's comparative advantages were balanced against those of her competitors? Graham repeated the comparative estimate of weaving costs submitted to the Royal Commission on the Employment of Children in Factories in 1833 by James Kempton, a cotton manufacturer from Norwich, Connecticut.[6] These annual operating costs were as follows[7]:

	United States £–s	England £–s
Interest on Dressing Machine	2–11	1–12

[2] *CSMA* 2, 289, 293.
[3] Graham, *Impolicy*, 22–30, Appendix, 11–34.
[4] Ibid., 46; also 43–48.
[5] Ibid., 30–43.
[6] GB, *PP (Commons)* 1833 (450) XX, "Report of the Royal Commission on the Employment of Children in Factories," Section E, 23.
[7] Graham, *Impolicy*, 42.

	United States £–s		England £–s	
Interest on 12 Power Looms	8–6	⎫ 6	4–10	⎫ 5
Cost per annum of one horsepower	3–10	⎬ per ⎭ cent	12–10	⎬ per ⎭ cent
Cost of dressing 3,756 pieces	23–9		46–18	
Cost of weaving 3,756 pieces	125–3		156–10	
	£163–00		£222–00	

10.50d per piece   14.25d per piece

This indicated that American manufacturers could make a piece of cotton goods for 4.375p, (or 10.5d) compared to a British cost of 5.94p (or 14.25d), a difference in favour of the American producer of 36 percent. The omissions from the calculation are obvious enough: the costs of spinning the yarn; total fixed costs; the overheads of transportation, insurance, depreciation, and maintenance; any difference in total horse power employed to make 3,756 pieces and differences in overall power costs; and the impact of British import duties. But it was at least a hard quantitative estimate of some comparative costs. And it invited further attempts to reach comprehensive comparisons.

The next came from Lancashire where the threat of American competition was used by the mill masters opposed to factory reform, in particular the Ten Hours Bill, to argue against state-directed change in their work practices. Robert Hyde Greg (1795–1875), whose family operated five mills (employing over 2,000 people) in Cheshire and Lancashire, advanced the argument in his pamphlet *The Factory Question* (London: James Ridgeway & Sons, 1837). After rehearsing the relative advantages of Britain and the USA, following Kempton and Graham, he presented his quantitative estimate of comparative costs. It came from an anonymous correspondent of the *Manchester Guardian* who asserted that the Lowell mills made 18–20 percent more yarn per spindle per week than did British mills, even allowing for differences in working hours.[8] The correspondent felt that a proportion of British

[8] 'G,' letter to the *Manchester Guardian*, "To the Manufacturers of Manchester and the Neighbourhood," reprinted in Greg, *Factory Question*, 146–148. For Greg see Mary B. Rose, *The Gregs of Quarry Bank Mill: The Rise and Decline of a Family Firm, 1750–1914* (Cambridge: Cambridge University Press, 1986).

manufacturers were apathetic towards the American threat: he noted that a copy of the published *Lowell Statistics* for 1836 and an engraving of Lowell had been hanging up in the Manchester Exchange for some months without occasioning much comment.[9] Greg agreed and, to rouse his fellow manufacturers, published the statistics again in his pamphlet.[10]

Still there was no definitive analysis of British and American comparative manufacturing costs. But at this point Montgomery was just settling into his new situation as manager of the York Manufacturing Company at Saco in Maine: a position which, with his practical experience and technical publications behind him in Scotland, uniquely fitted him to prepare reliable and comprehensive estimates of British and American manufacturing costs in comparable mills. This he had already begun to do.

In the first edition of his *Cotton Manufacture*, Montgomery advanced the technique of technical writing which he evolved over the three editions of his first book, the *Carding and Spinning Master's Assistant*.[11] Starting as a practical manual based on Scottish practices, this became by its third edition a handbook not only explaining how to operate individual machines and a whole mill production line, but also describing best practice manufacturing with examples drawn from and comparisons made between Lancashire and Glasgow. No doubt the popularity of that book and the opportunity presented by his emigration pointed towards the compilation of another survey embracing American manufacturing practice. Besides the anxieties of British manufacturers about their American competitors and besides Montgomery's earlier writings, other influences contributed to the writing of the *Cotton Manufacture*.

Before he left them, Montgomery's Scottish friends, eager to learn all they could about their rival manufacturers in world markets, induced him to prepare some kind of survey of the American cotton industry.[12] To them he must have appeared an ideal reporter: his technical knowledge, his writings, his record and connections among Glasgow cotton mill managers, all inspired confidence in his competence and trust in his reliability. And from Montgomery's viewpoint there were strong reasons why he should aim at the strictest veracity and impartiality. If his American po-

[9] Greg, *Factory Question*, 146.
[10] Ibid., 144–145.
[11] See the first section of this introduction.
[12] JM, *Boston Courier*, 27 March 1841; *CM* 2, Preface.

sition fell short of its promise he might wish to return to a Scottish or English situation where his most recent publication could promote or damage his prospects. If, on the other hand, he wanted to remain in America he would hardly want to alienate potential employers. So in writing his *Cotton Manufacture* Montgomery would best serve his own interests if he avoided passing prejudice and sought only factual information, as he claimed in his Preface to have done.

Apart from the urging of his friends, Montgomery's other motivations for writing his study can only be guessed. Presumably the book, published by John Niven, Jr., his first and only publisher, brought some financial return. His publicly professed opposition to any monopoly on knowledge may likewise have prompted him to make the study.[13] And, in view of his American employer's (Samuel Batchelder) interest in discovering the differences between American and British manufacturing methods,[14] it is not unlikely that Montgomery may have seen his book as lending some assistance to his chances of promotion in America.

Montgomery completed his *Cotton Manufacture* between June 1836, when he reached the United States, and September 1838, when he sent his manuscript to his Glasgow publisher. It was, he admitted in 1841, prepared in some haste.[15] He had little opportunity to visit all the American manufacturing districts on which he commented, which meant that his views of American cotton manufacturing were most clearly focussed on northern New England and the so-called Waltham system, a weakness he strove to surmount by quoting secondary sources for other districts. The limitation of Montgomery's perspective should not be exaggerated, however. The size, organisation, and technology of the northern New England cotton manufacturing companies following the Waltham model, the Boston Manufacturing Company, made them by far the most formidable rivals to British coarse goods manufacturers.[16] And precisely because Montgomery could offer an insider's view at a management level of such mills, his view assumes unrivalled importance and authority. A more serious limitation of Montgomery's work is that its objective treatment of

---

[13] *CSMA* 1, Preface.
[14] Batchelder to Treadwell, 13 February 1835, Treadwell Papers.
[15] JM, *Boston Courier*, 27 March 1841.
[16] For the Waltham system mills and their technology, see Caroline F. Ware, *The Early New England Cotton Manufacture. A Study in Industrial Beginnings* (Boston: Houghton Mifflin, 1931), 60–118; *TIR*, 92–103, 180–203.

cotton manufacturing at the purely technical and commercial levels left little room for any discussion of the human aspects of the
early factories, but as a manager he was not likely to know a great
deal about the experiences of American operatives unless members of his own family worked in the mills; further, the interests
of his own position could well have dictated a prudent silence.

As may be seen from the text, Montgomery's *Cotton Manufacture*
was most original for its thorough descriptions of Waltham system
technology and for its comprehensive and exact comparative estimates of British and American manufacturing costs. Yet his style
of writing was more relaxed (though in places still tediously technical to a modern eye) than the nuts and bolts descriptions found
in Rees's *Cyclopaedia* or Ure's *Dictionary*, the standard accounts of
British textile technology of the time.[17] It was closest in approach
to Ure's *Cotton Manufacture*.[18] Where a machine or component was
likely to have been unfamiliar to British readers, Montgomery
provided drawings and detailed explanation of how it was constructed and how it worked. Where American differences were
less extensive or less complex, he simply identified the essential
features concerned.

His comparative estimates made the earlier ones of Kempton
and the *Manchester Guardian* correspondent look very jejune.
Every possible operating cost, down to tallow and sweepers, was
accounted for. The one slip he apparently made was to make
insufficient allowance for the cost of capital, insurance, and wear
and tear, as was later observed. The structure for calculating these
costs he had already established in the third edition of his *Carding
and Spinning Master's Assistant*.[19] The example used in that structure he revived, updating overhead costs from information provided by his Glasgow correspondents.[20] The American model and
data he derived from one of the York Manufacturing Company
mills which he was currently managing.[21] Montgomery's conclusions about relative British and American cotton manufacturing
costs were radically different from those already publicised in
Britain. Costs within the mill, he reckoned, showed a differential

---

[17] Abraham Rees (ed.), *The Cyclopaedia: or Universal Dictionary of Arts, Sciences and Literature*
(45 vols., London: Longman et al., 1802–1820); Andrew Ure, *Dictionary of Arts, Manufactures, and Mines: Containing a Clear Exposition of Their Principles and Practice* (London:
Longman et al., 1839).
[18] Ure, *The Cotton Manufacture of Great Britain* (2 vols., London: Charles Knight, 1836).
[19] *CSMA* 3, 248–256.
[20] JM, *Boston Courier*, 27 March 1841.
[21] Below, *CM* 2, "Comparative Estimates," Note 2.

in favour of the British factory to the extent of 19 percent over a fortnight's production. Taking all costs into account, that is with the addition of transportation and import duties primarily, the differential swung in favour of the American factory, to the extent of 3 percent.[22] On the face of it, these figures lent little substance to British manufacturers' squeals against American competition: the differential was not 36 percent in favour of American manufacturers, as Kempton alleged, but less than a tenth of that. To an extent this eroded the British manufacturers' case against the Ten Hours Bill and further reforms, so it was not surprising that Montgomery's estimates received much less notice in Britain than in America. In the USA, as the second section of this edition shows, Montgomery's uncertainty about power costs led to a vigorous, and at times slanderous, debate in the Boston newspapers.

(b) Continuing importance

The debate over British and American manufacturing costs has re-emerged in the past thirty years among historians seeking to explain the distinctive and diverging courses of technical change in the two countries during the nineteenth century.[23] In an intricate and well-known theoretical study, Habakkuk maintained that two influences shaped early nineteenth century technical change in the USA: a higher ratio of labour (wages) to capital (machine prices, multiplied by the rate of interest plus rate of depreciation) costs than in Britain; and a more inelastic supply of labour in the USA than in Britain, which led to higher wages and lower profit margins, in turn countered by capital intensive techniques.[24] Pressure towards higher wages, despite population growth, derived from high returns in American agriculture or, if Lebergott is right, from a higher level of social mobility.[25] It is hard to establish labour–capital ratios but the experience of the Boston Manufacturing Company, the fountainhead of northern New England cotton manufacturing technology, does support Habakkuk's second point. The Boston Manufacturing Company met and overcame

---

[22] *CM* 1, 124–125.

[23] A good introduction to this debate is S. B. Saul (ed.), *Technological Change: The United States and Britain in the Nineteenth Century* (London: Methuen, 1970).

[24] H. J. Habakkuk, *American and British Technology in the Nineteenth Century: the Search for Labour-Saving Inventions* (Cambridge: University Press, 1962).

[25] Stanley Lebergott, *Manpower in Economic Growth: the American Record since 1800* (New York: McGraw-Hill, 1964).

high wage bills by replacing highly paid and skilled adult male operatives with lower paid and minimally skilled teenage female operatives working with a modified production technology.[26] The possibility that this substitution of capital for labour was hampered by high labour costs in capital goods manufacture seems to be mitigated by the recent finding that higher proportions of machine makers than of skilled industrial operatives or managers arrived from Britain during the 1820s.[27] The records of some of the early northern New England textile companies attest to the force of the inelasticity of the labour supply and in one case show that one firm adjusted its manufacturing equipment as a direct result of the problem.[28]

In a rider to Habakkuk, von Tunzelmann has argued that cheap power was a major influence in American technical change because it spurred manufacturers to run their spindles faster and led to the development of the cap and ring spindles.[29] Certainly American manufacturers did not have to worry about water power costs until the 1820s and 1830s when, starting with the Lowell promotions, power was sold to each mill by carefully calculated units of power *produced* at the waterfall. But American manufacturers apparently did not have the instruments to measure the power requirements of individual machines, gearing or belting, until the late 1820s or early 1830s.[30] If power was rarely sold in units *consumed,* nor even measured exactly until the late 1820s, it is hard to explain how it could have acted as a positive spur to American manufacturers to innovate before that date—and all the major Waltham system innovations were in use by that time. The universality of cheap water power does not explain the marked differences between the technologies of the Waltham and Rhode Island systems of manufacturing.[31] Even within those systems manufacturers inconsistently took advantage of their cheap power supplies. Higher spindle speeds ought to have been accompanied by faster production lines, if von Tunzelmann is correct. Yet Montgomery specifically noted the slower speed of American carding machines: Oldham machines were much larger than and

[26] *TIR*, chapter 10.
[27] Ibid., chapter 8.
[28] Ibid., 108, 201.
[29] G.N. von Tunzelmann, *Steam Power and British Industrialization to 1860* (Oxford: Clarendon Press, 1978), 274, 276.
[30] Below, *CM* 2, "Notices of the Various Machines," Note 31 and "Miscellanies," Note 1.
[31] *TIR*, chapters 10 and 11.

driven at twice the speed of American cards.[32] And, contrary to von Tunzelmann's theory, one of the virtues of both the Waltham spindle and the Waltham dresser was that they needed less power than pre-existing alternatives[33]—which did not necessarily imply that they would therefore be driven faster.

A more reasonable case seems to be that relatively cheap water power, just like relatively cheap, high quality raw cottons, simply extended the range of optional techniques open to American manufacturers by allowing not only faster but also cruder machines, such as the double speeder or the powerloom at Waltham, to be developed.[34] But faster and cruder technology came in response to other sharper pressures imperfectly experienced between regions and districts in the USA before 1830. One came in product markets with the demand for hard, high twisted yarns used to make cotton goods for mass consumption; high twist yarns were spun with high speed spindles.because the spindle, rather than the drafting rollers or any other part of the spinning frame, inserted twist.[35] Another pressure, perhaps the fiercest, came from the labour market, as noted earlier. A more relatively unskilled workforce, by which the factory wage bill could be further reduced, could only be employed on a production line more mechanized than the lines existing in Britain. And in the Waltham system mills mechanisation reached its greatest extent before 1840 in order to allow this substitution of capital for labour.

Montgomery describes the results: large scale, vertically integrated manufacturing units making one or two products for mass markets by means of a standardised and minimally complex technology; flow production within the mill; the integration of machines (as with the lap spreader or the railway card); self-acting movements within machines (like the rotary temple or the cloth take-up motion in the powerloom); higher speeds (as in feeding cards or driving spindles); less thorough maintenance (as in card stripping); stop motions to halt a machine when the material being processed was broken, impeded or exhausted (as in a variety of machines;[36] the lap doubler, drawing frame, and warper stop motions received most attention by Montgomery). The relatively elaborate fire precautions taken by the northern New England mills,

[32] Below, *CM* 2, "Notices of the Various Machines."
[33] *TIR*, 191; below, *CM* 2, "Notice of the Various Machines."
[34] *TIR*, 98–99, 181–182, 186.
[35] Ibid., 182, 184, 186.
[36] Ibid., 198 and passim.

and described in detail by Montgomery, may also be seen as part of the concern to use a combination of unskilled labour and additional capital to cut overheads.

Montgomery's comparative estimates of these operating costs in Britain and America disclosed two different results, both favorable to the American manufacturer. In the first edition of his *Cotton Manufacture*, the estimates showed the American mill manufacturing coarse goods on net cost at 3 percent less than the British mill. When he came to revise his book, Montgomery narrowed the margin further, bringing it down to 1 percent.

If interest on capital is introduced at the levels suggested by one of the reviewers of the *Cotton Manufacture*, then the differential between British and American costs moves against the American manufacturer, as the following table demonstrates.

COSTS PER YARD OF GOODS IN COMPARABLE AMERICAN AND BRITISH MILLS PER FORTNIGHT
(based on *CM* 2 data)

	AMERICAN MILL (51,300 yards a fortnight)		BRITISH MILL (35,200 yards a fortnight)	
	cents	% of total	cents	% of total
LABOUR (L)				
Preparation	0.4883		0.27	
Spinning	0.43664		0.35272	
Weaving	1.43265		1.35710	
Other	0.31384		0.28909	
Total	2.67143	25.61	2.26891	23.01
CAPITAL (K)				
Interest on fixed K (6% in USA; 5% in GB)	0.46575		0.23256	
Insurance, depreciation power*	1.i3847		0.28897	
Total	1.60422	15.38	0.52153	5.29
COTTON				
Cost	5.544	53.15	5.544	56.23
Transport	0.6098	5.85	1.5246	15.46
Total	6.1538	59.0	7.0686	71.70
TOTAL COSTS	10.42945	100	9.85904	100

Differential in favour of the British manufacturer is now 0.57041 cents a yard or 5.47 percent of the British manufacturer's total costs.

*This is calculated as 1/26th of the total "On-Costs" given by JM for a year; division by the corresponding yardage yields the cost cited here. Transport Sc is 11 percent for the American manufacturer, 27.5 percent for the British one.

Both labour and capital costs formed higher proportions of total costs in America than in Britain. American manufacturers therefore faced continuing incentives to reduce both sorts of cost through technical or organisational change. Crude capital–labour ratios indicate a much greater application of capital per unit of labour $\frac{(K)}{(L)}$ in America (0.6) than in Britain (0.23), which reflects the American substitution of capital for labour. The major comparative advantage held by American manufacturers, as was well known, lay in their lower transportation costs.

Montgomery's results, like those reworked, were for costs per yard of goods and on this basis all three estimates based on Montgomery's data indicate a marginal difference between British and American coarse goods manufacturers' costs. Much more ominous for British producers was the far higher throughput of the American mill. Even after compensating for the shorter hours worked in the British mill, the American mill processed 27 percent more yarn by weight and 14 percent more cotton goods by square yardage, as the following calculation shows.

COMPARISON OF FORTNIGHTLY PRODUCTIONS IN A BRITISH AND AN AMERICAN COTTON MILL (COARSE GOODS) EACH RUNNING 128 POWERLOOMS, ca. 1840

I   *Cloth Output per Fortnight*

British mill: 35,200 yards of 35″ wide cloth; 2,000 warp ends of No. 16 yarn; 63 picks per inch of weft, No. 18 yarn.

American mill: 51,300 yards of 30″ wide cloth; 2,400 warp ends of No. 18 yarn; 62 picks per inch of weft, No. 18 yarn.

II   *Formulae for Calculating Weight of Cotton*

$$\text{Warp weight} = \frac{\text{No. of warp ends} \times \text{length of cloth}}{840 \times \text{yarn No.}} \text{ lb.}$$

Weft/filling weight
$$= \frac{\text{Picks per inch} \times 36 \times \text{length of cloth} \times \text{width}}{840 \times \text{yarn No.}} \text{ lb.}$$

III   *Weights of Cotton per Fortnight*

British Mill:

$$\text{Warp} = \frac{2,000 \times 35,200}{840 \times 16} = 5,238.09$$

$$\text{Weft} = \frac{63 \times 36 \times 35,200 \times 35}{840 \times 18 \times 36} = 5,133.33$$

$\left.\right\}$ 10,371.4 lb.

American mill:

$$\text{Warp} = \frac{2,400 \times 51,300}{840 \times 18} = 8,142.86$$

$$\text{Weft} = \frac{62 \times 36 \times 51,300 \times 30}{840 \times 18 \times 36} = 6,310.71$$

$\left.\right\}$ 14,453.57 lb.

IV   *Relative Efficiency of British and American Coarse Cotton Goods Factories, as measured by Weights of Cotton Yarn processed in a Fortnight.*

British mill processed   10,371.4 lb. of yarn
American mill processed   14,453.57 lb. of yarn

But American mills worked a 75½ hour week and British mills a 69 hour week, i.e. in two weeks the British mill ran 91.39 percent of the time of the American mill. Therefore 8.61 percent is deducted from the American production figure. Revised American production figure is 13,209.12 lb.

By this measure the American mill was 27 percent more efficient than the comparable British one.

V   *Relative efficiency as measured by Square Yardage of Cotton Goods produced in a Fortnight*

British mill produced $35,200 \times \dfrac{35}{36}$ sq. yds.

$= 34,222.2$ sq. yds.

American mill produced $51,300 \times \dfrac{30}{36}$ sq. yds.

$= 42,750$ sq. yds.

Revised American production figure (8.61 percent deducted) = 39,069.2 sq. yds.

By this measure the American mill was 14 percent more efficient than the comparable British one.

Source: Montgomery, *Practical Detail,* 123, 174, and author's annotated edition (Kress Library, Harvard Business School) 79.

Since the effect of longer working hours in the American mill has been removed from the comparison, American efficiency must be attributed to a faster production line, achieved by a combination of re-organised work methods, technical equipment, the use of more spinnable cottons, and access to cheap water power, on the supply side. British coarse goods manufacturers met this challenge by using an increasingly efficient steam engine to power production lines which were further mechanised, by self-actor mules and more automated powerlooms, for example, and then driven at higher speeds, as von Tunzelmann has shown.[37]

(c) Editions

Partly to remove "some little inaccuracies"[38] which crept into the first edition (and are identified in places in the explanatory notes to this edition), partly no doubt because of the proof-reading arrangements, partly in response to the Justitia controversy (for which see the second part of this edition), Montgomery prepared a second edition. He worked on it between 1840 and August 1843, when he wrote the last of his dated revisions. Perhaps he expected sales to justify a further edition, as had happened with his first book. But for reasons unknown the second edition was never published. It is not even certain that he completed his intended revisions. He did not, for example, add first hand accounts of the manufacturing districts south of New England though he had the opportunity if, as reported, he moved into New York state in 1843.[39] The revisions were preserved in an interleaved copy of the first edition which eventually came into the possession of the

---

[37] Von Tunzelmann, *Steam Power,* 209–225.
[38] JM, *Boston Courier,* 27 March 1841.
[39] Graniteville Manufacturing Company records.

Kress Library of Business and Economics in Harvard's Baker Library.

What then does the second edition of Montgomery's *Cotton Manufacture* add to the first? Besides the correction of those typographical errors which crept into the first edition, the changes comprised two sorts of substantive alteration. Firstly, of course, Montgomery reacted to the Justitia debate. He adjusted his estimates of water and steam power costs in the USA so that the case for water power looked stronger than before: his new figures ranged between $16 and $61 per horse power per annum for water and $71 and $94 for steam. Surprisingly perhaps he made no comment on the water turbine experiments then in progress in Philadelphia.[40] In his estimates of comparative British and American manufacturing costs he raised Scottish mill construction costs marginally. At the same time he lowered his prices of British machinery (now freely exported, after August 1843, following wide relaxations the previous year)[41] by about 8 percent. As a result of all his changes, Montgomery's final net difference in manufacturing costs between coarse goods mills in Britain and the USA narrowed from 3 to 1 percent in favour of the American manufacturer. Presumably also in response to the Nativist and personal attacks made on him in the Justitia debate, Montgomery toned down certain emphases which were likely to irritate his American readers. Frequently he removed the epithet from Great Britain, deleted the first person singular from his remarks, and checked any tendency to make negative or patronising remarks about American circumstances.

Secondly, Montgomery updated much of his technical and economic information. This kind of change is scattered through the book and can be identified through the textual footnotes. It spans comments about developments in equipment (like fly frames or self-acting mules) and manufacturing methods (like American failure to improve preparatory processing techniques) to updated commercial information about America's manufacturing districts (such as the Lowell statistics for 1843, American census of manufactures statistics for 1840, and notes on the location and size of new mills in the South).

---

[40] Louis C. Hunter, *A History of Industrial Power in the United States: I. Water-Power in the Century of the Steam Engine* (Charlottesville, Virginia: University of Virginia Press, 1979) 322–324.

[41] David J. Jeremy, "Damming the Flood: British Government Efforts to Check the Outflow of Technicians and Machinery, 1780–1843," *Business History Review* LI (1977) 32.

A final note in defence of printing the second edition seems due, not least because the flood of British and American reprints of the 1960s and 1970s saw the first edition of Montgomery's *Cotton Manufacture* reprinted at least twice. The primary justification for publishing this second edition derives from the significance of the book. It is the single most important contemporary source for the technology and economics of American cotton manufacturing prior to 1850. The author's revisions therefore command commensurate attention. Rather than publish a long list of revisions in a scholarly journal, it was decided to produce the kind of edition Montgomery had in mind, indicating through textual footnotes divergencies from the first edition. The interposition of the Justitia controversy between the two editions also suggested an appropriate opportunity to republish the major documents in that debate, together with an account and discussion of the controversy.

It may be noted at this point that the first edition of the *Cotton Manufacture* was rapidly translated into German and appeared as *Die Baumwollen Manufaktur der Vereinigten Staaten von Nordamerika,* translation by Friedrich Georg Wieck (Leipzig: Robert Binder, 1841. Pp. 88 and three plates of illustrations). Wieck, an engineer of Chemnitz, the long-established centre of coarse calico manufacture in Saxony, made comparisons, in footnotes and in some of the tables, with German cotton manufacturing experience. Of particular note are the set of German manufacturing costs on pp. 41–48 of the German edition and the list of Chemnitz cotton machine prices on p. 88.

(d) Editorial methods

Many changes were contemplated by Montgomery as he revised his *Cotton Manufacture* for a second edition. Some were stylistic, some substantive. Where no alterations in meaning, either in factual information or in emphasis, appear, the differences between the two editions have been ignored. Rare instances of material transposed from one page to another have likewise been passed over. This present edition thus closely proximates to Montgomery's intended second edition; his spelling and typographical slips have been silently corrected, though not his punctuation (with its addiction to the comma).

(i) Textual notes

All changes between the two editions deemed worthy of notice on the foregoing principles have been indicated by alphabetically ordered superscript letters, commencing with each new chapter: a convention chosen because it seemed minimally disturbing to the reader's eye.[42]

Omissions from the first edition
One superscript letter, e.g. [a], signals an omission. The textual note to that letter provides both the location and the missing word or passage, intact or summarised. Summaries, printed in italics, were preferred if they saved space without losing any changes in information or emphasis. Several omissions from one revised passage are indicated by a shared superscript letter and sequential roman numerals, e.g. [aI, aII, aIII] & c.

Additions and alterations made to the first edition
Words or passages placed between two identical superscript letters, e.g. [a] nineteen thousand throstle spindles[a], mark interpolations or substitutions made by Montgomery. Where a long passage in the first edition received a mixture of substantive and stylistic changes, the limits have been indicated by two identical superscript letters, as before, and substantive changes within the passage identified by identical superscript letters and sequential roman numerals, e.g., [a] contained [aI] twelve [aI] mills and a total of [aII] nineteen [aII] churches, all of which are agreeably situated.[a]

Locations in the first edition
These are given in the textual note by citing the page and line number where they occur in the first edition. Thus 82, 24 means page 82, line 24; 82, 24–84, 2, refers to the passage between page 82, line 24 and page 84, line 2. Where an emendation occupies two adjacent lines only, the location of the first of the lines only is given. In counting lines in the first edition, title lines, but not the running head, have been regarded as lines of the text.

---

[42] I am very grateful to Kenneth E. Carpenter, Research and Publications Librarian of the Harvard University Library, for suggesting this method to me. It was directly derived from R. H. Campbell, A. S. Skinner, and W. B. Todd (eds.), *Adam Smith's An Inquiry into the Nature and Causes of the Wealth of Nations* (2 vols., Oxford: Clarendon Press, 1976).

Deletions of revisions
These are placed between % marks, as are replacements which were subsequently abandoned or deleted.

Manuscript illegibilities and lacunae
Illegible words or an occasional word missing from Montgomery's manuscript are indicated by square brackets [............]. Words appearing in these square brackets are the editor's insertions, designed to lend sense to the revision.
Only two sections appear, from internal evidence, to be missing from Montgomery's manuscript as he left it: one on the history of Waltham, the other from the section on power costs. This and any physical damage to the manuscript are noted in upper case type.

(ii) Explanatory notes

These are signalled in the traditional manner with superscript arabic numbers. While it would be possible to write very lengthy explanatory notes to Montgomery's text, I have confined mine primarily to the technical aspects of cotton manufacturing which, with estimates of costs, form the most original portions of the book. Even in this I have restricted the notes to the following objectives:

1. To provide a full citation for books or other sources cited by Montgomery.
2. To identify persons, firms, mills, machines, or patents.
3. To comment on general technical or economic statements known by this editor to be questionable.
4. To give meanings for archaic words.

A

PRACTICAL DETAIL

OF THE

COTTON MANUFACTURE

OF THE

UNITED STATES OF AMERICA;

AND THE STATE OF THE

COTTON MANUFACTURE OF THAT COUNTRY
CONTRASTED AND COMPARED

WITH THAT OF

GREAT BRITAIN;

WITH

COMPARATIVE ESTIMATES

OF THE COST OF MANUFACTURING IN BOTH COUNTRIES

ILLUSTRATED BY
APPROPRIATE ENGRAVINGS.

ALSO,

A brief Historical Sketch of the Rise and Progress of the Cotton Manufacture in America, and Statistical Notices of various Manufacturing
Districts in the United States.

BY JAMES MONTGOMERY,

SUPERINTENDENT, YORK FACTORIES, SACO, STATE OF MAINE;
AUTHOR OF "THE THEORY AND PRACTICE OF COTTON SPINNING," AND
"THE COTTON SPINNER'S MANUAL."

GLASGOW:

JOHN NIVEN, JUN., 158, TRONGATE;
WHITTAKER & CO., LONDON; J. & J. THOMSON, MANCHESTER;
OLIVER & BOYD, EDINBURGH; D. APPLETON & CO., NEW YORK.

MDCCCXL.

Preface

The writer of the following Details, upon leaving Scotland in the beginning of 1836, was[a] requested[a] by his friends to communicate some account of the practical state of the Cotton Manufacture of the United States, so far as it might fall under his observation. In complying with that request, he found that[b] general statements would not suffice, and was obliged, in order to fulfil his promise, to enter somewhat into details. In doing so, his materials accumulated to an extent greatly above what he originally anticipated;—they at least became too bulky for mere epistolary correspondence. Being also aware of the interest felt by the British regarding every thing connected with America, and, knowing the vague opinions which prevail regarding the practical state of the Cotton Manufacture[c] he was led to believe that there might be something in these details not altogether uninteresting to many employed in the Cotton Manufacture in Great Britain. Under these impressions he has been induced to lay them before the public, especially as the most contradictory reports have been circulated in that country, by many who have visited America.

[d]It has been the chief study of the Author to collect such information as might be most interesting to his friends and to state facts without depreciation or exaggeration on either side. He is not conscious of having made a single statement that may not be implicitly relied on. Each machine employed in the cotton manufacture has been noticed in its order, and any difference between them and those with which he[d] was acquainted in Great Britain, has been described. The practice of this country is contrasted with that of Great Britain, and the advantages or disadvantages of either pointed out. Drawings are given of some of the most important machines which are constructed in a different form from those employed for the same purposes in Britain.

[e]The speeds at which the machines are generally driven has

49

been stated[e]; the amount of work produced, the number of hands employed, the hours of labour, and the ordinary rates of wages; so that a fair estimate may be formed of the actual difference of the cost of manufacturing in the two countries. These, it is believed, will be interesting to those employed in the manufacture; while, it is hoped, that the Historical Sketch of the introduction of the Cotton Manufacture into the United States, and the Statistical Notices of various Manufacturing Districts, will be equally interesting to the general reader.

To ensure correctness, the various statements have been submitted to the inspection of several gentlemen in both countries, in whose judgement, experience, and practical knowledge of the Cotton Manufacture in all its details, the author has the utmost confidence.

Such is the origin and design of the present work; and the object of the author will be fully attained, if it shall in any manner contribute to disabuse the public mind, and assist manufacturers in Great Britain to understand correctly, the present position of both countries with regard to the Cotton Manufacture. It must be confessed that the most formidable rivals with whom the British have to compete in this important manufacture, are the Americans; their immense water power, together with their being the growers of the raw material, giving them advantages which no other nation possesses. So long as the British can manufacture cheaper than the Americans, just so long will they retain a monopoly of the trade. But every step the latter advance in reducing their expenditure, the nearer do they approach to an equality with the former. The Factories in Great Britain are already conducted in general with the most rigid economy, so that their only chance now, is improvements in their machinery, by means of which the processes may be expedited, and the cost of manufacturing reduced. But the Americans may also make improvements on their machinery, so as to derive similar results; and certainly they have more resources to which they can apply. By committing the charge of their Factories to competent persons, immense savings might yet be effected, of which they seem in general not to be aware. These savings too, might be effected without waiting new inventions or improvements in machinery, or resorting to a reduction of wages. These, together with the reasons already stated, operated as inducements to lay the following details before the public, most of which have been written at different times, and in the midst of other engagements, which is the only apology the author has to offer for their many imperfections.

CONTENTS

LIST OF PLATES.

WOOD ENGRAVINGS.

THE COTTON MANUFACTURE OF THE UNITED STATES CONTRASTED AND COMPARED WITH THAT OF GREAT BRITAIN.

PLAN AND ARRANGEMENT OF THE MILLS.

The Cotton Factories in America are scattered over a vast extent of territory. But there are three particular divisions which may be denominated the principal manufacturing districts; the first of which is the Eastern, comprehending Maine, New Hampshire, Vermont, and the Eastern parts of Massachusetts. The second, or middle district, includes the Western parts of Massachusetts, Rhode Island, and Connecticut. The third, New York, New Jersey, Pennsylvania, &c. &c. In the above districts the principal manufacturing towns and villages are,—in the first, Lowell, which is decidedly the largest and most important in the United States;[a] also Waltham, Taunton, Fall River, and Newburyport, all in Massachusetts. Dover, Great Falls, New Market, Nashua, and Amoskeag in New Hampshire,[a] and Saco in Maine. There are besides, a number of insulated Factories that do not require particular notice. All the manufacturing establishments in this district belong to joint stock companies, and, in general, they follow the Lowell plans in the form and arrangement of the Mills, as well as in the style of their machinery.

The principal manufacturing towns and villages, in the second or middle district, are—Providence and the vicinity, which, within a circuit of 30 miles, may comprehend from 70 to 80 Mills, including Pawtucket, Smithfield,[b] Blackstone, Woonsocket,[b] Lonsdale, Coventry, Cumberland, Cranston, Warwick, Scituate, Johnston, &c. together with Newport. In Connecticut there are Greenville, Cabotsville, Williamantic, Norwich, Jewitts City, &c.

Many of the Cotton Factories in this district belong to corporations, but the greater number are the property of private companies or individuals. As the cotton manufacture first commenced
here a vast portion of the machinery is old, and exhibits all the
different stages of improvement. But the best and newest Mills
being in or near Providence, the others generally copy their plans
and style of machinery.

The principal manufacturing towns and villages, in the third
or Southern district, are the towns of Paterson in New Jersey,
(which is next to Lowell in regard to the number of ᶜfactoriesᶜ;
Matteawan, New York; Manayunk near Philadelphia; Baltimore,
&c. &c. The Factories in this district generally adopt the plans
and improvements of Paterson and Matteawan, and these latter
obtain machines or models of all the newest improvements from
Manchester and Glasgow, which they put in operation in this
country. Their style of machinery is therefore a little different
from that of the other two districts. The Rhode Island machinery
also variesᵈ from that of Lowell or the Eastern district.

The principal Machine Manufactories are at Lowell, Massachusetts; Providence and Pawtucket, Rhode Island; Paterson, New
Jersey; Matteawan, New York.

The plan of the Mills is nearly the same in the different districts.
None that I am aware of exceed five stories in height, except two
at Dover, which are six stories on one side and five on the other.
The general height of the Mills in this country is three or four
stories with an attic.—*See Plate I. Fig. 2d.*—But the Mills recently
built at Lowell are five stories high, with a plain roof, such as Fig.
1st; from which it seems probable, that the double roofᵉ will here-
after be abandoned, as it is certainly the most expensive, nor does
it give so much room for machinery as the other.[1]

The general height of Cotton Mills in Scotland is six stories with
a plain roof. Those in England are from six to eight stories high;
Stirling and Beckton's Mill, lower Mosely Street, Manchester, is
nine stories.[2]

There are a few Mills in this country driven by high pressure
steam engines. There are four in Newport, and one in Providence,
Rhode Island; and three in Newburyport, Massachusetts.[3] The
coals used, whether anthracite or bituminous, cost fromᶠ five to
seven dollarsᶠ per ton.ᵍ The Mills throughout the United States
are generally driven by water and in consequence of the severity
of the cold in winter which sometimes descends to 30 degrees
below zero the water wheels require to be under cover and kept

in an atmosphere[gI] above the freezing point. For this reason they are generally placed in the basement storey which besides the water wheels may contain the mechanics shop and Cloth room or be filled with other machinery.[gII] All the large establishments have their cloth room, counting house and the mechanics shop in separate buildings.[gII]

The second flat[4] contains the carding engines, &c; the third the spinning; and the fourth and attic, the weaving,[gIII] warping,[gIII] dressing, &c. Such is the general arrangement of the Mills, in the eastern district. And here it is proper to remark, that the spinning is[gIV] all done by throstle frames,[gV] both warp, and weft.[gV] But in the Middle and southern districts, Mules are generally employed for spinning the weft; and in these places, the arrangement may be different, as the Mules are frequently placed in the upper stories.

Many of the Cotton Factories in Britain, have their picking or scutching rooms within the mill; but in this country they are separate buildings, erected for the purpose and standing[gVI] about 20 or 30 feet from the main building the connecting passages being secured by iron doors to prevent the communication of fire to the loose cotton in the picking-house.[g]

[h]Some of the Mills lately built at Lowell, have iron shutters outside the windows, to prevent the communication of fire from one to another; and each Mill has expensive apparatus fitted up for extinguishing fires, such as forcing pumps for raising water to a cistern at the top of the Mill, from which pipes descend into every apartment; and these not only serve to deluge the Mill in case of fire, but to supply each room with water for washing, as every apartment has its water trough, or what is denominated a sink, for the workers to wash their hands and face in; a most healthy, as well as cleanly operation, which is punctually attended to before every meal, soap being supplied for this purpose by the proprietors.[i] Besides the water pipes and forcing pumps in the inside, many of the large factories have horizontal platforms on the outside of both ends of the building; which are so arranged, that one is on a line with each floor, and connected with a perpendicular ladder, ascending from the ground to the top flat.[iI] Every Mill in the eastern district is likewise furnished with a watch clock, for the purpose of keeping the night watchman on the alert. These clocks resemble a common time piece, with a circular revolving dial: And surrounding the dial, about half an inch from the perimeter; there are a number of small pins which the watch-

man is required to shift but the clock is so constructed that only one pin can be shifted at certain intervals of time.[iii] It also contains a number of springs each of which must be lifted before one pin can be shifted but as the clock is all enclosed except the dial, it is necessary to have wires connected with the springs; these wires are likewise all enclosed in a wooden case except at their extremities in the different apartments so that before any one pin can be shifted the watchman requires to go into every room in the[i] Mill, for the purpose of pulling each wire separately, and this he must do at the end of every half hour; for if the pins are not shifted at the proper time, they cannot be shifted at all: and the superintendent of the works carefully examines these clocks every day to ascertain whether all the pins have been shifted; by which means he knows perfectly when the watchman neglects his duty. Some clocks are so constructed, that one wire only can be drawn at the end of every five or six minutes; so that when the watchman draws the wire in one room, he must wait some time before he can draw the next. This keeps him continually travelling from one room to another.

It is somewhat remarkable, that, in general, no such provision is made in the Cotton Factories of Great Britain for the prevention of fire. Except in a few instances, there are in that country neither forcing pumps and water pipes inside, nor platforms or ladders outside the Mills. Indeed there are a number of Mills in country places in Scotland that have no night watchman either in summer or winter.[5]

The method of conveying motion from the first moving power to the different departments in the Factories of Great Britain, is by means of shafts and geared wheels; but in this country it is done by large belts moving at a rapid speed, the breadth of which is 9, 12, or 15 inches, according to the weight they have to drive, and passing through a space of from 2500 to 3600 feet per minute. A belt 15 inches broad, moving at the rate of 3000 feet per minute, is considered capable of exerting a propelling force equal to 50 horses' power. All the recently built Mills are belted, whilst many of the older ones have had the shafts and gears removed, and belts substituted in their place; indeed, belts are generally preferred even by those who have had sufficient experience of both.[6] There are various opinions regarding the best plan of fitting up the drums and shafts so as to apply belting to most advantage. Plate I. represents two different plans; a greater number might have been given, but these, it is presumed, will be sufficient for our present purpose.

Plate I. Fig. 2d. [Fig. 7] represents a plan of driving the whole machinery by one large belt. B is the basement story or wheel room; C is the carding; S the spinning; and W W the weaving rooms. A is the water wheel; D the main drum driven by and geared from the water wheel, and is generally from eight to twelve feet in diameter; *e e e e* represent the lines of drums and shafts in the carding room, which also drive the spinning frames by means of belts passing up through the floor; *i i i i i* represent the lines of drums in the first weaving room, from which motion is conveyed to the second by belts passing up through the floor. The dotted line represents the main driving belt, which gives motion to all the drums; *a a a a* are the belt binders, which are suspended on springs or swivels, so as to bind or take up the slack on the belt, and keep it always at a proper degree of tension. The large belt, as here represented, would be between three and four hundred feet long, from twelve to fifteen inches broad, and would require from 600 to 700 lbs. of good belt leather to make it. Such belts are always made from the centre of the back of the hide, so that they stretch equally at both sides.

Belts upon this plan are bulky, ponderous, and unmanageable; and when they break, (an accident to which they are very liable on account of the great weight they have to drive,) run off the drums, and cause a hinderance to the whole work; besides, it takes five or six men nearly half a day to prepare them again for operation. They cause a great strain upon the journals of the shafts, thereby increasing the power required to operate the Mill;[j] the journals also heat and wear beyond the power of any lubrication to prevent, hence this plan of belting is not[k] approved of.[k]

Plate I. Fig. 3d. represents another plan of gearing with belts, which is generally adopted, about Lowell, and considered the most unobjectionable of any that has been tried; B is the basement story; C the carding; S the spinning; and W W the weaving rooms; E E the water wheels, F the main driving drum in carding room; H the main drum in first weaving room; A A A A the lines of main shafts in carding and weaving rooms; I I the two main driving belts; G the second driving belt. There are binders at I I, which exert a pressure of about 30 or 40 lbs. against each belt, so as to keep them always at a proper degree of tension.

Two such belts from 12 to 15 inches broad, are capable of

PLANS of DRIVING MAIN SHAFTS

IN COTTON FACTORIES

WITH BELTS INSTEAD of SHAFTS & GEARS.

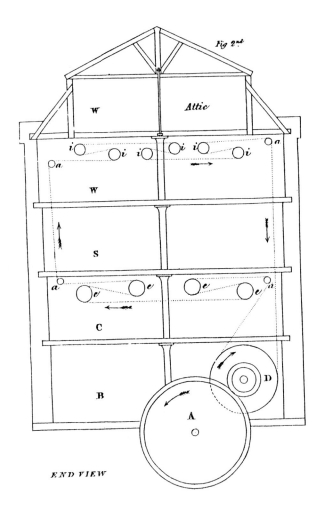

Fig. 7. Belt transmission system in American cotton factories, plate I, *CM*1.

58

Fig 1.

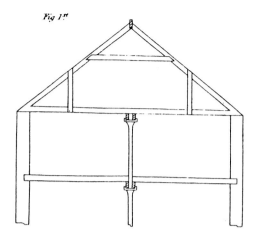

Fig 3.

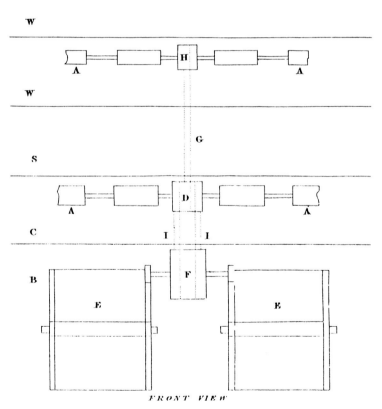

FRONT VIEW

Glasgow Published by John Niven, Jun.   *Swansea*

59

operating[l] 6000[l] throstle spindles with the necessary preparation and weaving for coarse heavy goods, which require a propelling force equal to between 80 and 90 horses' power. The spinning frames being constructed on the plan of what is called the dead spindle (known in Scotland as the Glasgow patent throstle)[7] they require much more power to operate them than the common throstles.[m]

The two plans of gearing with belts, represented in Plate I., will be interesting, the one from its novelty, the other from its being considered the best now in operation. However partial manufacturers in this country may be to this mode of conveying motion to the different apartments, those who have been accustomed to the neat manner in which Factories are geared in Great Britain, must regard the above as heavy, clumsy, and inconvenient, as well as more expensive. As all these large belts have to be enclosed, they occupy a large portion of the rooms they pass through; which, besides interrupting the view, gives less space for arranging the machinery; they are likewise very liable to stretch, and when too slack, they will slip on the drums: and owing to their breadth, it requires a considerable time to cut out one joining and sew it up again; but to prevent them from slipping, they are generally well soaked with currier or neat's foot oil, or the following composition which is much recommended, viz, two pounds of common tallow, one of bayberry tallow, and one pound of bees' wax: these are melted until they are completely incorporated; and, while boiling, applied to both sides of the belt with a brush; and in order to make the composition strike into the heart of the leather, the belt is then drawn slowly over a hot furnace, by which means the wax is completely decomposed, and with the tallow, penetrates every pore of the leather, until the whole belt is fully saturated. After being prepared in this manner, nothing more is required, than to lay on a thin layer of the same composition, at the end of every five or six months. The drums are also covered with leather, prepared in the same manner, and fastened on with wooden pegs, such as shoemakers use for fixing the soles on boots or shoes.

Though the Mills in this country are not so high as those in Britain, they are generally very strong and durable. Instead of joists for supporting the floors, there are large beams about 14 inches by 12, extending across from side to side, having each end fastened to the side wall by a bolt and wall plate: these beams are between five and six feet apart, and supported in the centre by wooden pillars, with a double floor above, consisting of planks three inches thick and covered with one inch board. Some have the planks dressed on the under side, others have them lathed and plastered: the floor being in all four inches thick, is very strong and stiff. The average thickness of the side walls may be from twenty to twenty-four inches and they are generally built of bricks. There are very few stone walls, free stone being scarce in this country.

The preceding embrace the principal things which a stranger, on visiting the Cotton Factories of the United States, is most likely to notice as differing from the general plan of those in Great Britain. It will be observed, that the only particulars in which those of the former differ from the latter country are, *first*, the height of the Factories; *second*, the double roof is peculiar to the American Factories; *third*, belts are employed for conveying motion to the various apartments; *fourth*, the floors are laid on large beams instead of joists; and *fifth*, the arrangement of the machinery.

In Great Britain, the weaving is generally in the lower stories, and the carding and spinning above; but in the States, the weaving is contained in the upper stories, with the carding and spinning below. Instead of large beams laid across the house for supporting the floors, the Factories in Britain have joists about three inches by ten; these are laid on their edges about twenty inches apart, with one inch flooring above, lathed and plastered beneath, or sheathed with thin boards. The joists are also supported in the centre by a beam about eleven inches by six, running from end to end of the building; the pillars are of cast iron, and placed right under this beam: the beam does not rest on the pillar, but on a cast iron case, which passes up on each side of the beam, and meets together above; so that, whilst the under part of this case rests on the top of the pillar below, the upper part supports the pillar above; thus leaving the beam entirely free of the pillars in the rooms above; by which means the uppermost floors are supported on columns of cast iron from the foundation; consequently there is no danger of such floors sinking in the centre. But in this country where the cross beams rest on the top of the pillars, and the pillars above rest again upon the beams, the floors in the upper stories sink down in the centre, in consequence of the shrinking of the timbers, and the pressure of the ends of the pillars into the beams."

The Factories in Great Britain are generally moved by steam engines placed outside of the main building, from which motion is conveyed to the different apartments by shafts and geared wheels. Some of the factories in this country have the water wheels outside of the main building, but generally all the new Mills use belts for gearing; but whether belts require more or less power than the other mode of gearing, the writer has not been able to ascertain satisfactorily.° I am inclined to believed that if shafts and geared wheels are fitted up on the newest and most improved principles, they will be both cheaper and run lighter than belts. Some object to the grating noise of wheels, but this in Great Britain is no objection at all, as the wheels are mostly outside the building; besides, geared wheels, when properly made and fitted up, will run as silently and smoothly as belts.[8]

NOTICES OF THE VARIOUS MACHINES

THE WILLOW.

The Willows[a] that were formerly used in this country were[a] generally made in the form of a cone enclosed within a concentric case with one row of spikes on each side of the case, and four rows on the cone, placed at right angles to each other. The cotton is put in with the hand, by an opening right above the smaller end of the cone, and carried rapidly round until it is thrown out at the larger, by the centrifugal force.

The self-acting conical Willow, made by Mr. Lillie of Manchester,[1] has been introduced into this country, and put in operation at Matteawan, New York; and Fall River, Rhode Island; but is not likely to be generally adopted; it altogether appears to be a heavy, clumsy machine.[b]

A new machine called a whipper has lately been introduced which the writer considers the best of the kind he has yet known either [in] Britain or America. *See Plate II. Fig. 1st and 2d.* [Fig. 8]—A B are two parallel shafts about $2\frac{1}{2}$ inches diameter; *a a a a*, &c. are arms, or spikes, about six inches long, and fastened into the shafts. The shaft A is surrounded with a gird or harp from *c* to *c*, and the shaft B has a harp from *e* to *e*. The gird has several bars containing spikes pointed inwards; see *s s s s s*. The front of the machine is open from *b* to *b*; all the other parts of it are enclosed, except a small opening above, represented by the dotted lines *n n*: this opening is about $2\frac{1}{4}$ inches, extending across the top, by which the cotton is introduced, when the revolving arms

PLATE II.

MASON'S WHIPPER.

Fig 2ᵈ

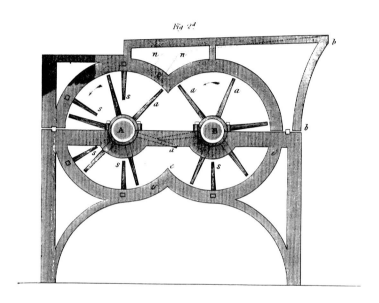

Fig 1ˢᵗ

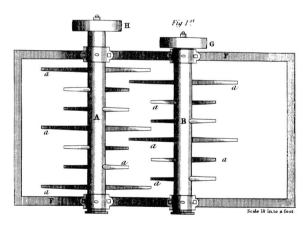

Scale 1½ in.to a foot.

Glasgow Published by John Niven Junʳ   Swan Sc.

Fig. 8.   Mason's whipper, plate II from *CM*1.

of shaft A immediately take hold of it and carries it rapidly round, by which means it is agitated and torn against the spikes *s s s;* but as it proceeds round with the arms of the shaft A, it is met by the arms of the shaft B, which clear it off, and throw it out by the mouth *b b.* The belt pullies G H are of different diameters, so as to make the shaft B revolve faster than A, by which means it has more power, and frees itself more perfectly of the cotton that becomes entangled between the arms of the revolving shafts. The diameter of the pulley G is six, and H seven inches, or the driving drum may be of different diameters, to effect the same variation in the speed of the shafts. The speed of the shaft B ought to be 1800 revolutions per minute, and A 1600; and as the shaft A has to carry round a[c] greater weight[c] of cotton, it is generally stronger than B.

As the chief use of the willow is to open or separate the clotted tufts of cotton, so as to make it spread at the following machine; the tearing process it must pass through to accomplish this is very liable to injure and break the tender staples; therefore every machine that has been employed for this purpose, is liable to many objections. The writer has been acquainted with almost all the different machines that have been in *general* use for the last thirty years, and he considered the whipper, represented in Plate II. as decidedly the best which he has seen. It is called Mason's Whipper, from the name of the inventor,[d] Samuel P. Mason Esq. of Newport R.I.,[d] and though of small dimensions, being only three feet high, and two and a half feet broad, it is capable of willowing one bale of upwards of 400 lbs. in an hour and a half. It occupies little room; is easily managed and kept in order, and costs 75 dollars = £15.15.6.[2]

SCUTCHING AND SPREADING MACHINE.

In the Cotton Factories of Great Britain, the above are generally two separate machines; but in this country they are combined into one, denominated the lap spreader. In any of the British Factories where the two are combined into one machine, they have generally four or five beaters or scutchers; but here they have only one, two, or at most three.[e] In the preparation department of the cotton manufacture there are three most essential processes which are generally carried to greater perfection in the British than in the

American factories. These are the MIXING, CLEANING, and CARDING the cotton. The first is done previous to its being willowed where a number of bales are opened and the cotton spread out layer above layer and trampled down firm and compact. This heap or mixture is called a "bing" and may contain from 20 to 40 bales. The utility of this mixing will be obvious to every practical manufacturer.[cl] As in every lot of cotton there are a great variety in the qualities of the different bales. But when a number are thus intermixed so as to incorporate these various qualities a large quantity of cotton is thereby obtained of an equal and uniform quality.

When this cotton is to be used it is pulled out from the side of the bing, equally from top, to bottom, so that a portion of all the different bales may be used at once.

To select such cottons as are best suited for making a given quality of goods and especially such as will unite or incorporate together so as to make superior yarn requires on the part of the British Manufacturer a degree of skill and practical experience that is altogether unknown to the American. The manager of a British factory carefully examines every bale that is carried into the picking house and makes it his particular study to have as perfect uniformity as is practicable in the quarters of each successive bing. Every mixture must if possible be equal to the preceding unless some variation in the quality of goods is intended. By a steady uniform equality in the cottons there will be the least possible variation in the sizes of the yarn: and the nearer we can attain to a perfect uniformity in the quality of the yarn there will be less waste made: a much larger quantity of work produced: and fewer imperfections in the goods.

But in the Americans factories so far as I have observed there is no selection whatever. The cotton used may perchance be New Orleans or Uplands. It is piled up in the cotton store just as it is received and afterwards carried into the picking house in the same order in which it was piled. And there being only a few bales given out at one time it often happens that the bales used one day are inferior to those of the preceding, this cannot fail to cause a variation in the yarn: imperfections in the goods: and a deficiency in the quality of work. Such management may pass in the manufacturing of coarse heavy goods from superior cottons but for fine goods or coarse goods of the best quality a very different system must be pursued.

It is therefore of the utmost importance to have the cotton

mixed in large quantities, the larger the better as without this it is impossible to keep the quality of the yarn regular and uniform. And in order to do so there must be sufficient space in the picking houses. But in the American factories generally there is barely room for mixing from five to eight bales at one time. Manufacturers however are now fully aware that this is an error. Some have already enlarged their picking houses. And the new mills that have been lately erected have these departments built of such dimensions as to afford every facility for mixing, assorting and cleaning the cotton.[3]

To have the cotton thoroughly cleaned, is of equal if not more importance to its being properly mixed. To effect this the scutching process is chiefly designed viz. to clean the cotton from sand, seeds and other vegetable substances as well as to open out the tufts more perfectly than can be done at the willow and thereby make it spread into a more uniform lap for the cards.

A large portion of the American manufacturers have no steam pipes nor other means of heating the picking houses which is certainly a serious error. Steam is as necessary in this as in any other apartment of the works not so much for the purpose of drying the cotton as to equalise the temperature of the room. It is well known that cotton wool has a great affinity for dampness and during wet weather it imbibes a large quantity of moisture which easily evaporates in dry weather especially if accompanied by a light drying wind. And as cotton that is full of moisture weighs heavier than when it is very dry; if the weights at the spreading machine are kept always the same, the variations in the temperature, by affecting the weight of the raw cotton, cannot fail to produce variations in the sizes of the yarn—a consequence which is always injurious to the work and productive of serious derangements in the various processes.—Steam heat in the picking houses is especially required to absorb the superabundant moisture in wet weather and to render the apartment sufficiently comfortable for the hands employed there during the winter.

Some manufacturers have supposed that steam pipes or hot air ought to be introduced into the picking houses for the purpose of drying the cotton but such an idea is absurd. If the cotton has been wetted by rain or otherwise it certainly ought to be dried before using it but the best method of doing so is by exposing it to the heat of the sun or to a cold drying wind in the open air. There is a moist glutinous matter however which is naturally inherent in the cotton and ought never to be dried out because when cotton is divested of its natural moisture it becomes weak, short, and brittle. It adheres to the leather rollers, breaks down and makes waste in every department of the process. Practical Manufacturers know that cotton always works better when it is a little moist and providing the sizes of the yarn could be kept uniform no evil whatever would arise from any degree of humidity which the cotton [illegible line] much more liable to injury from being too dry than too moist. In very droughty weather it is frequently necessary to sprinkle the floors in the picking house and carding rooms with water—in consequence of the dry husky nature of the cotton—sometimes a quantity of steam is allowed to escape and fill those apartments with a moist humid atmosphere to supply the cotton with an equal quantity of moisture to what may have been absorbed by the drought.

An equal and uniform temperature being the great desideratum for the picking house, whatever means can be adopted to supply this, will effect the greatest benefit to the cotton. If a communication could be formed between the picking house and the flumes or water race-way and constructed so as to throw in a stream of humid air among the loose cotton when necessary, particularly in very dry weather it would be far more beneficial to the work than steam in wet weather. But perhaps the best means of regulating the temperature would be steam heat for drying the superabundant moisture when the weather is wet and a communication with the water race-way for supplying the lack of moisture when it is very dry.[e] When the cotton is uniform in quality, well opened and cleaned previous to its being put through the cards, the latter process will be more perfect. The great object to be principally attended to in spinning cotton yarn is, to make it equal in its grist, clean, and smooth; but that can never be attained without a proper system of mixing, cleaning, and carding. In these three processes are to be found the principal deficiencies of the Cotton Factories of America, as compared with those of Great Britain.

Mr. Whitings of South North Bridge, Massachusetts, is supposed to make the best lap spreaders, or scutching machines now made in this country.[f] [g]They are the same in principle as those generally used in Britain. The only difference is, that in Whitings' machine, the cotton is carried forward from one beater to another by means of wire cylinders, instead of revolving aprons. The cylinders are made with iron heads (like the heads of a bobbin) between two and three inches broad. These iron heads constitute an improvement for which Mr. Whitings has taken out a patent.[4]

Mr. Leonard of Matteawan, Pitcher and Brown of Pawtucket make spreading machines equally as good as Mr. Whitings' without the patent head on the wire cylinders.[5]

There is a material defect in all these machines which cannot be too soon remedied. That is, the space from beater to beater is not sufficient to allow the cotton to clear the beaters perfectly. Whenever the cotton does not clear the beaters freely it is liable to be carried round with the revolving arms: a very common defect in many picking machines, and one which is most injurious to the cotton, as it strings it, and beats it into naps, and thereby puts it in such a state that the cards cannot tease it out so perfectly as to make good work. If the above machines were lengthened out, and sufficient space allowed for the cotton to clear the revolving arms of the beaters, and expand freely before it is carried through the second feeding, or the calender, rollers, it would greatly improve the work and add to the merits of the machines.

Some of the most improved spreading machines have lately been imported from England. One known by the name of "Hetherington's spreading machine" has been put in successful operation in Lowell and some other places.[6] Mr. Whitings has likewise adopted Hetherington's principle and makes his new machines upon a similar construction, some of which that have been put in operation make superior work and give complete satisfaction to those who have introduced them. Indeed the whole process of picking, mixing, and cleaning the cotton, has undergone an entire change in this country within a few years back, which, with the enlarged picking houses, will doubtless bring up this department of the business to as high a state of improvement as in the cotton factories of Great Britain.[g]

CARDING ENGINES.

That the remarks upon this head may be as explicit as possible, it has been deemed necessary to give a sketch of the different forms of Carding Engines now in *general* use. *(See Plate III.)* Fig. 1st [Fig. 9] is a common carding engine with a licker-in A, and 12 tops or flats B B. Fig. 2d has 13 tops, but no licker-in. These two represent the general form of carding engines in this country. Fig. 1st is the breaker, and Fig. 2d the finisher. The former is generally furnished with a licker-in, but not the latter. There are but few Mills in the United States that use single carding; mostly

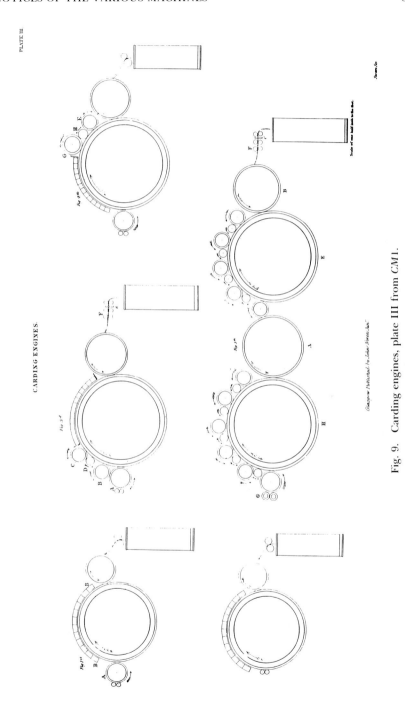

Fig. 9. Carding engines, plate III from *CM* 1.

all have breakers and finishers, even those that manufacture the coarsest goods. In the Eastern district the general breadth of the cards is 37 inches, and diameter of main cylinder 36, doffing cylinder 13, feeding rollers 1¼ inches. The main cylinders are made of cast iron, and covered with broad filleting instead of sheets.[h]

In the middle, or Rhode Island district, the cards are made of various breadths, from 18 to 36 inches, and are mostly wooden cylinders covered with sheets: indeed, sheets are generally recommended even by those who have used fillets for the main cylinder.

[i]In the southern district the general breadth of cards is [ii]30 inches[ii] and in each district the average speed of the main cylinder is about 110 revolutions per minute.

I have neither seen nor learned of any carding engines in this country that are driven so fast or make work equal to those used in England. With single carding the English manufacturers produce superior work to any double carding I have seen here.[i]

Fig. 2d represents the general form of carding engines used in Scotland, and for very fine numbers in England. In the former the breadth of cylinder is 24 inches, and diameter 36; the average speed of cylinder 120 revolutions per minute. When used to card for fine yarns, either in Scotland or England, the breadth of cylinder is 18 inches, and speed from 90 to 110 revolutions per minute.

Fig. 3d represents the general form of carding engines used in England for middle and coarse numbers. The breadth of main cylinder is 36 inches, and diameter 42; diameter of doffer 18 inches; speed of main cylinder, from 130 to 150 revolutions per minute. The cotton is taken in from the feeding rollers by the licker-in roller A, from which it is transferred to the main cylinder, and again carded between the cylinder and the carding rollers B and C, both of which revolve with a slow motion. Whatever cotton adheres to B is cleaned off by A, which acts both as a licker-in and cleaner to B. The carding roller C is cleaned by D. These are technically denominated carders and cleaners: in this country they are called workers and cleaners. The relative surface motion of the main cylinder, and A and D, should be as three of the former to two of the latter, that is, the surface of the main cylinder should move through a space of three inches, in the same time that A and D move two. The English carding engines are generally mounted with a drawing head called a drawing box, represented at F, figures 3d and 5th, the two under back rollers of which are slightly fluted, and the upper ones covered with a ply of cloth and leather. The two front ones, being the delivering rollers, are quite smooth, without any covering. The cotton, as it is delivered from the doffer in a thin fleece, is conducted into the back rollers, where it is drawn between that and the middle rollers, about $1\frac{1}{2}$, or 2 to 1, and again compressed by the conductor e, from which it is delivered by the front rollers into the can. A number of the English Factories have the card ends delivered into cans, others have the ends from several cards wound on large wooden bobbins, with tin-plate ends: these bobbins when full, are carried direct to the first heads of the drawing frame. Those who use two sets of cards generally retain the old lap-drum in front of the breakers, from which the lap, when sufficiently thick, is broken off, and placed to the back of the finisher. In Scotland, as well as in this country, it is common to have a separate machine, between the breaker and finisher cards, for the purpose of forming laps for the latter: this is called a lapping machine or lap doubler.

[j]Fig. 4th represents a particular plan of the carding engines, which are used in various places in Britain; chiefly for single carding. And [jI]with the exception of the Matteawan cards,[jI] are the best the writer is acquainted with.[jII] E and H are a carder and cleaner, G is a fancy roller. These being placed next to the doffing cylinder, allow the tops to be arranged so that they can be stripped with the greatest facility.

The advantages of this arrangement will be more apparent, when we consider that all the impurities that may pass in with the cotton; are likely to be detached from it, and held fast by the first tops: and the oftener these are stripped; will prevent those impurities from again intermixing with the cotton; which being thus freed from seeds and dirt, will be better carded and straighted, by the cleaner and carder H and E, before it is thrown on to the doffer.

The roller G, denominated the fancy roller, is driven somewhat faster than the main cylinder; by which means the cotton that adheres to the surface of the latter, is slightly raised up, and again straightened by the carder E, before it is laid on the doffer.[j] This continual operation of the fancy roller keeps the cylinder always clean, and thereby supersedes the necessity of stripping it more than once every second day or so. This not only saves time and waste, but makes better work, as every practical carder knows when the cylinder is full of cotton, the carding is not so clear and uniform as immediately after stripping it; therefore, by keeping it always clean, the carding is uniformly good. The fancy roller requires to be covered with filleting, having longer and finer wires than is used for the others: the wires are likewise[k] nearly straight, having only a slight bend, so as to prevent them from carrying round the cotton by their accelerated motions. And the back of the wires on the fancy roller being towards the back of those on the cylinder, very little cotton adheres to its surface; besides, the cleaner H may be placed so as to act as a cleaner to both E and G, and thereby prevent the cotton from collecting on either.

Fig. 5th represents a species of carding engines used about Oldham, (England), called double carding engines, which are certainly the most powerful machines of the kind[l] in operation.[18] They are similar to those used in Woollen Factories, and found equally applicable to the cotton manufacture. This machine, as may be seen by the sketch, consists of two complete carding engines combined, the main cylinders of which are surmounted with small cylinder cards instead of flat tops. These latter are known amongst practical manufacturers by various names. They are called urchins, squirrels, carders and cleaners, or workers and cleaners. The breadth of cylinder is 48 inches, diameter 42, and the speed at which they are driven, is from 160 to 180 revolutions of main cylinders per minute. This may appear incredible to some, but the writer had full opportunity of ascertaining the fact by personal observation. The intermediate doffer A is about 28 inches diameter, and revolves once for every ten revolutions of first main cylinder H. The second doffer B, is about 22 inches diameter, and revolves once for every twelve revolutions of second main cylinder E. The quantity of yarn produced is from eight to nine cwt. per week of 69 hours, from each of these double engines, No. of yarn 36; equal to 155 lbs. of No. 36 per day, of $11\frac{1}{2}$ hours. The feeding rollers G, instead of being fluted, are covered with half inch broad filleting made of strong wires formed with diamond points. Indeed carding engines broader than [m]30 inches[m] ought to have a pair of small cylinder cards, from two to three inches in diameter, instead of fluted feeding rollers. The teeth of these small cylinders should be pointed inwards, so as to operate as lickers-in; and instead of feeding the cotton immediately on to the cylinder, they should be surmounted by a larger cylindrical card, for the purpose of transferring the cotton from them to the main cylinder. Every practical carder knows that when weights are suspended on each end of a long feeding roller, it will spring up in the centre so much, as partly to lose hold of the lap, and instead of the cotton being fed on to the cylinder in single filaments, it will be pulled in by the card teeth in large tufts, thereby producing bad carding, and of course inferior yarn; to enlarge the diameter of the feeding rollers with a view to prevent the spring in the middle, is attended with an equally bad effect "as it removes the bite of the feeding rollers too far from the surface of the cylinder. The distance from the points of the card teeth to the bite or centre of the feeding rollers should be rather less than the length of the staples of cotton; as by that means the wires take hold of each filament separately and carry it up to the tops, as it escapes from the rollers. To accomplish this in the best manner, various contrivances have been attempted but the following only is deemed worthy of notice in this place.

See the annexed figures representing the feeding apparatus of carding engines. A, A, are cotton laps from the spreader. B, B, are the feeding rollers. Fig. 1st has two rollers B & C. Fig. 2d has one only. The diameters of B, B, are 2 to $2^{1}/_{4}$ inches. The diameter of C is only $1^{1}/_{4}$ inches. D, D, represent the main cylinders, which may be 36 or 42 inches in diameter represent cast iron shells operating on the upper surface to fit the circular form of the roller. The use of these shells will be easily understood from their position. They are made fast to the framing of the card, and as near to the main cylinder as from experience is found necessary. The incumbent roller B, lies in the trough of the shell, and being weighted, bears down upon the cotton so as to carry it into the cylinder. The feeding rollers B & C fig. 1st are fluted. B fig. 2d may have coarse fluting or covered with half inch filleting having strong wires diamond pointed. The shell fig. 2d may be moved round behind the roller B, so far as to suit any length of staple. With this explanation it is presumed that the principle of the whole contrivance will be fully understood. Though this improvement has been in operation for some years in Britain it has not been introduced in this country except as a mere experiment so far as known to the writer. It is however a very important improvement and when properly fitted and adjusted produces superior work. "As the principal use of the carding process is to separate or divide the fibres of the cotton, and straighten them to a certain degree, so as to form an evenly sliver, it is essential to the attainment of good carding, that a proper method of first introducing the cotton into the main cylinder be well understood; and here it may be proper to notice the difference between the practice of the United

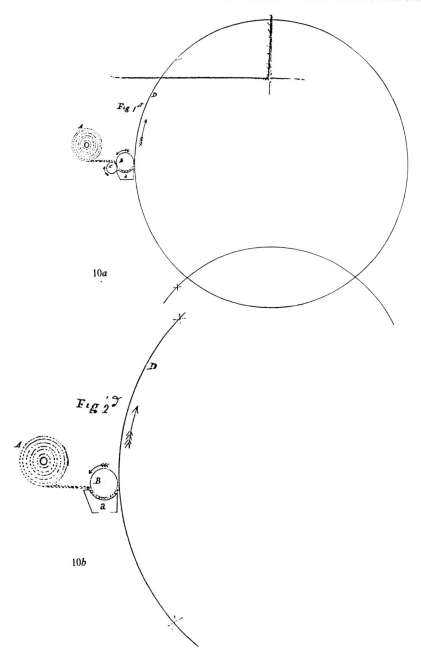

Fig. 10. British feeding arrangement for finisher carding engines.

States and that of Great Britain. In the former it is common to crowd the cotton on the cylinder so rapidly, that, instead of being taken away from the feeding rollers in single filaments, it is dragged in by the slow motion of the revolving cards in large flakes, which are not allowed to remain long enough under the operation of the tops, to be sufficiently teased out, the doffing cylinder being also driven too fast in proportion to the speed of the main cylinder. Now the practice in [o]Britain is directly the reverse: there the cotton is led into and delivered from the cards, by a very slow motion; that is, the motion of the feeding rollers and doffing cylinder, are comparatively slow in proportion to the speed of the main cylinder; for example, the mode of regulating the motions of carding engines is as follows: in Britain a main cylinder 36 inches diameter, will revolve between 70 and 80 times for one of the feeding rollers: in this country their motions are as 35 of the former to one of the latter. The proportion between the revolutions of the main cylinder and doffer are in Britain as 25 of the former to one of the latter. In America, it is 17 to 1.

From the above it may be easily perceived how the American manufacturers do not generally produce yarn of equal quality to that of the British. In the first place, they do not mix a sufficient quantity of cotton at the first process: and secondly, they do not clean it so well at the scutching; [p]and their carding is very defective.[p] The British manufacturer makes the cotton undergo a greater amount of operation, both at the scutching and carding, he is thus enabled to produce a cleaner, smoother, and more evenly thread of yarn. In order to make smooth level yarn, it is absolutely necessary to have the cotton well cleaned and carded: if the tufts or knots of cotton, are not perfectly teased out, and the fibres well separated the fleece delivered from the doffing cylinder will exhibit inequalities, or appear[q] *"clouded,"* and from card ends of that texture, it is impossible to make good yarn. I am aware that bad carding will frequently arise from inaccurate adjustment of the operating parts of the machine, but at the same time, if the *principle* upon which the various processes are conducted be wrong, the most perfect adjustment of the machinery cannot remedy the evils arising therefrom. It has been already stated, that the English spinners make their cotton undergo more operation at the carding process than the Americans; and in order to produce a sufficient quantity of work, they drive their cards at a much higher speed. While a cylinder of 36 inches diameter moves at the rate of 100 to 110 revolutions in the American Factories; in England a cylinder 42 inches diameter, and of the same breadth, revolves from 130 to 160 times; which, taking into consideration the difference of their diameters, is *nearly* double the speed of the former.

Before leaving this subject, it may be proper to notice, that the manner of attending and managing the carding engines in this country, is different from that of Great Britain. Here the cards, &c. are not divided into what is called systems, or preparations; nor is the mode of stripping and grinding the same. In Britain, one person strips the cylinders, and another the tops: and a regular system of stripping is of the utmost importance, the top stripper is kept constantly going round a certain number of cards, of which he or she may have the charge, which number generally constitute a system or preparation: but if these be too many, they may be divided, so that one person may strip the breakers, and another the finishers: and suppose the cards to have twelve working tops, the stripper proceeds in the following order. Beginning at the one end of the range of cards, the first four tops are stripped all round the system or preparation: again, commencing at the first card, the second four, or the 5th, 6th, 7th and 8th tops, are stripped all round in like manner; this completes two courses, wherein the first eight tops have been stripped once: in the third course, the first four tops are again stripped, and at the fourth round, the last four, or the 9th, 10, 11th and 12th, are stripped: thus the stripper takes four courses to go round all the tops; during which the first four have been stripped twice, and the other eight only once. The preceding is the order of stripping the tops for coarse numbers; for middle numbers, two, or at most three tops, should be stripped at each course, carefully observing to strip the first two oftener than the others: for fine numbers not more than two should be stripped at each course. In some Factories the whole of the tops are stripped in three courses, in the following order. In the first course, the stripper cleans the 1st, 4th, 7th, and 10th: second course, the 2nd, 5th, 8th, and 11th: and lastly, the 3rd, 6th, 9th, and 12th. If there are more than twelve tops, the order of stripping is so arranged as to secure uniformity of work. But it is an invariable rule, that as soon as the whole of the tops have been stripped all round, the stripper immediately goes over the whole series of cards in the same order as before; thus keeping up a constant and uniform system of stripping during the whole time the machinery is in operation. The cylinder stripper has also a certain number of cards assigned him, which he is careful to strip at least every two hours, so that the top stripper is left at liberty to proceed with his work without interruption, and is paid by the weight of the strippings taken off; which strippings are weighed and examined every day by the overseer, or some person appointed to do so.

In the American Factories the order of stripping is very different from that described above, particularly in the Eastern district; where the stripper having charge of a certain number of carding engines, begins at the first card in the series, and strips every second top all over, that is, the 1st, 3rd, 5th, 7th, 9th, and 11th; and having gone over the whole of the cards in the set in this manner, he is allowed to rest a certain time, say fifteen minutes, before he begins his second course, which embraces the 2d, 4th, 6th, 8th, 10th, and 12th tops: thus the whole of the tops are stripped in two courses, and at the end of each, he is allowed to rest about fifteen minutes: but as it is left to the stripper himself to notice the expiration of the fifteen minutes, the utmost punctuality is not expected, for the overseer, however attentive, cannot always have his eyes upon him. The top stripper likewise strips the cylinders, and during this process he must either stop all the cards under his charge, or allow them to run without stripping, which is too frequently the case. And supposing him to have the charge of ten cards, it will sometimes take 30 minutes to strip the whole, the tops being stripped all over with the cylinders as he goes along: and if the cards have been kept running all the time, they will thus have been working full 30 minutes without stripping: it certainly cannot be expected that perfect work will be produced from such a process. The strippers are generally paid by the day, not by the weight of strippings taken off; as owing to the frequent changes amongst the hands it is difficult to establish a system of piece work in some departments, which can be done with the greatest convenience in Britain.

The mode of grinding or sharpening the cards in this country, is very different from that of Great Britain. In the latter country, when a carding engine is first clothed with new sheets, fillets &c the practice is to put the cylinders in motion the right way; and a light emery board about four inches broad, is traversed over the top of the cylinders with a very delicate hand: this is called facing up the teeth, because the points of the wires are running against the board, and is intended to cut down any single wires that may be too long. After running the cylinders in this way for about fifteen minutes, their motions are reversed; and small cast iron cylinders coated with No. 4 emery, are mounted on the top of each cylinder, that is, one above the main cylinder, and one above the doffer: these are denominated fast grinders; and after being properly set, are caused to revolve in an opposite direction to the card cylinders. This operation is continued until the whole of the teeth on both cylinders are ground to a uniform length; but during the process of grinding, the emery cylinders are made to traverse a little each way, so as to grind the wires to a round point, and prevent them from being hooked or barbed. After being sufficiently ground, the cards are then dressed up, first with a brush dusted with chalk, and then with emery boards called straikes or strickles; this latter process is called sharpening, and is afterwards continued, at least once every day to the breakers, and every second day to the finishers. The fast grinders are not applied perhaps above once a year, or only when the cylinders are found what is called "off the truth;" that is, when some part of their surface may have become higher than others; then the grinders are employed to reduce all to the same level. By this method of grinding the cards only when necessary, and sharpening them every working day, they are always kept in good order, and consequently produce more perfect work: besides, when the practice of sharpening is continued daily, it can be done in much less time. Two hands are quite sufficient without much exertion, to sharpen thirty carding engines in four hours, the card belts being all fitted with buckles, no time is lost in shortening or lengthening them for the purpose of reversing the motion of the cylinders. The tops are also brushed out and sharpened once every week.

The practice in this country differs from the above in this respect, that the cards are never sharpened except when they are ground; or rather, the grinding and sharpening constitute the same operation, which is repeated only once every two, three or four weeks, and is done in the following manner. One fast cylinder grinder is placed between the main cylinder and the doffer, so as to grind both at the same time; and after being allowed to operate in this way for one or two days, it is removed, and the cylinders dressed off,—not with emery boards, but strickles made of belt leather, coated with emery in the same manner as the boards;—when the card is in this way sufficiently sharpened, it is put in operation, and allowed to run a week, a fortnight, or a month, and perhaps longer, before the same operation is repeated; the tops are also brushed out and sharpened at the same time. ˢThe preceding describes the general practice of[s1] the eastern factories;[s1] and as for the south, they are still more deficient, in all that is essential to the production of superior work.

The sheets and fillets for clothing cards, are all made by machinery, and in general are as good as those used in Britain, only the leather is not so well tanned. From five to six years, is about the average time allowed for one clothing. Many of the manufacturers in Britain, renew their card clothing every[s11] four[s11] years, especially those who spin fine numbers; the old sheets are sold to those who manufacture coarse goods.[s]

Plate IV. [Fig. 10] represents a system of carding engines introduced at Matteawan, in the State of New York, the chief peculiarities of which are the following: the feeding rollers A, and the doffing cylinder B, are[t] set somewhat[t] below the centre of the main cylinder; so that a greater portion of the latter is in actual operation. In the common system of carding, a little more than one-third of the surface of the main cylinder may be said to be in operation at one time; but in the Matteawan cards about two-thirds. D D are not doffing cylinders, but carding rollers, at the back of which there are small cleaners, one end of which is seen at E.—F F are standards supporting a drawing head. C is a horizontal box, called a railway, along the bottom of which the belt G moves with a slow motion, proportioned to the delivery of the cards. The fleece of cotton delivered by the doffers B B descends down to the railway box C, where it is compressed by the small rollers H H, and carried forward by the belt G, until it reaches the back of the standards F F, where the railway terminates, and from which the fleeces delivered from each card in the set, being compressed into slivers, are drawn up in regular order to the drawing head J, where they undergo a draught regulated according to circumstances, and then delivered into a can. In the plate only four cards are represented in connection; but it is obvious that any number from two to sixteen might be connected in the same way, and the drawing head, instead of standing in front of the cards, could be placed at one end of the range; so that the railway might run along in front of the whole.[u] Having examined the above cards, and minutely inspected their various operations I have no hesitation in saying that the work produced by them (on single carding) was equal if not superior to the best double carding I have seen in the country. To trim, adjust and keep them in good working order will no doubt require more skill and attention than the common kind but the expense of attending is very little.[u9]

PLATE IV.

MATTEAWAN RAILWAY CARDING ENGINES.

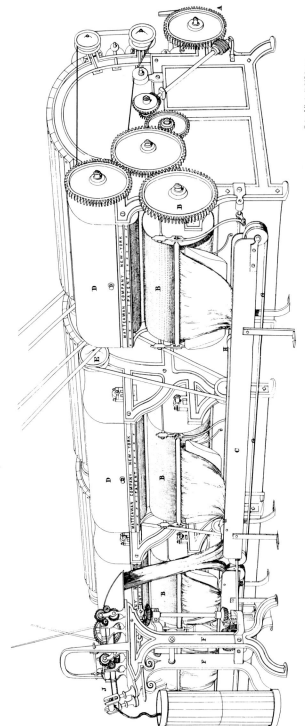

Maclure & Macdonald lith. Glasgow

Glasgow; Published by James Niven 1876'

Fig. 10. Matteawan railway carding engines, plate IV from *CM1*.

The railway system is extensively adopted about Rhode Island and other parts of the middle and Southern districts. The writer has seen sixteen cards in all in one line delivering into a railway, having the whole doffing cylinders and feeding rollers connected, and driven by the same motion, whilst the main cylinders were driven separately; so that the whole feeding and delivering motion could be instantly stopped, or put in operation, without affecting the motion of the main cylinders.

The above, or railway style of carding, was put in operation in Glasgow, by Mr. Neil Snodgrass, in 1835; and although the writer has seen it in a number of Factories in various parts of America, he has no hesitation in saying, that the manner in which Mr. Snodgrass put it in operation, was altogether the neatest and most perfect of any that he has yet seen.[10] He had eight carding engines, all delivering into one railway, one or more of which could be stopped when it was necessary to strip or grind, and as soon as the broken off sliver entered the drawing rollers at the end of the range, the person attending them, by means of a small shifting lever, threw one pair of pinions out of gear, at the instant another pair was put in operation, which diminished the draught exactly in proportion to the number of cards stopped. All this was done without interrupting the progress of work: the sliver delivered into the can from the drawing head, was of the same grist as when all the cards were running. The Matteawan carding engines have the same means of shifting various pinions in and out of gear, for the purpose of increasing or diminishing the draught in the drawing head J, to suit the number of cards stopped or in operation. In some Factories this is accomplished in a very imperfect manner by means of a pair of cones: in others one or two spare carding engines are kept standing all the time, except when an equal number are stopped for being stripped or ground [Fig. 11].

[v]The railway style of carding is liable to various objections but when connected with other arrangements in the drawing process it greatly abridges the expenses of the carding department.[vi] As for example. If single carding is adopted and six cards delivered into one railway with which a drawing head is connected this will give six doublings and at the same time reduce the produce of the six cards to a sliver sufficiently fine for being put up at the drawing frame. If eight of these slivers are put through one head of the drawing frame and two heads made to deliver in one can, this will make sixteen doublings at the drawing process; with six at the former, being 96 in all; which are amply sufficient for

common American cottons, and for all numbers of yarns under 30's. This is properly only twice drawing. First at the drawing head connected with the railway, and second at the drawing frame. But as there is only one operation at the drawing frame, the cotton being put once through; two girls are able to attend as many heads, as will pass through between two and three thousand pounds of cotton per day; and at the same time, attend the foreside of as many cards as necessary to supply this amount of carding. The drawing process is generally the most expensive in the carding department, but by this arrangement, it can be made the cheapest and suit any goods manufactured in this country.

The chief objections to the railway style of carding are the liability of the slivers to break down in front of the doffers in consequence of interruptions at the feeding—laps running out &c. &c.[vII] When this occurs it causes a deficiency in the grist of the sliver as it is carried forward to the drawing heads and ultimately a weak part of the yarn. There is also much defective work made while the card cylinders are being stripped or ground all of which tends to make imperfections in the goods.

When each card delivers into a separate can, inequalities in the sliver may be easier removed and more perfect work secured. The manner of pressing the can should also be carefully attended to otherwise the work may be greatly injured.[v] A method of pressing the card ends into the cans has been introduced into various Factories in Scotland with very good effect.[11] It consists of having the can placed with the bottom upwards,[w] and the mouth resting[w] upon a round plate of iron, with a small hole in the centre; this is placed about six inches off the floor, having a pair of calender rollers right under it, which are moved by an upright shaft; this shaft is driven by small bevel pinions on the delivering shaft of the card, so that the calender rollers of the card have the same motion as the under ones; and the card end, as it is delivered from the calender, or delivering rollers, descends to those under the iron plate, and being directed by a tin conductor, is made to pass up through between the under rollers, which are pressed together with a spring, and by them through the hole in the centre of the plate into the can. Thus the can is filled upwards: and being furnished with a loose or false bottom, which lies down upon the plate when the empty can is placed there to be filled, but rises gradually as the sliver is pressed up into the can; and when the can is full, the loose bottom has reached the top, where, by disengaging a catch, it causes a spring to ring a bell for the purpose

NOTICES OF THE VARIOUS MACHINES   83

of apprising the card tenter, who immediately removes the full
can, and replaces it by an empty one to be filled in like manner.
Those cans that are pressed in this way contain a great[x] length or
weight[x] of card ends, and from the gradual manner in which they
have been filled, are easily drawn out without stretching.

[y]The plunger motion for pressing cans now in operation in
Britain has been lately introduced into this country and into successful operations in various places.

When the cans in front of either cards or drawing frame are
very large in diameter they are liable to fill unequally and if hard
pressed the slivers become so entangled that it is impossible to
pull them out without stretching or breaking. To prevent which
it is very common in this country to make the cans revolve while
they are being filled by which means the slivers are distributed
equally all round within the cans and though hard pressed they
are not so liable to be entangled and consequently may be drawn
out without injury. But this though regarded as an improvement
by many is subject to serious objections. As the revolving can
imparts a twist to the sliver which renders it incapable of being
drawn equally and uniformly at the drawing rollers of either the
drawing frame or speeder. The same objections—to a certain degree—apply to the practice of compressing the sliver into the form
of a round cord or rope at the calender rollers of the drawing
frame.

To make the sliver draw freely and equally, it ought to be thin
and flat like a ribbon. If it is twisted and compressed like a rope
the centre will form a body too thick for being drawn, but the
edges being thin, will draw freely and unequally; and so become
ragged and broken into loose fibres, which adhering to the leather
rollers will be carried up to the cleaners, and there forming into
large tufts will occasionally pass through the rollers and greatly
injure the rovings and consequently the yarn.[y]

As already stated, the preparation machinery in the American
Factories, is not divided into systems, or preparations, as in those
of Great Britain. In general, every Factory in this country is fitted
up for making only one kind of goods, the warp and filling* being
made from the same cotton, undergoes the same operation;
whereas in the British Factories, there are generally various qualities manufactured at the same time; so that the waste made from
the finer, may be used in the coarser; the warp and weft are always

* Weft in this country is always called filling.

made from different qualities of cotton. Weft does not require
the same quality as the warp; neither does it require the same
expensive process: it would be more profitable to the manufacturers of this country, if they had sufficient room in their picking
houses to make at least two mixtures of cotton, one for warp and
another for filling, by which means all the waste used could be
put into the filling, thereby leaving nothing but good cotton for
the warp; as the latter requires to be strong, smooth, and wiry;
the former soft and woolly, so as to fill up the cloth, and give it
a more rich and full appearance. When waste is put into the filling,
it imparts to it this rough and woolly quality; and when prepared
in a separate system of cards, drawing, &c. the process could be
abridged and better adapted to the quality of yarn required.

The manufacturers of this country use up a considerable portion of the inferior waste into what is called batting, that is, after
being spread into a card lap in the usual way, it is put through a
breaker card, which is mounted with a lap drum; and when the
carded lap has acquired a proper thickness, it is broken off from
the drum, and rolled up in paper[z] in which state it is sold, and[z]
afterwards sewed between two plies of cotton cloth, and used
instead of blankets. These are called comforters, and are extensively used in this country both by rich and poor; one good one
is certainly superior to a pair of Scotch blankets: and when neatly
covered with printed calico, quilted, and bound round the edges,
they appear extremely neat and cleanly upon a bed. It is somewhat
surprising that these comforters are not,—at least so far as known
to the writer—used in Great Britain; as poor people might thus
have good warm bed clothing much cheaper than woollen blankets.[12]

Double carding, or breaker and finishers cards, are generally
used in all the American Factories, as formerly stated, and between
the breaker and finisher, there is a separate machine employed
for making laps for the latter, denominated a lapping machine,
or lap doubler, which, though used for the same purpose, is dif-
ferently constructed from those used in Scotland, but is not
deemed of sufficient importance to merit the expense of a separate plate. All that is considered necessary in this place, is to
endeavour to describe the difference between the two.

The lap doublers used in Scotland, require to have double or
treble rows of cans crowded behind the machine; and the card
ends, from these being guided by proper conductors, are directed
through between a pair of calender rollers, from which they are

wound on to the lap roller for the finisher cards. Those used in this country, have a long frame stretching out about seventeen or eighteen feet behind the machine: this frame is mounted with wooden rollers for the purpose of carrying forward the card ends from the cans, which are stretched out in one row along the angled side of the frame; the card ends are brought forward in regular parallel rows between the wooden rollers, till they reach the cal- ender rollers, where they pass through the same operation as the above. But each end as it comes out of the can, passes through a guide or conductor,[a] so constructed, that when an end breaks or runs out, the latch drops down; and as it drops, it touches a pin fastened into a horizontal rod lying along one side of the frame, and causes the rod to turn round so far as to disengage a catch at the head of the frame: the disengaging of this catch allows a spiral spring to operate upon the belt lever, so as to shift the belt from the fast on to the loose pulley, by which means the machine instantly stops. Thus when an end breaks or runs out, the machine stops instantly, until the attendant repairs the broken card end, and lifts the latch to its proper place. The writer is not aware of any such machines in Great Britain being fitted up with a similar contrivance; there it is wholly left to the attendant to stop the machine, or piece up the broken end while running: and practical carders know the difficulty of always obtaining correct work at this process.[13]

DRAWING FRAMES.

The Drawing Frames used throughout the Eastern district are made on the same plan as those generally used in England, and consist of a single beam with three pairs of rollers.[b] The draught on each head is divided between the middle and back, as well as between the middle and front rollers. Some of those frames which are made about Providence in Rhode Island, have double roller beams; but none that I have yet seen, are equal to those now made in Glasgow, (Scotland).[14]

The drawing is a most important process in the art of cotton spinning, and requires to be arranged with great skill and atten- tion, so as to give the slivers the necessary doubling and drawing that is, and no more; at the same time the slivers ought to be kept as heavy as possible, until they pass through the fly frame where they are twisted. So far as I have had an opportunity of ascertaining by visiting Factories or otherwise, I do not find that the manufacturers in this country double the slivers so much as is done in Britain. In many Factories the slivers are as fine as in those of Manchester, that manufacture from Nos. 150 to 200. This must be attended with very injurious effects; as owing to their extreme softness and delicacy, they require to be very tenderly handled; it is almost impossible to pull a fine sliver out of the can without stretching it; hence the necessity of keeping them as heavy as can be done with safety, at every department of the process.

The drawing frames of this country, with the single beam of three pairs of rollers, though made upon the simple plan, are not so well adapted for making good work as those with a double roller beam; because when the sliver is drawn between a pair of rollers, it spreads out, and becomes thin and broken at the edges. Now, in a three roller beam, where the draught is distributed between the middle and back, as well as the middle and front rollers, the first draught spreads out the sliver to a certain extent, whilst the second causes it to spread still farther: and when delivered from the front rollers, the edges or selvages of the sliver are so thin, that they become ragged and broken, whilst the fibres or filaments losing their hold of each other double up, and adhere to the upper front rollers, by which they are carried up to the cleaners, and there collect in large tufts, which, besides making waste, causes a great deal of very imperfect work.[c] This objection is equally applicable and perhaps more so—to those drawing frames with four rollers, having three separate spaces for a draught. Such drawing frames, made in the best style were introduced many years ago and (so far as known to the writer of this) disapproved of by all practical men who have had them under their own charge.[c] I have never known the drawing frames in[d] Britain cause so much trouble as those[e] in this country, in consequence of the slivers adhering to the upper front rollers. If the same evil is general in all the Factories,[f] [g]it is not surprising[g] that the manufacturers of this country have not as yet attained to great perfection in the *quality* of the goods manufactured. Besides the cause above stated, viz. the double draught in the single roller beam, there is another, which[h] no doubt operates in some measure to produce the same effects, that is, the quantity of electricity generated in the carding rooms. It was formerly stated that the spinning frames were generally driven from the carding room, by means of belts passing up through the floor; this, of course, requires a great number of carrying belts in the card room[i] which

by the rapidity of their motions and consequent friction generates
a greater amount of electricity in this apartment of the work than
any other.[ii] The climate of this country is certainly not so moist
as in Great Britain and in dry weather here, the air becomes at
times so charged with electricity; and the work in the carding
room so affected by it, that not a single drawing frame would
operate freely; all more or less lapping up on the leather rollers;
retarding and spoiling the work, besides making a great quantity
of waste.[i] But this may be greatly remedied by the double roller
beam, such as that represented in the annexed figure. Here the
slivers in passing through the back beam A, undergo a draught
of 2, or $2\frac{1}{2}$ to 1; and as it, (the fleece or reduced slivers) is thereby
caused to spread out, it is again gradually contracted as it proceeds
from the front roller of the back beam, to the back roller of the
front beam B, by which means the thin edges of the fleece are
doubled over; and entering the front beam in that state, where it
undergoes a draught of 3 or upwards, it is again delivered from
the front rollers with smooth unbroken selvages, being equally as
thick at the edges as in the centre; and the fibres having hold of
each other, are not so liable to double up, or lap on the front
rollers [Fig. 12].

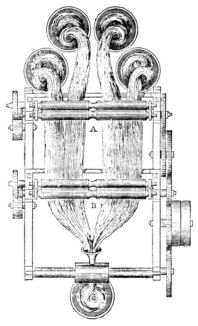

Fig. 12.   Drawing frame, wood engraving from *CM*1, p. 55.

[j]The above frames are now made at Saco[15] and Matteawan and I have not seen any drawing frame in this country, producing work equal to those from which the above sketch was taken. Although only 4 slivers are presented as passing through the head, it is obvious that six, or eight, might be put up if required.

In general the cotton is passed three times through the drawing frame each frame having three single heads, or six heads paired two and two.[j] In[k] Britain the drawing frames are made with six, eight, or ten drawing heads, according to the number of times the slivers are to be drawn. For three times drawing, the frame will have six heads paired, two and two: for four times drawing, there will be eight heads, and ten heads for five times drawing. Five times drawing, however, is only used for very fine yarns; four times for middle numbers; and three times is deemed sufficient for all low numbers, whilst the number of doublings is often a mere matter of opinion: in all Factories there are more or less, according to the peculiar notions or experience of the superin-tendents, or the quality of the yarn wanted, and the cotton from which it is to be made. I have often found that too many doublings and drawings had a greater tendency to injure the yarn, than improve it; and I believe it is generally admitted, that the less operation the cotton passes through at this department of the process, it is the less liable to injury; hence every experienced[l] manufacturer[l] studies to give just what is necessary and no more.

[m]The principal use of the drawing frame is to straighten the fibres of the cotton and lay them in a longitudinal direction par-allel with each other which may generally be accomplished with three times drawing provided the cotton has been properly carded and cleaned at the previous processes. And when once the fibres are sufficiently straighted it is unnecessary to continue the process of doubling and drawing any farther. For long stapled cottons when prepared for fine goods four or five times drawing is some-times necessary as it smooths down the fibres and makes a more glossy and silky thread of yarn. But for all kinds of American cottons three times drawing with from 100 to 250 doublings are amply sufficient. The writer has made excellent water twist No. 24's, from common New Orleans and Upland cottons, with single carding, twice drawing, and only 64 doublings in the whole pro-cess. The number of doublings and drawing should always be adapted to the quality of the cotton. Every unnecessary doubling requires an extra draught and when the cotton is too much drawn the fibres become so disconnected and disengaged—especially

short stapled cotton—that they are weakened and losing their hold of each other by too frequent separation that they (the fibres) are apt to fly off in dust or make weak soft and tender yarn.

The drawing frames in this country are mostly all mounted with a self-acting stop-motion which has not so far as I am aware been extensively introduced in Britain, nor is it very[m] necessary that it should; because the helps* in that country are very different from this. Here they are constantly changing, old hands going away, and new ones learning. The great majority of girls employed in the American Factories are farmers' daughters, who come into the Factory for, perhaps, a year or two, and frequently for but a few months, until they make a little money to purchase clothes, &c. and then go home. In consequence of this continual changing, there are always[n] a number[n] of inexperienced hands in every Factory: and as the drawing process requires the utmost care and attention to make correct work as well as to prevent waste, it is necessary to have the most expert and experienced hands attending the drawing frames; but such hands cannot always be obtained here as in[o] Britain; hence it is necessary to have some contrivance connected with the machinery which will, to a certain extent at least, prevent the work from being injured by inexperience on the part of attendants. All the drawing frames, therefore, which I have seen in this country, are mounted with a self-acting stop-motion, so that when an end (sliver) breaks, or runs out, that head with which it is connected instantly stops [Fig. 13].

The annexed figures represent the stop-motion; A A is the wooden beam, supporting the stand B, with the drawing rollers: L L, called the latch, is supported by a pin P P, upon which it is so nearly equipoised, that it requires only the weight of the sliver H to make it stand upright; but as soon as the sliver breaks, it falls down, and the under point $n\ n$ touches a small pin projecting from the rod $c\ c$, which causes it to turn partly round, and in turning round, another projecting pin presses up the catch i; this catch turning on the stud s, the point o is brought down, and thereby relieves the rod X X, which is immediately forced forward by the spiral spring d; and striking against the belt lever h, causes it to shift the belt from the fast on to the loose pulley; by this means the head is instantly stopped before the end of the broken sliver enters into the back rollers. The small steel spring a presses

* The work people in Cotton Factories, as well as house servants in this country, are *always* called "helps;" upon no occasion do we ever hear them designated *workers* or *servants.*

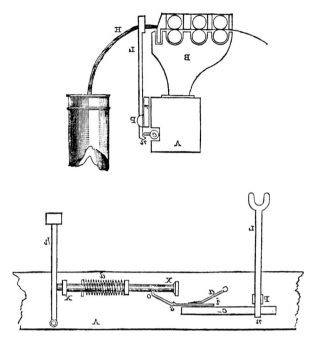

Fig. 13.   Stop motion for drawing frame, wood engraving from *CM*1, p. 58.

slightly on the catch *o s i* at *i*, and thereby keeps the point *o* in its place. The whole of this apparatus except the latch L L, is concealed when the frame is in operation, between a piece of wood *r*, and the beam A.

The stop-motion is[p] an American invention, and is particularly necessary in this country for the reasons already stated. But besides the expense of fitting up, it is very liable to get out of order, and when that is the case, it makes a great deal of imperfect work, as it usually begets carelessness on the part of the attendants, who depend almost entirely upon the latch for stopping the head when the work goes wrong; and when one latch falls, it frequently brings down others, and breaks or injures the slivers, besides causing some trouble in putting them all right before the head can be again started. Now in Great Britain where there is always a command of experienced hands, the introduction of this stop-motion would be attended with no advantage, as in my opinion, two active girls by close attention, would do more and better work on a drawing frame having no self-acting stop-motion, than any I have yet seen with it, even the most improved.[16]

SPEEDERS

The first machine which follows the drawing frame is in this country usually denominated a "Speeder," and of these there are some variety; as the Taunton Speeder, the Double Speeder, the Eclipse Speeder, and the Plate Speeder. All these are[q] American inventions. The first, *viz.* the Taunton Speeder takes its name from the place where it was first put in operation, Taunton, in the State of Massachusetts, a place where there is a great deal of very superior machinery made. This machine was patented in England in 1825, by Mr. Dyer of Manchester, and is known by the name of Dyer's Frame.[17] They are also made in Glasgow, by Mr. Holdsworth and Messrs. William Craig & Co. (late C. Girdwood & Co.) where they are known by the name of tube frames.[18] Though the principle of these frames is the same in both countries, at least in as far as the tube is concerned; those that are made in Britain are so much improved in their general form and construction, as to render them altogether a very different machine from any of the kind I have yet seen in this country; indeed, so far as I have learned, the Taunton Speeder is rather an unpopular machine[r] and is likely to be[s] superseded by others of a much superior character.

The tube frames made in Glasgow have generally two rows of bobbins in one frame; those of this country have only one row: and in the various Factories which I visited in England in 1836,[19] I do not remember having seen any of Dyer's frames with two rows of bobbins. The Glasgow frames [t]may therefore be considered the best now in operation in either country[t] and are well adapted for spinning any numbers of yarn under No. 36.

The Eclipse Speeder, or, as it is sometimes called, the Bellows Speeder, was introduced into Manchester in 1835, where it is known by the name of the Eclipse Roving Frame; a short description of which is contained in the "Theory and Practice of Cotton Spinning."[20] The rovings produced from this frame are in every respect the same as those from the tube frame, having no twist, and built on the bobbins, or spools, with conical ends in the same manner. Instead of the roving passing through a tube, as it descends to be wound on the bobbin, it passes through between two opposing surfaces of a travelling endless belt, which produces the same effect. In the quality of the roving, it has no advantage over the tube frame, but it is much more simple in its construction, occupies very little room, and requires much less power to work it; whilst, at the same time, its power of production is astonishingly great. The front roller, being $1^{1}/_4$ inches in diameter, may be driven at the rate of from 700 to 750 revolutions per minute; and a machine of ten bobbins, the size generally made, will produce 40 to 50 hanks of roving per hour, allowing for time to remove and replace the bobbins or spools as they are filled.

The Eclipse roving frames are made Messrs. Sharp, Roberts & Co., of Manchester, by whom they have been much improved, and made altogether superior to those used in this country.[21]

The Plate Speeder was introduced into Glasgow in 1835, by Mr. Neil Snodgrass, who imported one direct from America.[22] The general form or construction of this frame is similar to that of the tube frame, only in place of tubes, there are friction plates; and each ply of roving, as it proceeds from the front rollers to the bobbin, passes through between one pair of these plates, which, revolving rapidly in opposite directions, twist and untwist the roving in the same manner as the tube. The whole surfaces of the plates are not in contact but only about one-half inch from their periphery, which part is beveled off so as to make those sides of the plates nearest the front rollers stand about $1^{1}/_2$ inches apart, in order to bring the two surfaces of the beveled portion exactly parallel to each other. The plates are therefore placed so as to form an acute angle; the acute point of which, pressing on the bobbin upon which the roving is wound, makes it wind on as firm and compact as if done by the tube. That portion of the inner surfaces of the plates which are in contact, are slightly grooved or fluted, so as to operate more effectually on the roving, and they may be set closer or farther apart, as may be thought necessary, by which means they are equally adapted to coarse or fine roving, as well as dirty or clean cottons.

"The Eclipse and Plate Speeders are very unpopular in this country and not likely to be renewed by those who have once had a trial of them. The Taunton Speeder was likewise going down but an alteration has lately been made upon it by Mr. Danforth of Paterson, New Jersey which may perhaps protract its existence for a short period longer.[23]

Mr. Danforth removed the tubes entirely and substituted a rapidly revolving strap of leather or cloth in their place. And as the rovings descend from the rollers instead of passing through a tube they pass through between the opposing surfaces of this strap which produces the same effect. The strap being about $1^{1}/_2$ inches broad and lying parallel with the bobbins but placed so that one edge of it comes close up to the bite between the bobbins and the friction cylinder. The strap is kept in its place by tin plates or guides. This improvement is properly a combination of the Eclipse and Taunton Speeders & as thus united seems to suit the purpose better than either.[u]

The Double Speeder is properly a fly frame, but differing very materially in its construction from those generally used in Great Britain. It is considered the best in this country, and is more extensively used than any of the three already mentioned. I am not aware, however, that it has ever been introduced into[v] Britain, nor do I think it would succeed there; as, with all its merits, it cannot be compared with the fly frames now made in Manchester and Glasgow. As regards the quality of the roving produced, it is equal to any fly frame[w]; but it is heavy, clumsy, and extremely complex. The fly frames of Great Britain have all two rows of spindles, whilst the double speeders of this country have only one; yet one of the latter with twenty·spindles will[x] make more waste[x] and consume double the quantity of oil required for one of the former containing forty-eight spindles: at the same time, the English fly frame will produce much more than double the quantity of work.

The double speeder is made in two different forms: the first receives the cans direct from the drawing frame, and has only one row of spindles on one side of the frame; these are usually called speeders[24]: the other receives the bobbins from the speeder, and reduces the rovings to a finer texture, and is denominated a stretcher or extenser: these have one row of spindles on each side of the frame similar to that of the common throstle or spinning frame.[25]

The figure [Fig. 14] represents the exact position of the spindles and bobbins in the extenser. The bobbins A A A A having been filled at the speeder, are here mounted on a common creel containing four rows; and one ply of roving from the upper and under bobbins being doubled, they pass through the beam rollers B B, where they undergo a draught of from 5 or 6 to 1; from these the reduced rovings are delivered to the flyers, and entering the funnel at *a a*, they descend the tubes *c c c c*, from the bottom of which they are wound on to the [Fig. 14] bobbins D D, which, when full, are removed and carried to the spinning frames. By examining the figure, it will be seen that the front rollers are lower than the back ones; this is called the bevel of the rollers. All the frames of this kind, as well as the spinning frames, have the rollers more or less beveled, as by that means the twist from

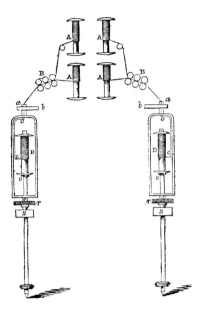

Fig. 14.   Stretcher or extenser, a form of double speeder, wood engraving from *CM*1, p. 64.

the flyers proceeds direct to the bite of the front rollers. It likewise shortens the distance between the top of the flyers and the rollers, so that the yarn or roving may be as softly twisted as may be thought necessary.[y] This method of beveling the beam rollers is certainly an improvement, as applied to throstle spinning frames; but of very little consequence in any other machine.[y]

The fly frames used in Britain have the flyers balanced on the top of the spindles, in the same manner as in the common throstle frames; here they are altogether disconnected. The tops of the flyers run in collars *b b*, fastened to the framing of the machine. The under part of the flyers rest upon *e e* of the wheels *r r*, these are again supported on the wooden beams *s s*. The bobbins D D, rest on the rings *o o*, which are connected with the spindles. The bottom rail upon which the spindles rest, ascends and descends by a regular motion, and thereby carries the spindle and bobbin up and down with it. As the bobbin rests upon, and is moved by the spindle, the speed of the latter varies according to the increasing circumference of the bobbin, whilst the motion of the flyers is equal and uniform, giving a regular twist to the roving.

The double speeder has only one beam, containing three pairs of rollers; the extenser has two, one on each side; therefore the slivers, from leaving the drawing frame until delivered from the extensers, pass through two beams of drawing rollers: now one English fly frame mounted with two beams, will do the work of both speeder and extenser to much better purpose, and at a great saving of room, power, and oil, as well as expense of attendance.

The newest fly frames from Manchester, made by[z] Mr. Higgins & Son[z] containing their improved spring presser, have already been introduced into this country, and put in operation at Matteawan, in the State of New York; and I have no doubt that as soon as their merits become known, they will be generally adopted, and entirely supersede all the other roving machines mentioned in the preceding pages.[a] Fly frames are better adapted to this purpose than any other [machine?] with which the writer is acquainted. They are superior to the double speeders and extensers so much admired in this country. First they are less complex and consequently easier adjusted and kept in order. Second they condense the process and thereby save waste and expense for attendance. Third they save room, power and oil.[a1] This however only applies to fly frames containing two roller beams, by which the slivers may undergo the same amount of draught as when passing through both speeder and extenser.[26]

It must be admitted that some of those fly frames lately imported into this country from England do not justify the above statement—as regards their superiority over the speeder and extenser—When last seen by the writer, they did not seem at all equal the same machines as operated in England nor did they equal those now in operation at Matteawan & Manayunk in this country.[a]

Fly frames cannot be driven at so high a speed as the others. The revolutions per minute of the spindles in fly frames are from 800 downwards; while those of the speeder and extenser may, with equal safety, be driven at the rate of from 900 to 1000. The speed of the front rollers of the latter are from 150 to 200 revolutions per minute, diameter $1\frac{1}{4}$ inches.

The double speeder and extenser are almost universally used throughout the whole of the Eastern district. The eclipse and plate speeders are used chiefly about Paterson in New Jersey, and some other places in the Middle and Southern districts. Mule stretching frames are employed for finishing frames about Rhode Island and Connecticut, particularly for fine spinning, that is, from No.

40 to No. 50. The finest yarn spun in the United States, of which I have been able to obtain any account, is No. 60, except that which is made for sewing thread, some of which is as fine as No. 110.

Twenty finisher cards, 36 inches broad, supply six drawing frames, containing three heads each; six double speeders, containing eighteen spindles each; and seven extensers, containing eighteen spindles on each side, = 252 spindles in all on the extensers.

Seven extensers, containing 36 spindles each, supply two-hank roving for 4,416 throstle spindles, producing six hanks each per day of[b] No. 20[b] yarn = 26,496 hanks. Therefore each extenser spindle produces 105 $\frac{1}{7}$ hanks of yarn per day; each speeder spindle 245 $\frac{1}{3}$ hanks; each drawing frame of three heads 4,416 hanks; each finisher card 13,248 hanks, or '66 $\frac{1}{5}$ lb' of yarn per day.

The preceding may be regarded as the average produce of these machines: some may produce more, but there are many that produce less. It is necessary to mention, however, that the throstle spindles referred to as producing six hanks per day, are what in this country are called the *dead spindles*, known in Great Britain by the name of *Montgomery's Patent Throstle Spindle*, or the *Glasgow Patent Throstle*.[27]

To attend six drawing frames, the front of twenty finisher cards, and the back of six speeders, will require twelve girls receiving[d] 42 cents each per day = 1/9 Sterling[d]—six speeders require three girls at[e] 46 cents per day = 1/11 Sterling[e]—seven extensers require five girls; four attending three sides each, and one attending two sides, at[f] 46 cents per day = 1/11 Sterling.[f] Girls attending the spinning frames are paid at the rate of [g]44 cents per day = 1/10 Sterling.[g] One girl can attend four sides of 48 spindles = 192 spindles of the above-named throstles, spinning No. 18, and producing six hanks per spindle per day. In general, the spinning girls, and in some Factories, the speeder and extenser girls, are paid by the quantity of work done. This is ascertained by small clocks mounted at one end of each frame, and moved by the back rollers which properly indicate the revolutions of the back roller, and from these, calculations are made to find the length taken in, according to which a scale is made out, by which each girl is paid a given rate for a given quantity of work.

Turning to the figure on page 94, it will be seen that the spindles are driven by bevel wheels at the bottom, like the fly frames of

Great Britain, but the flyers in the speeders and extensers, are driven by spur gears. (*See the figure.*) The wheels *r r* drive the flyers, and instead of the teeth of these gears being cut right across, they are cut in an angular or slanting direction, denominated spiral gears. Small gears or pinions of this kind, are cut in a cutting machine, with a very fine pitch, and are extensively used in this country on the drawing rollers of spinning frames, speeders, and extensers. For the drawing rollers, the pitch is from 23 to 32 teeth to the inch. They run perfectly smooth and free, without the least vibration, the whole breadth of the teeth not being in contact at one time.[h] The friction of the teeth in the two opposite gears, is not so apparent as in those cut in the common way; for that reason the spiral gears are much better adapted for the drawing rollers of all roving and spinning machines.

SPINNING MACHINES.

Throstle Spinning Frames are universally used for spinning warps in the American Cotton Factories, with the exception of a very few where a superior quality of fine goods is manufactured, from No. 40 yarn and upwards. The few Factories in which these fine goods are made, use mules for both warp and filling. Some of the ⁱMills at Lowell spinʲ No. 40ʲ yarn, both warp and filling, on throstle frames; but the cloth made from such yarn cannot be compared with that spun by mules [Fig. 15].

Mules are generally used for spinning, filling, or weft, about Rhode Island, Connecticut, and the Southern Factories.[k]

In Lowell, and generally throughout the whole of the Eastern district, throstle spinning frames are used for spinning both warp and filling; and in this district too, the particular kind of spinning frames employed, are almost entirely the *dead spindle,* mentioned at page ninety-six. In the Middle and Southern districts the *live spindle,* or the old common throstle frames are generally used.

ˡThe Cap Spinner known in ˡˡBritain as the Danforth Throstle [Fig. 16] is used in some factories at Paterson and a few others about Rhode Island [Fig. 17] butˡˡˡ ˡˡˡˡis not a very popular machine. ˡˡˡˡIn this country various attempts have been made to improve and render it more efficient, or less objectionable: none of which have been very successful. Its very limited success in Britain is no doubt owing to the improvements made upon it since its first introduction there.[128]

Fig. 15. Warp spinning frame (64 spindles) built by the Locks & Canals Company of Lowell, ca 1830s, in Old Slater Mill Museum, Pawtucket, Rhode Island. Author's photo, courtesy of the museum.

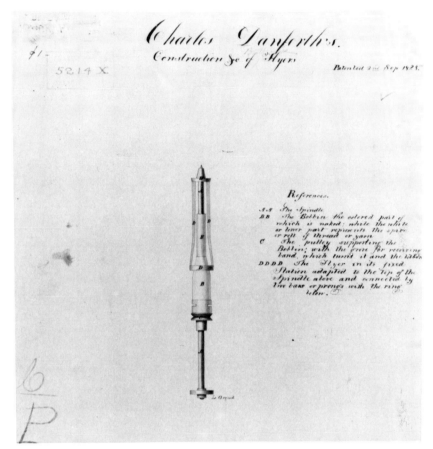

Fig. 16.   U.S. patent drawing of cap spindle patented by Charles Danforth, 2 September 1828, in U.S. National Archives, RG 241, patent drawing no. 5214. Courtesy of the National Archives.

The Ring Throstle, or, (as it is sometimes denominated) the Ring and Traveller, has been put in operation in various Factories about Rhode Island and other parts; and I am told it has given great satisfaction wherever it has been tried. Having had one under my own charge working along with others of the dead spindle kind, I have no hesitation in saying, that so far as I have been able to test the two, the ring throstle is[m] superior to the other. It requires much less power, and may be driven at a[n] higher speed, and makes a better quality yarn.[29]

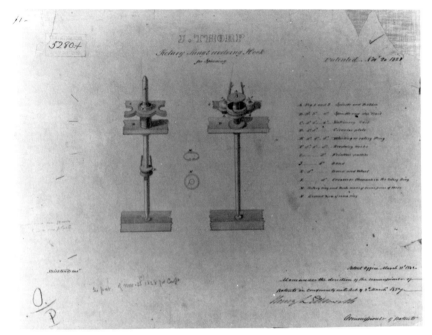

Fig. 17.   U.S. patent drawing of ring spindle patented by John Thorp, 20 November 1828, in U.S. National Archives, RG 241, patent drawing no. 5280. Courtesy of the National Archives.

Mr. Gore's patent throstle spindle has been introduced into this country, and put in operation at Taunton, Massachusetts, and is regarded by competent judges as a most important improvement.[30] Having had an opportunity of seeing this throstle in full operation in various parts of Great Britain, I have no hesitation in pronouncing it the best that has yet come under my observation. I found the current opinion of practical men both in Scotland and England strongly in favour of the old common throstle. But Mr. Gore's patent partakes of all the merits of the other, while at the same time it can, with safety, be driven at a much higher speed. Although driving machinery at a high speed, does not always meet with the most favourable regard of practical men in °Britain; because in that country where *power* costs so much, whatever tends to exhaust that power, is a matter of some consideration: but in this country, where water power is so extensively employed, it is of much less consequence. Besides, the expense of labour being much greater in this country than in ᴾBritain, the American manufacturers can only compete successfully with the British, by producing a greater quantity of goods in a given time; hence any machine that admits of being driven at a higher speed, even though it should exhaust the power, if it does not injure the work, will meet with a more favourable reception in this country. For these reasons alone, can I account for the extensive use of the *dead spindle* spinning frame in America; as the same spinning frame was tried in Glasgow under the most favourable circumstances, and never realized the high expectations entertained regarding it upon its first introduction. In fact it may be said to have been a total failure in that country, and yet it has been very successful in this. The chief objections to it in Glasgow were its immense weight and consequent exhaustion of power, together with the great quantity of oil it consumed. The same objections would prevail against it here, but for the reasons already specified.

The weight of the dead spindle spinning frame compared with the common throstle, is considered to be as 80 of the former is to 100 of the latter; that is, one horse power is deemed sufficient to move 100 spindles of the common throstles with the necessary preparation, while the same power is required for 80 of the other.[31]

Both Mr. Gore's patent spindle and the ring throstle obviate, in a great measure, the objections urged against the dead spindle. They require much less power and oil, and may with safety be driven at[q] the same high speed.[q]

To ascertain the average speed and produce of the dead spindle spinning frames of this country, I have access to memoranda, containing an account of them in a number of Factories in various parts of the United States; and without specifying any particular Factories, it may be stated in general, that the speed of the front rollers of these frames, which are one inch in diameter, will range from 60 to 100 revolutions per minute for all numbers between 12's and 40's; but some of the fine spinning frames are even below 60. Some Factories in Lowell had the speed of the front roller as high as 110 revolutions per minute for No. 14, but it was afterwards brought down to 100, as being considered more profitable. The average speed of the flyers according to their diameters, will range from 3,800, to 4,700 revolutions per minute; and their average produce is from $4\frac{3}{4}$ to $6\frac{1}{2}$ hanks per day of $12\frac{1}{4}$ hours. The ring throstle can produce one-fourth more than the dead spindle. Their relative speeds, as I have had them working, are as 80 of the latter to 100 of the former. Dr. Ure gives the following as the produce and speeds of the English throstle frames.

"The quantity turned off is about 24 hanks per spindle of 30's twist in 69 hours. . . . In some Factories, with new throstle-frames, fully 30 hanks* of 34's or 36's may be turned off.

"In spinning 32's, the front rollers of the common throstle make 64 revolutions per minute, and the spindles 4,500. For the spinning of lower numbers, the rollers go quicker; thus, the 28's to 30's they make from 68 to 70 revolutions. . . . In Mr. Orrell's Factory, for 36's the front rollers make 72 revolutions per minute, and the spindles 4,000. . . . . . At Hyde, where excellent throstle-yarn is spun, $3\frac{1}{2}$ hanks of 36's are the average daily quantity per spindle, or about 21 hanks in 69 hours.

"I visited a great Factory at Stockport, where the throstle-spindles revolved 5,000 times in the minute; and the front rollers 90 times, in spinning 36's. These machines were constructed by Mr. Gore of Manchester. I was informed that Mr. Axton, of Stockport, had contrived a modification of the throstle-spindle, in consequence of which he could give the front rollers a speed of 80 turns in the minute, and the spindles 7,000 turns, in spinning 24's."[32]

From the above it will be seen, that the common throstle in England is capable of producing nearly as great a quantity of yarn in a given time, as the dead spindle of America: and that Mr. Gore's improved spindle will even exceed the latter. Hence from all I have seen or can learn regarding the different throstles that have been brought into general use, I consider the ring throstle and Mr. Gore's improved-throstle, as superior to any others that have yet been introduced.

The ring throstle, owing to the vibration it gives the yarn, tends to start the fibres, and make the thread more rough and spongy; but Gore's throstle gives the yarn all the wiry smoothness of the common spinning frame. Each of these will be appreciated by manufacturers according to the particular purposes for which the yarn is to be appropriated.

Throstle spinning frames upon the principle of Mr. Gore's patent, are made by Messrs Rogers, Ketchum, and Grosvenor, at their large Machine Making and Engineering establishment in

* If such is the produce of the common throstle in England, it is more than I was accustomed to witness in Scotland, even with the best constructed machinery. I am not aware of any in that country that have been able to keep up more than 24 hanks per week on any numbers.

Paterson, New Jersey, where machines of every description are manufactured.[33]

The principle of Gore's patent spindle has likewise been applied to the bobbin and fly frame by Mr. William B. Leonard of Matteawan New York.[34] And as applied to this latter machine it has effected a very material improvement by obviating the vibratory motion to which the spindles in the fly frame are peculiarly liable when driven at a high speed.[r]

Good throstle frames cost about 10/- per spindle in Glasgow and Manchester. In Lowell they cost from 4 to 4½ dollars, or about 18/- Sterling per spindle.

[s]The best hand mules I have seen in this country are those made by Pitcher and Brown, Pawtucket.[sI] [35]Those made by the Providence Machine Co. and by Crocker & Richmond, Taunton are very excellent machines except in some of the minor modifications they are equal to those of Manchester and Glasgow.[36]

Mule spinners are generally paid from[sII] seven to ten cents per 100 skeins; only a few are paid as high as ten cents; eight and a half to nine may be the average = 4¼d to 4½d sterling.[sII] [sIII]In some few small mills they are paid one and a half dollars per day or where each spinner is allowed to supply filling for a certain number of looms they are paid at a fixed rate per cut of cloth, produced by these looms.[sIII]

[sIV]The several Self-Acting Mules now in operation in Britain have been introduced into various places in the United States but generally on a small scale, few mills having more than four pairs in operation. At the Rocky-Glen Mill near Matteawan in the State of New York there are 16 pairs of Smith's Self-Acting Mules in operation containing 6,144 spindles.[37] And this so far as known to the writer is the only establishment in the country where the yarn both warp and weft is entirely spun on Self-Acting Mules.[38] The results of this mill have been very satisfactory to the owners. The yarn counts No. 36 and is woven into printing cloth. The goods are so far superior to any others made in the country of a similar texture that they are greatly preferred by calico printers and generally sold at a higher price.

Besides the British Self-Acting Mules that have been imported another has been invented in this country by Mr. William Mason of Taunton in the state of Massachusetts,[39] and considered by those who have had experience of both to be fully superior to Sharpe and Roberts mule. It contains a beautiful assemblage of mechanical contrivances exhibiting great ingenuity on the part of the inventor but rather too complex for general adoption yet it performs all its operations with mechanical precision and accuracy.

Having been but lately introduced it has not yet been extensively adopted. It is likely however that Mr. Mason will simplify some of its moving parts after which there is every probability that it will be more extensively used as all those who have had trial of it express themselves entirely satisfied with its operations.

The Great Falls Manufacturing Company having determined to introduce Self-Acting Mules into one of their factories but being unable to satisfy themselves regarding the merits of the different mules now in operation they resolved by actual experiment to test the capabilities of each and only adopt that which after a fair trial proved to be the best. Accordingly one pair of Sharpe and Roberts mules[40] and a pair of Mason's were obtained and placed side by side so as to give them equal advantages; at the same time the respective machinists were required to provide the best hands they could procure to take charge of them during the trial to which they were to be subjected. The following are the results obtained after a fair and impartial competition during four months viz July, August, September, and October of 1842. One pair of Sharpe and Roberts's Self-Acting Mules containing 840 spindles.

One pair of Mason's Self-Acting Mules containing 704 spindles Produce of four months

	lbs	size	hks pr spindle pr week	waste per cent	cost of piecing cents pr 100 hanks	
Sharpe & Roberts	11.042 ..	28	$21^{30}/_{100}$	$1^{3}/_{4}$	$34^{3}/_{10}$	= $1s.5^{1}/_{3}d$
Mason's	9.845 ..	$28^{64}/_{100}$ ..	$22^{86}/_{100}$	$2^{1}/_{4}$	$37^{1}/_{2}$	= $1s.6^{3}/_{4}d$

  Average hours of labour .... 75 hours pr week

The above mules had been in operation about seven or eight months when one of Smith's was procured and put into the apartment for the purpose of being tested with the others. But after being in operation a very few weeks its superiority was so apparent that the company came to a very speedy conclusion regarding the respective merits of the different Self-Acting Mules which had been thus subjected to a fair and impartial trial. An order was immediately given for sixteen pairs of Smith's Self-Acting Mules containing in all about 10,000 spindles, and which are now (August 1843) making at Lowell and expected to be in operation in the course of 1844. This is the first large order for Self-Acting

Mules that has been executed in this country. And the very favourable circumstances under which they will be introduced at this place will furnish a fair opportunity for proving the capabilities of these machines upon their adaptation to the American factory system.

The results of Smith's Self-Acting Mules at the Rocky Glen Mill are as follows
6,144 spindles average produce *pr week* 9.450 lbs of No. 36 yarn = 151,200 hanks. Cost of labour per week 40 dollars = £8.6.8 including overseers' wages which is something under 3 cents or 1½d per 100 hanks.

The following statements with which I have been kindly favoured will show the results of Self-Acting Mules in Great Britain.[5]

Cost of Mr. Smith's Mules per spindle when new, ... 8/-
—of adapting the self-acting motion to Hand Mules,
  pr spindle 3/6
—of Messrs. Sharp, Roberts & Co.'s Mules, per spindle,
  new, ... 9/6
—of adapting their self-acting motion to Hand Mules, ... 4/1
—of Mr. Potter's Mule,[41] per spindle, new, ... 7/6
—of adapting his motion to Hand Mules, ... 3/3
—of Common Hand Mules in Glasgow, per spindle, new, . . . 5/6

From the above it will be seen, that for Mr. Smith's Mules, about £75 of extra capital is required to fit up 600 spindles, self-actors, and about £100 to adapt the motion to hand mules; perhaps £10 may be added to this last sum for sundry articles that may be required. Keeping this in view, the following will shew the cost of working a pair of self-actors of 600 spindles for two weeks, producing 21 hanks per spindle per week of No. 36 weft.

Wages to Piecers, 25,200 hanks @ 1/9 per 1,000 hks.	£2	4	1
Extra attendance of Spinning Master and Mechanics, ...	0	5	0
Interest and charge for tear and wear on £110 extra Capital, @ 7½ per cent, ...	0	6	3
Extra charges, including Power, Oil, and Banding, . .	0	5	0
Insurance on extra Capital, ...	0	0	8
Nett charges for two weeks, ...	£3	1	0

being 2/5.05 per 1,000 hanks.

The lowest price paid for hand-spinning in Glasgow at present is 2/10¾ per 1,000 hanks; this is what is technically called the 2¾d. per shilling rate.*

At the termination of a strike in Glasgow, in April 1838, an agreement was made between the operatives and their employers, by which the rate was fixed at 3/5 per 1,000 hanks for No. 40's, with a discount of 1 per cent off the gross wages for every 24 spindles which any spinner had under his charge above 600.[42] In consequence of this clause, many of the manufacturers got two 'mules' coupled together, and gave 1,200 spindles and upwards to each of their spinners. The following small table is intended to shew the annual difference on every 1,000 spindles of each of the three methods,—common, hand-coupled, and self-acting mules; the produce supposed to be 21 hanks per week, No. 36's weft, and the rate allowed for the common mules is the lowest at present paid in Glasgow, viz. 2/10¾ per 1,000 hanks,—that for the coupled mules 3/5, with the discount allowed for 1,200 spindles,—and the rate for the self-actors is that found in the preceding table.

	Hand Mules.			Coupled.			Self-acting.		
	£	s	d	£	s	d	£	s	d
Price per lb. for No. 36's,			1.241			1.096			1.038
Cost of spinning 30,333 lbs. No. 36's from 1,000 spindles in 12 months,	156	16	11	138	15	6	131	3	10
Gain per annum on 1,000 spindles of coupled & self-acting over hand mules,				18	1	5	25	13	1

In the above estimate no allowance is made for the difference of produce. It is likely to be greater from the self-acting, and less, with the yarn somewhat inferior, from the coupled mules.

The following statements have been obligingly furnished by gentlemen in and around Glasgow, all of whom have had extensive acquaintance with self-acting mules; and having had ample opportunities of judging of their merits, their statements will be valuable to all who are anxious to obtain correct information regarding self-acting mules in Scotland.

---

* The above rate refers to an old list of prices, in which No. 100's was paid @ 2/- per lb. but in the progress of the manufacture, the price has gradually fallen to 5½d., or 2¾d. per shilling.

The first relates to a Factory having a large number of spindles fitted up with Smith's self-acting motion.

"Speed of our spindles, for No. 36's warp, 6,400 revol. per minute.
.... for No. 18's weft, 4,800 ....
Produce .... of No. 34's warp, 18½ hanks per week
.... of No. 40's do.   17   do.   do.
.... of No. 18's weft, 23   do.   do.
.... of No. 34's do.   22   do.   do.
Wages of Piecers, ...for No. 34's to 46's 1/9 per 1000 hanks.
.... for No. 20's   1/11½ do.   do.
.... for No. 18's   2/   do.   do.

"There is considerably more trouble with the self-acting, than with the hand mules, arising from the extra machinery required to perform all the different movements: our spinning master attends upwards of 10,000 spindles, but we find he has rather too much to do."

The next relates to the same description of mules, and contains a comparison of the price paid per lb. for self-acting, and that paid for hand spinning, in the immediate neighbourhood, the rate being 3/1¾ per 1,000 hanks; but it must be observed that the price noted for self-acting is the nett amount paid for wages, and does not include anything for extra capital, charges, &c. These will amount to 20 per cent on the wages, yet, even taking this into account, the difference is still very considerable.

No. of Yarn.	Self-acting.		Hand Mules.
	Produce, Hanks.	Cost per lb. (d.)	Cost per lb. (d.)
30	24*	.5625	1.125
40	24	.75	1.5
50	19	.7813	2.125
60	17½	1.047	2.75
70	16½	1.313	2.375

"*This is the produce per spindle per week of 69 hours."

"We allow one spinning master for every 8,000 spindles, and our piecing costs about 1/5 per 1,000 hanks."

The next statement relates to a work in which several pairs of Mr Smith's self-acting mules have been for a considerable time in operation along with hand mules, and the statement shews the actual cost and produce from 1,872 spindles, each, self-acting, and hand mules, during the period represented.

Time taken by Hand Mules to produce
    359,424 hanks, No. 41's, ... 114½ days.
Produce per spindle per week, ... 22 hanks.
Cost of spinning @3/1¾ per 1,000 hanks, ... £56   10   8

Time taken by Self-actors to produce
    359,424 hanks, No. 41's, ... 96 days.
Produce per spindle per week, ... 24 hanks.
Piecers' wages, 1/9 per 1,000 hanks, ... £31   9   0
Extra attendance of spinning master & mechanic,   4   4   1   35   13   1

*Gain by Self-actors, ... £20   17   7
*It must be observed that no allowance has been made in the above statement for extra capital, charges, or insurance. From 20 to 25 percent should be added for these, which would still leave about £13 in favour of self-actors.

The two statements which follow refer to Messrs. Sharp, Roberts & Co's. self-acting mules, and exhibit the produce, &c. of two Mills in which they have been extensively adopted.

      1st. "Produce, No. 50's, 19 hks: cost of piecing, 1/6 per 1,000 hks.
         .... No. 60's, 17⁹⁄₇ do. .... 1/7   do. do.
         .... No. 70's, 17   do. .... 1/9   do. do.

"It takes two men two weeks to fit up the head-stocks of a pair of mules. A spinning master may attend 10,000 spindles."

2*d.* "The speed of our spindles runs from 4,800 to 5,500 revolutions per minute.

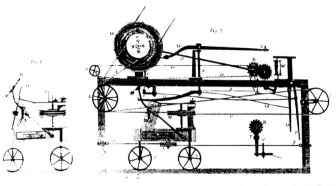

Fig. 18. Hand mule. Illustration from Abraham Rees (ed.), *The Cyclopaedia* (45 vols., London, 1802-1819) s.v. "Manufacture of Cotton" Plate XI.

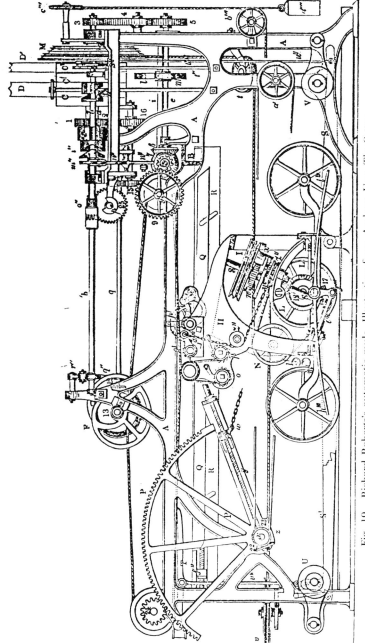

Fig. 19. Richard Roberts's automatic mule. Illustration from Andrew Ure, *The Cotton Manufacture of Great Britain* (2 vols., London, 1836) II, 174.

"Produce, No. 20's & 24's, 22 hks.:   cost of piecing,1/8 per 1,000 hks.
. . . . No. 36's   21 do.   . . . . 1/9 per 1,000 do.

"One spinning master may attend from 6 to 7,000 spindles. A mechanic is required for the same number; and the mules are expensive to keep up."

I have made particular inquiry regarding the self-acting mules said to be invented by Dr. Brewster, and I find that these were not mules, but rather a species of self-acting common jenny, invented, not by Dr. Brewster, but by a machinist of the name of Brewster.[43]

The chief peculiarities of Brewster's self-acting jennies, consisted in having the spindles placed in a horizontal position, about 12 or 14 inches off the floor; and the clasp, together with the creels, containing the rolls or coarse[v] rovings[v] rose and fell in a perpendicular direction. That is, the clasp ascended, while the yarn was being stretched and twisted; and this completed, the spindles backed off, and wound on the twisted yarn: while winding on, the clasp descended till within a few inches of the spindles, and again commenced another operation, and so on, alternately ascending and descending, whilst the spindle frame was stationary. These jennies were only used in Woollen Factories, but are now entirely abandoned, the same process being[w] performed by the common stretching mule, similar to those used in cotton spinning.

There is a species of self-acting mules used at Nashua, New Hampshire, which are said to be very imperfect and complex, and not likely to be adopted by any other manufacturing company, as not being considered equal to the common hand mules.[44]

SPOOLING OR WINDING MACHINES

Many of the old Factories of Great Britain are exclusively occupied with machinery for manufacturing cotton yarns, whilst others are entirely appropriated to weaving. In this country there are but a [x]few small Mills, chiefly about Rhode Island, that are employed for spinning only; a considerable portion of the yarn thus manufactured is used for sewing thread and stocking yarn. All the respectable establishments have both spinning and weaving conducted under the same roof.

In those departments of the cotton manufacture which relate
to carding and spinning,[y] this country is much behind[z] Britain,
especially in the carding. But in those which relate to common
power loom weaving the Americans have[a] equalled, and in some
things surpassed, anything I have yet seen, either in Glasgow or
Manchester. In fancy weaving, either by power or hand, this country[b] has made little progress.[b]

Amongst that series of machines which belongs to power-loom
weaving, the first in order is the Spooling Machine, by which the
yarn is wound off copes, or small bobbins, on to larger ones, called
spools, for the warping machine.

Plate V. [Fig. 18] represents the spooling machine used in this
country, which ꞌis[c] superior to the common winding machine of
Great Britain. It is certainly much more simple, and can be attended by girls from eleven to fourteen years of age, who, upon
the American spooler, are capable of doing as much work as a
woman of thirty, upon the British.

Fig. 1*st* represents a common cylindrical shaft containing sixteen
drums: A A A A are four bobbins, or spools, laid on two of the
drums, and driven by friction: B B represent cast iron arches
placed between each pair of drums, which serve to keep the spools
in their places (*see a a, Fig.* 2*d.*); each spool has small projecting
ends, which serve instead of skewers (*see Fig.* 1*st*): *c c* are the
bobbins from the spinning frame: *n n* are round pieces of iron
covered with cloth, lying on the moveable rails *o o.* Pieces of cloth
are also fastened on the rails beneath the cleaners *n n,* so that the
thread passes through between two plies of cloth, which partly
smooth down the fibres, and clean it from any loose specks that
may adhere to it: *e e* are guide pins fastened on the rail *o o.* The
pulley E, driven by a band from the cylinder shaft, is connected
with a heart motion, which moves the rails *o o* alternately in a
horizontal direction the full length of the spools, and by means
of the guide pins *e e,* causes the yarn to wind on equally from
end to end. Each drum is covered with cloth or leather, and requires to be perfectly true, as otherwise it would give a vibratory motion to the spools while the yarn is winding on.

The whole machine is extremely light, simple, easily attended
and kept in order. One of twenty drums may be attended by two
girls of twelve years of age, and is capable of winding 3,000 hanks
per day of $12\frac{1}{4}$ hours.

PLATE V.

SPOOLING or WINDING MACHINE.

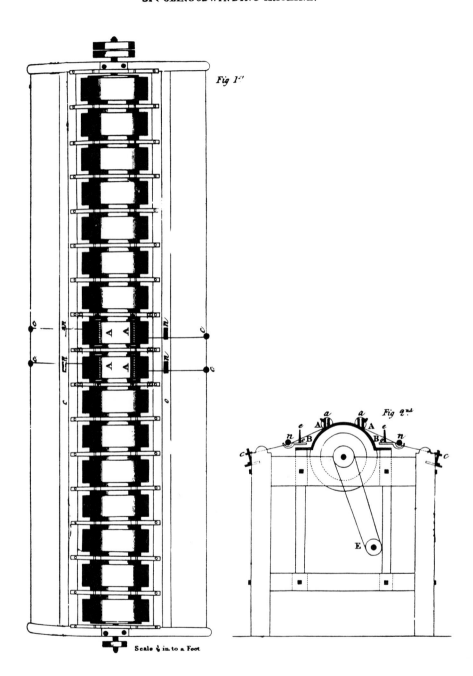

Fig 1st

Scale ⅛ in. to a Foot

Fig 2nd

E

Glasgow Published by John Niven Junr.

Swan S.

Fig. 20.   Spooling or winding machine, plate V from *CM*1.

Instead of winding the yarn off the small bobbins on to others of a larger size, it is common in a number of Factories to take the small bobbins direct from the spinning frame to the warping machine, which is mounted with a small rack or creel suited to the size of the bobbins. This creel, rack, or bobbin frame, is attached to the back of the warping machine, and lies in a horizontal position, but hollowed in the centre like a cradle; hence it is denominated the cradle warper. The girl who attends this machine stands with her face towards the back of the warper, having the bobbin frame intervening; she thereby has all the bobbins within her reach, so that whenever she perceives one about empty, she is ready to remove it, replace it with a full one, and tie the two ends of the thread, without stopping the machine. And owing to the number of bobbins in the frame, and the small quantity of yarn contained on each, they are constantly emptying, while the attendant is constantly supplying their places with full ones: but in order to prevent them from running out entirely, she requires to take out a considerable number before the yarn is completely wound off. The yarn, therefore, which is left on the bobbins, if not wound off at some other machine, is[d] made into waste. Hence the cradle warper has not been generally adopted, as it has been found that the loss from the quantity of waste made by it, is greater than the expense required for spooling, or winding the yarn from small bobbins on to others of a larger size, suited to the common bobbin frame of a warping machine.

WARPING MACHINE

[e]The next in the series, is the warping machine; and as[e1] those generally used[e1] in this country, contain a very ingenious invention, by means of which, when a thread breaks, the machine stops instantly. Two views of this warper are given in Plate VI. chiefly for the purpose of illustrating this rather curious contrivance,[e11] invented by Mr. Paul Moody,[e11] and not by Mr. Perkins as stated in a former edition of this work [Fig. 19].[e45]

Fig. 2d, Plate VI. [Fig. 19] is a side view of the warper: A A, *figs. 1st and 2d,* the framing of the machine, which, being made of wood, gives it a heavy appearance in the engraving: B, *figs. 2d and 4th,* are the threads of yarn proceeding from the bobbin frame to the iron plate *c,* where each thread is separated; the plate *c* being perforated with small holes corresponding to the number

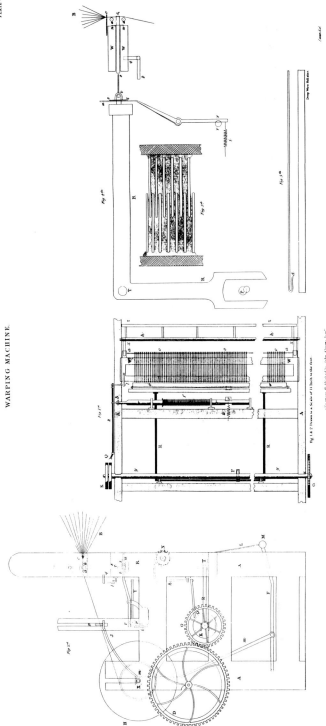

Fig. 21. Warping machine, plate VI from CM1.

of threads to be wound on the beam. Passing the plate *c*, where all the threads are brought into one horizontal plane, they thence pass over the rods *a a;* from these through the guide reed *p*, and on to the beam H, which is represented as containing only the first round of yarn. The belt pullies K are on the same shaft with the wheel G, which drives E; on the same axis with E, is the drum D, which drives the yarn beam H. The iron axis of the yarn beam rests in two slots of the framing at X, and is pressed down upon the drum D, by the stirrup *m m*, which is also weighted down by the cross lever F. From the top of the stirrup *m*, an arm J extends to the guide reed *p* so that as the yarn fills on the beam H, it gradually rises, and the arm J presses up the guide reed with the same gradual motion, so as to keep it always in a proper position in relation to the increasing diameter of the yarn beam: L, *fig. 2d*, is a strap attached to the weight M, and which winds round a small shaft, on the end of which the ratchet wheel N is fastened. When the yarn beam is sufficiently full, the strap L is wound up by means of a wrench attached to the ratchet wheel, which thereby lifts the weight M, the lever F, and the stirrup *m*, until the hook on the axis of the yarn beam at X, is so far relieved as to be pressed back: the full beam is then removed, and an empty one put in its place—the stirrup is brought forward till the hook is right above the axis of the beam—the catch of the ratchet wheel is lifted—the strap unwound—and the machine is then ready to warp another beam.

From the preceding description, it will be seen that this warping machine in its general operations, differs very little from those used in Great Britain.[46] In every respect it is equally as simple, efficient, and easily attended; besides having the advantage of the stop-motion, which is now to be described.

As the yarn from the bobbin frame enters the plate *c*, it passes over the rods *a a;* but between these rods, there is a drop-wire suspended upon each thread: these drop-wires are pieces of flattened steel wires, about four inches long, from $\frac{1}{8}$ to $\frac{3}{16}$ broad, and $\frac{1}{16}$ of an inch thick. (*See Fig. 5th.*) Some of these which I weighed, varied from 4 grains 4 dwts. to 4 grains 10 dwts. They are hooked at the top, and suspended by their own weight on each thread. (*See o o o, front view, fig. 1st, and o, fig. 4th.*) When the machine is in operation, the drop-wires are borne up by the tension of the yarn, but as soon as any one thread breaks, it slackens, and, of course, the wire drops down till the point of the hook at *d, fig. 5th*, rests on the plate *n n, fig. 4th;* and it is this dropping

Fig. 22.   See caption on next page.

down of the wire that stops the machine. The shaft $y\,y$ extending across the machine, has an eccentric at P, *figs. 1st and 4th,* which works into the fork of the lever R R. On the top of the lever R R, there is a small tumbler $s\,t\,s$ attached to the steel plate e, *figs. 1st and 4th.* The lever R R turns upon a journal at T, *figs. 2d and 4th;* and in consequence of the eccentric P working into the fork, the top of the lever, and with it the tumbler $s\,t\,s$, and the plate e, are made to oscillate right under the drop-wires; so when a thread breaks, the wire drops down, and retards the oscillating motion of the plate e, which immediately depresses either end of the plate $s\,s$ of the tumbler, which again presses down the lever $u\,v$ at u, and raises the other extremity at v. By lifting the lever at v, the rod $r\,r$, being then disengaged, is operated upon by the spiral spring f, *fig. 1st,* which causes it to shift so far as to act upon the upright rod k, and turn it round as far as to make the belt lever at Q, shift the belt from the fast on to the loose pulley. And as these various parts are fitted so as to operate all at once, the whole machine upon the breaking of one thread is instantly stopped.

When the broken threads are all tied, and the machine ready to be put in motion, the girl attending, lays hold of the rail $z\,z$, *fig. 1st,* and pulls it forward: $x\,x$ are straps of leather fastened to the wooden frame W W, containing the drop wires; therefore, by drawing down the rail $z\,z$, the shaft $h\,h$ turns round, and causes the straps $x\,x$, to raise the frame W W so far as to lift all the drop-wires above the top of the plate e, which keep their places by the tension of the yarn, as soon as the machine gets into full operation. In lifting the drop wire frame W W, it also draws up the point q, of the small lever $q\,l$, *figs. 2d and 4th,* which causes the other extremity l, to operate upon an arm of the upright rod k, and turn it round as far as to let the belt lever at Q shift the driving belt from the loose, on to the fast pulley: at the same time another arm Y, of the upright rod k, *fig. 2d,* also operates upon the rod r r at b, *fig. 1st,* and shifts it to the right hand, until the point v of

◄ Fig. 22. Stop motion in twisting frame similar, though not identical, to that in the warping frame illustrated in *CM*1. The photo shows the back of the twisting frame. The stop motion consists of a bolt held held back by a coil spring and resting on the axle of an oscillating steel bar (set about five inches above, but parallel to, the axle). When a yarn broke it released a drop wire which fell into the path of the oscillating bar. The sudden jolt released the spring bolt which then shot against the shipper (the fork guiding the driving belt), knocking the driving belt from a fixed to a loose pulley in the machine. The picture shows the Merrimack Manufacturing Company's twisting frame no. 3, built at Lowell in the period ca. 1830–1840 and presently in the custody of the National Museum of American History, Smithsonian Institution, Washington. Author's photo, courtesy of the Smithsonian.

the lever *u v*, drops into the square groove seen in *fig. 1st:* the lever or catch *v*, is kept in the groove of the rod *r r*, by the small spiral spring *i*. Thus, by pulling forward the rail *z z*, the drop-wires are lifted, and the whole machine is instantly put in operation; and, by lifting the catch *v*, the rod *r r* being operated upon by the spiral spring *f*, it is as instantly stopped.

Fig. 3d is a front view of the guide reed *p*, seen in *fig. 2d*, for directing the yarn on to the beam H. It consists of a piece of sheet iron cut into a number of slits, corresponding to the number of threads to be warped on the beam. By examining the figure, it will be seen that the slits are so contrived, that a lease may be formed on each beam if necessary.

The views of the warping machine given in Plate VI, are chiefly designed to exemplify its general movements, but more especially those of the self-acting stop-motion, which, by carefully examining the figures, and attending to the descriptions given, may be easily understood. *Fig. 4th* is also represented in its proper place in *fig. 2d*; the former being drawn to a larger scale, merely for the purpose of giving a more accurate representation of its operations.

Those unacquainted with this machine might suppose from the representation here given that owing to the number of levers, springs, &c. depending on each other, it would work inaccurately, and be difficult to keep in order. This, however, is not the case. Although acquainted with the most improved warping machines made in Glasgow up to the year 1836, and having had opportunities of observing the operations of the newest made in Manchester, particularly in Mr. Orrell's Factory at Stockport, and in other places[47]; I have not seen any that made more perfect work, or required less attention, than those just described. The warping machines used in Great Britain require the utmost attention on the part of the attendant to notice[f] when a thread breaks; as should her eye be diverted from her work but one moment, the end of a broken thread might wind round the beam so far, as to require five minutes or more to find it, and put the machine again in motion. But this is not the case with those used in this country; for while the machine is in operation, the attendant is frequently behind the bobbin frame, taking out empty spools, and supplying their places with full ones; nor could the cradle warpers of this country be used, except by being furnished with a self-acting stop-motion. This motion is, therefore, eminently entitled to the appellation of an important labour-saving improvement.

I omitted to state in its proper place, that the drum D, *fig. 2d*, on which the yarn beam rests, and by which it is moved, is exactly one yard in circumference, and upon the one end of its axis, there is a screw working into small geared wheels connected with an index, which indicates the revolutions of the drum during the warping of each beam, from which the length of yarn on each beam is ascertained, and the attendant is paid accordingly

Upon an average a smart girl can warp 13 beams per week of No. 12's, each beam containing 300 threads of 3,000 yards in length, for which she is paid about[g] 22 cents per beam, being 2 dollars 86 cents, = 11/11[g] for 13 beams, or, per week. Another ordinary hand can average 9 beams per week of No. 18's, each containing 300 threads of 5,000 yards in length, and for which she is paid[h] 30 cents per beam, being two dollars 70 cents, = 11/3[h] for 9 beams, or, per week. Besides finding the length of the yarn warped each beam is weighed and a correct account of the weight regularly kept.

DRESSING MACHINES.

The Dressing Machines used in this country, are made upon an entirely different construction from[i] those used in England and Scotland.[i]

[j]They are not so complex; and consequently, easier attended and kept in order. They require much less power and oil and might be made in Britain for about two thirds of the cost of those now generally used in Manchester and Glasgow.[j]

Plate VII. [Fig. 21] represents a side view of the dressing machines used at Lowell, and generally throughout the Eastern district. A A is the centre frame, supporting the centre beam *a*, containing the dressed yarn. The wheel E on the centre beam, is fastened with a set screw, so as to be easily taken off when a full beam is to be removed, and put on the empty beam which is to replace it. B B B to the right and left are the section frames, all made of wood, containing the section beams H H H H. The ends of the section beams are of cast iron, with a square groove for receiving a friction strap, and a weight represented at *v v v v*. The sizing rollers are represented at *y y y y*. The yarn as it leaves the section beams, passes through a raddle or ravel, made of small pieces of hard wood, and represented at *u u u u*,—from that, through between the sizing rollers,—again, through a brass wire

LOWELL DRESSING MACHINE.

PLATE VII

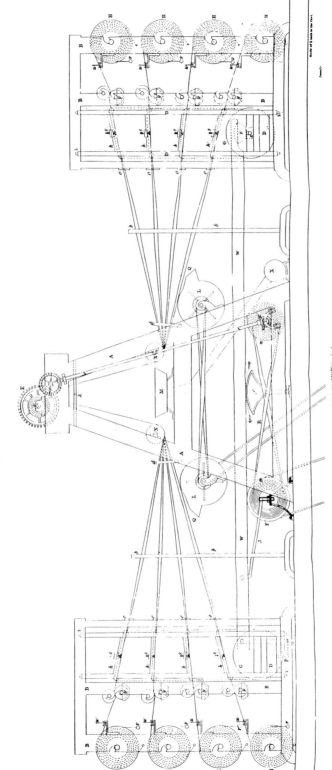

Fig. 23.   Lowell dressing machine, plate VII from CM1.

reed at *o o o o*,—through a copperplate supported by *b b*,—through another brass wire reed at *d d*,—and under the measuring rollers N N,—at which place the yarn from the four beams on each section are, for the first time, brought all into one horizontal plane. From the rollers N N, the two sections of dressed yarn pass up through heddles at *k*, called the lease harness; from that it is wound on the centre beam *a*, at the top. The lease harness may be shifted to either side of the frame by means of a screw at *k*, and only one section of the dressed yarn goes through heddle eyes in the lease harness; so that when a full beam is to be removed, a lease rod is introduced between the two sections of yarn above the harness, then, by drawing the harness to one side, another lease is formed, into which a rod is introduced.

The different operations of the machine are effected in the following manner. F are the fast and loose belt pullies, driven by a belt from the room below; *n n* are two cones, that to the right being driven from the one to the left, by the cross belt R. On the axis of the cone to the right hand, there is a small bevel wheel working into another on the bottom of the upright shaft *r r:* on the top of the upright shaft, there is a small bevel wheel working into the wheel *c*, and on the axis of the wheel *c*, there is a small spur gear, not seen in the engraving, working into the wheel E, on the centre beam *a*. Motion being thus given to the cone on the left hand (by the belt pulley F which is fastened on the same shaft,) it is next communicated by the belt R to that on the right, and from it to the beam on the top, containing the dressed yarn; hence the speed of the centre beam *a*, on the top of the centre frame, may be increased or decreased, by shifting the belt R, on the cones *n n*.

The brush motion is next to be considered: D D D are the brush racks, or brush frames; they are not fastened to the section beam frames B B B, but are fitted so as to move up and down, short spears *z z* being fixed to the top and bottom of each side of the brush frames, which slide into the eyes of studs, and serve to keep them in their proper positions, as well as to let them move freely up and down: *s s s s* are small blocks of cast iron, which are fitted to slide freely on the polished steel rods *h h h h:* the dotted lines represent straps or belts passing over small pulleys on each side, and descending down to the large wooden pullies G G, to the surface of which the belts are fastened: the blocks *s s s s* are fastened to the belts by a small nut and screw on the under side, whilst the brushes rest on the blocks above. The feathers represented on the blocks at *s s s s*, fit into slits in the ends of the brushes. W W represent two beams of wood, (one at each side of the machine,) about four inches broad, and three inches thick, called sweeps; these are supported in the centre at *f*, and at the end towards the left hand, they are attached to the lever P P, the under point of which supports the whole brush frame; the other end of the sweeps being attached to the block *i*, towards the right hand: the block *i* is a projection from a shaft, extending across the machine at each section, the axis of which is seen at *x*, and the pullies G G are fastened on each end of this shaft. By carefully examining the engraving of the various parts of the machine, the reciprocating motion of the brushes, together with the up and down motion of the brush frames, will now be easily understood. The lever J is connected with a sliding crank on the axis of the cone *n*, towards the left hand; consequently, the revolving of the crank moves the sweeps alternately, from section to section; and the end of the sweeps to the right hand, being attached to the block *i* by a strap of belt leather, the alternate motion of the sweeps moves the shaft *x*, and with it the pulleys G G, about one-fourth of a revolution each way: this reciprocating motion of the pulleys, draws each end of the straps represented by the dotted lines, and thereby produces the necessary reciprocating movements of the brushes upon the yarn; while, at the same time, the other end of the sweeps towards the left hand, by means of the lever P P, raises the brush frame, and with it the brushes, up and down at every alternate stroke. Thus, when the brushes are at *h h h h*, the frame is down, and they are then in a proper position for moving along the surface of the yarn, (which has just been coated with size, in passing through between the sizing rollers,) and having made one full stroke, they are then at the opposite side of the frame, which is immediately raised by the lever P P, connected with the sweeps; and in raising the frame, the brushes are lifted out of the yarn, until they return to their former position at *h h h h*. The whole movements of the machine commence at the cone *n*, towards the left hand. From it, motion is communicated by the belt R, to the opposite cone, and from it, to the centre beam *a* at the top: and from it (the cone *n*), motion is also communicated by the sliding crank, and the connecting lever J, to the sweeps W W, of which there are two, and the one end of each moves the brushes alternately from side to side, whilst the other end produces the up and down motion of the brush frame. The whole machine is extremely simple, and all its different movements so contrived, that they can easily be adjusted so as to operate with the most perfect accuracy.

L L represent the fanners enclosed in wooden boxes, open only at the centres for admitting a current of air, and at the mouths Q Q, for throwing it out: by this method of confining the air, it rushes out with much greater force, and the mouths Q Q are made so as to direct it right up amongst the dressed yarn. The fanners here represented have four wings each, but some have only two or three: that to the left hand, is driven by a belt from the room below, and from it, a cross belt communicates motion to the one on the right. X is a hot air pipe, with a branch extending up to the hot air box M, placed between the two rollers N N. The cover of this box extends till within one half-inch of each side, which leaves a small opening for the escape of air, which issues out at each side upon the yarn, and being entirely hot air, it has a peculiar effect in absorbing any remaining moisture upon the yarn, before it is wound on to the centre beam a, on the top. Instead of the hot air box M, some dressing frames have a centre fanner, similar to those used about Manchester.

The sizing rollers $y\,y\,y\,y$, are generally made of soap stone, with an iron axis; the under roller only is covered with cloth: one of these rollers, when finished, costs about eight dollars = £1 13s 4d.

The two sections of these dressing machines may be extended out as far as may be thought necessary. In order to diminish the size of the plate, the distance from the centre of the section beams to the centre of the frame, is represented as only $9\frac{1}{2}$ feet; some, however, extend to 17 or 18 feet. As the greater the distance from the sizing rollers to the centre beam containing the dressed yarn, more time will be gained for drying; but when the section beams are stretched out too far, the yarn is more liable to break with the drag of the centre beam.

On one end of the axis of the measuring rollers N N, there is a screw or worm working into gears, connected with an index which points out the number of yards of dressed yarn on the centre beam; every 33 yards is marked with paint, which allows 30 yards of cloth to each piece, the 3 yards, (equal to 10 per cent.) being allowed for shrinkage in the weaving, &c.

The measuring rollers, in general, are common wooden cylinders, about eight inches in diameter; some are revolving steam cylinders, which, when properly packed at the journals so as to prevent the steam from escaping, have the best effect of anything that has yet been tried for drying the yarn speedily.

It is impossible to give even the average produce of these machines; indeed, in all my experience, I have never known anything so variable as the produce of the dressing machines of this country. In some Factories, their average produce will be about[k] 18 pieces[k] per day to suit 900 reed; yarn, Nos. 14's to 18's. In other Factories, on the same kind of work, these machines will dress[l] 30, in others 40, 50, and 60 pieces[l] per day. Some of those fitted up with revolving cylinders, are said to produce even 70 pieces on the same kind of work, viz. coarse 900's, yarn No. 14's, and each piece is invariably intended to make 30 yards of cloth. One[m] cause of this difference in the quantity produced from the dressing machines, arises from the different temperatures in the apartments where they are in operation, and the mode of applying the heat to the dressed yarn. In some Factories the dressing machines are in the same room with the looms, where the temperature seldom exceeds 75°. Those mounted with steam cylinders,* in place of wooden measuring rollers, generally produce[n] a greater[n] quantity of work; next are those with the hot air pipes; and next to the latter, those with three fanners, that is, one at each side, and one in the centre. Those with only two fanners, produce the least quantity of work.

The prices paid for dressing, are as variable as the produce of the machines. Some Factories pay as low as $1\frac{1}{2}$ cents per piece, whilst others pay as high as[p] $2\frac{1}{2}$ cents = $3\frac{1}{4}$d.[p] sterling, for the same kind of work. In fact, the dressing, as well as some other kinds of work, are paid more or less, according to the quantity the machines are able to produce. The great object with all manufacturers in this country, is to pay their help just such wages as will be sufficient inducement for them to remain at the work. Hence the greater the quantity of work produced, the higher the profits, because paid at a lower rate of wages.

---

* The steam cylinders for measuring rollers were invented by Samuel Batchelder, Esq., Agent of the York Factories at Saco, State of Maine, for which he obtained a patent in 1835.[o48] It may also be remarked that this mode of drying makes a very material improvement on the yarn. As every practical manufacturer knows that the quicker the yarn is dried in the process of dressing it will work much better in the weaving and greatly improve the quality of the cloth.

Every dryer knows the sooner his yarn is dried after having passed the process of colouring that the colors will be brighter and purer. The advantage of quick drying holds equally good in both cases.[o]

The size used for dressing, is generally made of potato starch for coarse work; and of flour, for the finer, or such as are intended for printing. The mode of preparing the starch for size requires particular attention; and although different places may have different methods, the two following have been found to suit the purpose remarkably well.

First,—*Method of making Size from Potato Starch for coarse Goods.*

qFrom 80 to 85 lbs of dry potato starch, are mixed up with eight common ten quart pails full of warm water. And during the process of making up this batch, a person keeps stirring it with a poll or stirring stick, so as to make the water incorporate perfectly with the starch. The two last pails of water, should bring up the heat of the whole batch to 140°. It is then allowed to stand from 10 to 14 days in a temperature of 70°, until it is thoroughly fermented. After which the said water, and whatever has been thrown up by the fermentation, is skimmed off, and the whole after being passed through a sieve, is emptied into a boiler containing 150 gallons of water at a heat of 190°, where it is allowed to simmer about ten minutes; it is then emptied out, and fit for immediate [use?].q

The above makes a very superior size. It is smooth, clean, and entirely free from any offensive smell: and although about the same price as flour, it is found to answer the purpose much better for coarse goods; very little of it adheres to the yarn, yet quite enough to make it weave well.

Second,—*Method of making Size from Flour for the finer Goods.*

300 lbs. of flour mixed in 45 gallons of water, and allowed to stand for four or five days at bloodheat, until it is perfectly fermented; this is called yeast. To the above are added about 140 gallons of water heated to 180°. The whole is then boiled by steam from 30 to 45 minutes; at first it boils thick, but by continued boiling becomes thin in the middle, when it is considered done; after which it should stand over one week, and be reduced with cold water when used.

The following mode of making Size from Flour, is practised in Glasgow for various kinds of Goods.

One barrel of flour is soaked in water, which had been previously heated a little over 120°, and allowed to stand in this state about a week, or until it ferment thoroughly. It is then mixed with about 110 gallons of water in a copper boiler, with a cast iron casing; and by introducing steam into the boiler, as well as into the vacant space between the casing and the boiler, it is gradually heated until it boils; after which the steam may be admitted at any pressure, and the boiling process continued about an hour; during which an agitator, driven by the engine, moves round with a slow motion, until all the concretions or lumps are completely dissolved, when a wooden roller being dipt into it, if the small portion which adheres to the roller has a thick, smooth, glutinous appearance, it is then ready to be emptied out into narrow deep vessels to cool, in which it is allowed to stand for three or four days before using it.

It now only remains to be noticed, that one of these dressing machines when complete, costs in this country 400 dollars = £83 6s 8d, and upwards. And throughout all the Eastern manufacturing district, girls are employed to attend them. Men are employed as dressers about Rhode Island and some other places; but so far as I am informed, there are none in this country make so high wages, as the same class of workmen generally average about Glasgow, (Scotland.)

Besides the kind of dressing frame just described, there is another made at Providence, Rhode Island, and generally used throughout that part of the country, known by the name of Pitcher and Gay's dressing frame.[49] The principal difference between it and the one already described, is, that the former has four pairs of sizing rollers on each section, while the latter has only two; that is, the yarn from the two upper beams passes through between one pair of sizing rollers and that from the two under beams through another pair. By this and some other contrivances, it appears somewhat more simplified than the other, and, I believe, will be generally preferred by workmen, as being more convenient and easy to attend.[r]

[s]The English dressing machine for dressing the yarn in hot size has just been introduced into this country but has not yet been sufficiently tested to form a correct estimate of its merits.[50] The method of dressing and dying yarn in the chain has long been practised with great success.[s]

There is one peculiarity of the American dressing frames, viz. their requiring but little oil, so that the machines, and all about them, are so perfectly clean that the girls who attend them, have no appearance of being even in the vicinity of anything that could soil their clothes. This is not the case with those used in Glasgow, for there no class of workers employed about a Cotton Mill, seem to have dirtier work than those who attend the dressing machines. In some of the new Factories of this country, the floor under the dressing machine is painted a high cream colour, which by a slight washing over once in two weeks, has always a fresh cleanly appearance, not a drop of oil being seen all around it.

WEAVING BY POWER

'It has already been stated that the Americans have attained to as much perfection in all that relates to common power loom weaving as the British. The crank looms are now generally used and as regards steady smooth working machines or the quantity and quality of work produced they are in every respect equal to those made by the best machinist in either Manchester or Glasgow.

It is not to be understood that all the looms now made in this country are equal to those referred to in Britain[u] but the crank looms made at Lowell, Saco and by Pitcher and Brown [of] Pawtucket are certainly not excelled by any now in operation either in this country or elsewhere [Fig. 22].[u' t]

The looms[u] used throughout this country are generally[u] mounted with a letting-off motion as it is called in Britain; and with a very few exceptions, they have all self-acting temples. After witnessing the simplicity and efficiency of these contrivances[v] it is surprising that[v] they have not been more generally adopted in[w] Britain. The self-acting temples, besides saving a great deal of labour on the part of the attendant, make a much superior, and more uniform selvage,[51] whilst the letting-off motion equalises the number of picks in the cloth, by giving off from the warp beam, exactly as the cloth beam takes up [Fig. 23].

It is unnecessary to describe either of these improvements, as it is presumed they are already sufficiently known in[x] Britain. When in Manchester in April 1836, I met with Mr. Amos Stone from Rhode Island, who was then introducing his power loom for weaving silk, which contained the self-acting temples and letting-off motion, upon a much improved principle, for which he had taken, or was about to take out, a patent: and since that time,

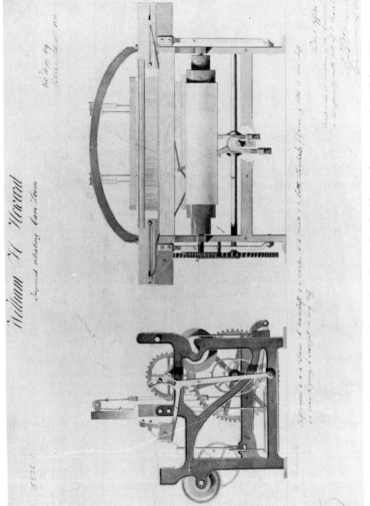

Fig. 24. Plain cotton powerloom, iron framed and with wooden accessories, of the 1830s: illustrated in the U.S. patent of William H. Howard of 12 February 1830 in the U.S. National Archives, RG 241, patent drawing no. 5826. Courtesy of the National Archives.

Fig. 25.   U.S. patent drawing of rotary temples patented by Ira Draper, 7 June 1816 in U.S. National Archives, RG 241, patent drawing no. 2608. Courtesy of the National Archives.

they have been partially adopted in various parts of^y Britain.[52] In this country there is some variety in the form or plan of these contrivances; yet I have not seen any superior to those of Mr. Stone. But there is another little improvement which has been added^z I do not recollect having seen on Mr. Stone's loom, viz. a self-acting apparatus for shifting the pace weight gradually towards the extremity of the lever, as the cloth beam increases in diameter, by the continued winding on of the cloth.

The ratchet wheel, which moves forward the cloth beam, is itself moved by clicks attached to a perpendicular lever at the side of the loom, which (lever) is operated upon by a stud fastened in the sword of the lathe; and on a horizontal arm of the lever, there is a weight suspended, denominated the *pace weight*. Now every practical man knows, that this weight acts with more power when the cloth first begins to wind on an empty beam; and as the beam gradually fills with successive layers of cloth, the effect of the weight diminishes in a corresponding ratio. The improvement here referred to, is a simple contrivance, by which the weight is shifted progressively towards the extremity of the horizontal arm of the lever, so as to keep the cloth always at a uniform degree of tension, which, together with the other contrivances, make a perfect uniformity in its quality.

[a]The mode of propelling the shuttles[a1] is the same as in the looms used in Scotland,[a] with this difference, that instead of shuttle cords made of coarse cotton yarn or roving, strips of leather are used, which have been tanned and prepared for the purpose. These are called picker-strings, and are made to pass up through a slit in the bottom of the shuttle box of the lathe, and fastened with nails to the under part of the shuttle drivers, which are made of Buffalo hides. The picker-stick, as it is technically denominated, is also fastened to a cross bar of cast iron, which connects the under ends of the two swords of the lathe, with which it likewise reciprocates, so that in every position of the lathe, the picker-stick is always perpendicular to it; and the treadles, by which the shuttle is propelled, are pressed upwards.[b] This connection of the picker-stick with the lathe gives the latter more of a trembling motion than if it was fastened either to the cross bar of the loom or on the floor.[b]

In order to ascertain how long one of these leather picker-strings might last, I have kept account of the number of sides of picker-string leather used during one year, in a weaving room containing upwards of 100 looms, together with the average number of picker-strings cut out of each side; and according to this, I find that each string upon an average, will last between eight and nine months; each side of leather will make between 80 and 90 strings. Picker leather costs about 50 dollars per dozen sides in Boston.

The heddles, or, as they are commonly called in this country, the harnesses, are generally made of cotton yarn: each Factory makes to suit themselves; but the yarn must always be of a good quality, and generally spun from double roving. Three plies of yarn No. 8's make very good heddle twine for weaving heavy goods, from yarn sizing No. 24, and downwards. Before using the harnesses, they receive two coats of varnish made of the following ingredients:—

'Varnish for harnesses
1 quart of the solution of Indian Rubber*
5 do   good copal varnish
4 do   shelac varnish
1 do   boiled linseed oil
3 oz gum of camphor
1½ oz bees wax

The above are mixed over a slow fire and well stirred until the wax is completely dissolved. After removing from the fire one quart of the spirits of turpentine is added. And when used, if found too thick, add a little more.' Before putting on the varnish, the harnesses ought to be brushed down with paste or size from the dressing machine: and after the varnish is thoroughly dry and hard, they ought to be brushed down with tallow, to smooth them well before they are put to use. Harnesses properly made in this manner, and perfectly dry before they are used, will generally last over one year.

A patent has been taken out in this country for a peculiar method of forming the eyes on heddles, which consists only of a double knot, one on each side of the eye [Fig. 24].

Fig. 26.   Patent knot for heddles, wood engraving from *CM1*, p. 105.

The one side A, is perfectly straight, whilst two single knots are formed on the side B, each of which encloses A, and when drawn tight, the heddle-eye is formed between the two knots, so that in the up and down motion of the harnesses, the warp threads are pressed against the knots, which being hard tied, the warp is not so likely to cut the eyes, as if they pressed against a single loop of the heddles.

These patent heddles are made or knit upon a revolving frame represented in the annexed figure [Fig. 25].

* The solution of Indian Rubber is made by soaking one pound of caoutchouc in four gallons of the spirits of turpentine until the whole is completely dissolved which may require four or five weeks. The solution is then to be strained clean and kept for use.

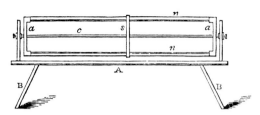

Fig. 27.   Harness knitting frame, wood engraving from *CM*1, p. 106.

The above figure represents what is denominated a harness knitting frame: A is a plain top of a common form, or stool, supported on the legs BB. The centre rod *c* is oval shaped, and made of polished steel or iron: *a a* are the two ends of the knitting frame fastened on the centre rod *c,* and as this rod revolves, the two ends revolve with it: *n n* are the two shafts of the heddles or harness, which fit into slits of the revolving end of the frame *a a,* and are fastened by set screws: *s* is a moveable binder fastened by screws to each shaft, intended merely to prevent them from twisting or bending, and can be moved from side to side, or taken off at pleasure. The eyes of the heddles are formed on the centre rod, at each side of which the knot is tied; the under shaft is then turned up, and the loop formed by the cord; then the other shaft is turned round, another loop formed, and then again the two knots, &c.&c.

One smart girl will knit harnesses on the above frame, to the amount of 120 beers (porters) per day of 12¼ hours, for which she is paid at the rate of[d] 75[dI] cents for 20 beers.[dI]

The average speed at which power looms are operated in this country may range from[dII] 115 to 125 picks per minute.[dII] [dIII]In Lowell the speed for all numbers of yarn was *formerly* from 120 to 140 picks.[dIV] But during the depression of 1842 it was found expedient to reduce that speed to from 106 to 114 and to give one girl the charge of three or four looms instead of two. By this arrangement a small saving was effected amounting to from ⅕ to ½ cent on the pound of cloth. It was doubtful however whether this arrangement would be carried out in the event of business reviving with an increased demand for goods. As it is acknowledged on all hands that two looms operating at 125 picks per minute is as much as any girl can attend to good advantage.[dIV] [53]

Shuttles are generally made of appletree and sold at [dV]43 cents each.[dV] Persimmon wood is sometimes used being superior to the apple and nearly equal to the box wood used in Britain.[dVI]

Shuttle drivers as already stated are made of Buffalo hides and cost in pairs from[dVII] 7½ to 8 dollars[dVII] per gross.[d]

Dr Ure[54] gives the following as the results of experiments which had been tried to ascertain the necessary power required to move certain machinery employed in the cotton manufacture.

	Mule Spindles with preparation.	Self-acting Mules with preparation.	Throstle Spindles with preparation.	Power-looms with dressing.	Power-looms without dressing.	Dressing Machine.
One horse power drives in the following Mills:						
Mr. Orrell's Mill, spinning No. 36's, ...	500	300	180	10		
Peil & Williams' Mill, ...				6		
Factory of J. A. Beaver, ...					15	
Clark & Son's Mill, ...					11	1
Average of the above, ...	500	300	180	8	13	1

The following are the results of experiments made at Lowell, in the State of Massachusetts.

	Throstle Spindles with preparation.	Power-looms with dressing.	Power-looms without dressing.	Dressing Machines	Throstle Spindles without preparation
One horse power moved as under:					
Spinning No. 14's, and producing 7 hanks per spindle per day, ...	77*	8.5	12	1.6	105
Spinning No. 40's, and producing 6 hanks per spindle per day, ...	127	11	12	3.8	{ 160 warp. 188 weft.

*The throstle frames used at Lowell are the dead spindle, which are supposed to require at least one-fourth more power than the live spindle.

ᵉThe preceding details embrace all that is deemed worthy of notice regarding the machinery employed and the mode of conducting the various processes in the cotton manufacture of this country. Their average produce and cost of attendance have been stated so that any person acquainted with the business may easily make comparisons. Something may have been omitted which ought to be noticed but nothing we believe has been introduced which upon examination will be found incorrect.ᵉ

From the number of American inventions that have been patented in Britain, many have supposed that the manufacturers of this country must have attained great perfection in labour-saving machinery. But it is not so.ᶠ For in regard to the most improved machinery, and the whole internal arrangement and management of the Mills, they are many years behind the British. The American machinery is generally neither so solid nor so well proportioned nor is it so easily adjusted and adapted to the particular uses for which it is employed. The British machinery is so constructed that all their moving parts may be adjusted with the greatest accuracy to suit the various qualities of cotton used or goods manufactured. The American machinery is greatly deficient in those little but necessary contrivances by which they might be altered and adapted to changing the system from one style of goods to another so as to take advantage of variations in the market or to change the system of working to suit any quality of cotton that may be used. Considerable improvements however are now being introduced. The late importations of English machinery will doubtless suggest many alterations that may be worthy of being adopted.ᶠ

In giving the preceding details of the various machines employed in the cotton manufacture of this country, I have endeavoured to point out their advantages and disadvantages, compared with the machines used in Great Britain. These will be interesting chiefly to practical men. To the minds of proprietors, an interesting inquiry will naturally occur, viz. *What may be the actual difference in the cost of manufacturing in the two countries?* In order to ascertain this as nearly as possible, I have endeavoured to give in the following pages correct estimates of the cost of erecting manufacturing establishments, and putting them in operation; together with their produce and expense of workmanship in both countries; from which a fair estimate may be formed of their comparative advantages, for manufacturing cotton goods.

COMPARATIVE ESTIMATES OF COTTON FACTORIES IN THE UNITED STATES OF AMERICA, AND IN GREAT BRITAIN.[1]

IN the following Estimates of the cost of Buildings, Machinery, &c. of Manufacturing Establishments in the United States of America and in Britain, each Factory is supposed to contain 128 power looms, with all the subordinate machinery required. The extent and form of the buildings, together with the number of machines, and the arrangement of the various departments are suited to the practice of each country. Likewise the number of hands employed—the rates of wages paid—the cost of materials—and the amount of goods produced—are also adapted to each country respectively.

In estimating the cost of building a Factory in the United States, the extent of the Mill is supposed to be four stories besides the attic. The first, or basement story, contains the water wheels, cloth room, and mechanics' shop; the second, contains the carding; third, the spinning; fourth and attic, the weaving and dressing; leaving the picking-house and cotton store for separate buildings.

The Factory in Britain is supposed to be five stories in height, and arranged in the following manner. The first and second stories contain the weaving and dressing; the third and fourth, the mule and throstle spinning; fifth, the carding; and the attic, the picking and scutching. At one end of the carding and spinning rooms, there may be small apartments partitioned off for a mechanic's shop, cloth room, &c. The engine and boiler house being situated at one end of the Mill, and the cotton store in a separate building.[2]

In the following, and all other computations of money throughout this work, the dollar is reckoned at par value, equal to 4/2 sterling, making the cent equal to one halfpenny; each dollar contains 100 cents, of 100 halfpennies, equal to 50 pence, or 4/2. When exchange is at par between Great Britain and the United States, 108 dollars here are equal to 100 in Great Britain, supposing the dollar 4/6.

COMPARATIVE ESTIMATES *of the cost of* BUILDINGS,
MACHINERY, &C. *and of the* EXPENSES IN WAGES,
&c. for a Cotton Factory in GREAT BRITAIN
and the UNITED STATES OF AMERICA.

BUILDING AND GEARING.

	AMERICA.				BRITAIN.		
	AMOUNT.				AMOUNT		
	Dollars.	Sterling.			Sterling.		
A Cotton Factory, built of Brick, 142 by 42 feet within the walls, four stories in height, besides an Attic, including all expenses for Materials, Labour, &c., ...	25,000	5,208	6	8			
Two Water Wheels, equal to 80 horses' power, including Gearing, Gates, Shafting, Belting, &c.	17,000	3,541	13	4			
A Cotton Factory, built of Brick, 90 by 38 feet within the walls, five stories in height, including all expenses for Materials, Labour, &c.	. . . . .	. . .			[b]1,000[b]	0	0
A Condensing Steam Engine of [a]35 horses' power,[a] with Boilers, &c.	. . . . .	. . .			660	0	0
Gearing, including Wheels, Shafting, Drums, Fitting up, &c., ...	. . . . .	. . .			290	0	0
Furniture, Gas and Steam Pipes, Lathes, Tools, &c. &c., ...	2,000	416	13	4	570	0	0
	44,000	9,166	13	4	[c]2,520[c]	0	0

MACHINERY
PREPARATION DEPARTMENT.

	AMERICA						BRITAIN				
	No. of Machines.	Rate.**	AMOUNT				No. of Machines.	Rate.	AMOUNT		
			Dollars.	Sterling.					Sterling.		
Willow, ...	1	100	100	20	16	8	1	£20	20	0	0
Scutching Machine, ...	1	600	600	125	0	0	1	58	58	0	0
Spreading Machine, ...	..	..	..				1	50	50	0	0
Carding Engines*, including Clothing,	40	210	8,400	1,750	0	0	64	28	1,792	0	0
Lapping Machines, ...	1	250	250	52	1	8	2	20	40	0	0
Drawing Frames of 3 Heads each, . .	6	200	1,200	250	0	0					
ᵈ4 Do. ᵈof 6 do. do.	..	..	..				ᶜ24hds ᶜ	9	162		
Double Speeders of 18 Spindles each,	6	660	3,960	825	0	0					
Extensers of 36 Spindles each, . . .	7	900	6,300	1,312	10	0					
ᵍ6 Fly-frames⁸ of 48 Spindles each, .	..	..	..				ʰ 288sp ʰ	381	542	2	0
Roving and Card Cans, ...	..	..	542	112	18	4			61	16	0
Top and Cylinder Grinders, Brushes, &c.	..	..	..								
. . . .			210	43	15	8			30	0	0
Miscellaneous Charges, ...	..	..	100	20	16	4			20	0	0
			21,662	4,512	18	4			2,775	18	0

N.B.—Some variations from the prices stated for the machines in the above and following Tables, may, no doubt, be found in the various places where they are made; but those assumed are believed to be a fair average for machines of good quality.

*In the above estimate the American Cards are understood to be 37 inches broad, and the British 24 inches. The diameter of Cylinder in both is 36 inches.

**'Rate' means cost of machine either as a single unit or pro rata per spindle, bobbin &c.

SPINNING DEPARTMENT.

	AMERICA.						BRITAIN.				
	Number.	Rate.	AMOUNT.				Number.	Rate.	AMOUNT.		
			Dollars.	Sterling.					Sterling.		
Throstle Spindles, ...	4,992	4.50	22,464	4,680	0	0	2,160	9/6	1,026	0	0
Mule Spindles, ...		. . . .	. . . .	. . . .		.	2,400	5/6	660	0	0
Machine for covering Rollers, ...		. . . .	. . . .	. . . .		.	1	£15	15	0	0
Rove Bobbins, ...	10,000	.6	600	125	0	0	74 gross	19/	70	6	0
Spinning Frame Bobbins, ...	12,000	.1	120	25	0	0	59 "	8/	23	12	0
Skewers, ...	6,000	1½	90	18	15	0	35 "	10/	17	10	0
Spools or Warpers' Bobbins, ...	6,000	.3	180	37	10	0	40 "	12/	24	0	0
Miscellaneous Articles, ...	. . . .	. . . .	100	20	16	8	. . . .	. . .	22	0	0
			23,554	4,907	1	8			1,858	8	0

DRESSING AND WEAVING DEPARTMENT.

	AMERICA.						BRITAIN.				
	No. of Machs.	Rate.	AMOUNT.				No. of Machs.	Rate.	AMOUNT.		
			Dollars.	Sterling.					Sterling.		
Spooling or Winding Machines, ...	6	70	420	87	10	0	3	£7	21	0	0
Warping Machines, ...	4	150	600	125	0	0	3	17	51	0	0
Dressing Machines, including Mounting,	9	400	3,600	750	0	0	6	49	294	0	0
Looms, including Mounting, ...	128	75	9,600	2,000	0	0	128	10	1,280	0	0
Twisting Frames, ...	. . . .	. . . .	. . . .	. . .			3	1	3	0	0
Miscellaneous Articles, Boilers, Tubs, &c.	. . . .	. . . .	100	20	16	8	. . . .	. . .	65	0	0
			14,320	2,983	6	8			1,714	0	0

COMPARATIVE EXPENSE *in Wages per fortnight in* BRITAIN *and* AMERICA.

PREPARATION DEPARTMENT.

	AMERICA.				BRITAIN.		
	No. of Hands.	Rate per week.	Amount per fortnight. Dollars.	Amount per fortnight. Sterling.	No. of Hands.	Rate per week.	Amount per fortnight. Sterling.
Attendants at Willow, ...	1	4	8	1 13 4	1	6/	0 12 0
Do. at Scutching Machine, . . .	1	5	10	2 1 8		6/	0 12 0
Do. at Spreading Machine, . . .	. .	. .	. . .		2	6/6	1 6 0
Do. at Cards, ...	2	3.50	14	2 18 4	4	4/	1 12 0
Do. at Lap Machine, ...	1	4	8	1 13 4	2	6/	1 4 0
m*					n*		n*
Do. at Speeders, ...	12	2.50	60	12 10 0			
Do. at Extensers, ...	3	3	18	3 15 0	p*		p*
o*	7	3.25	45.50	9 9 7			
Top Strippers, ...	4	4	32	6 13 4	4	7/6	3 0 0
Sharpers and Grinders, ...	2	4.75	19	3 19 2	2	7/	1 8 0
Overseer, ...	1	12	24	5 0 0	1	21/	2 2 0
Assistant for Overseer, ...	1	6	12	2 10 0			
	35		250.50	52 3 9	11		19 16 11

*Changes in *GM2* unclear here.

SPINNING DEPARTMENT.

	AMERICA.				BRITAIN.		
	No. of Hands.	Rate.	Amount per fortnight — Dol.	Amount per fortnight — Sterling. (£ s d)	No. of Hands or Quantity.	Rate.	Amount per fortnight — Sterling. (£ s d)
Hands attending Throstle Frames,	30	3	180	37 10 0	12	5/	6 0 0
Cost of spinning Mule Yarn No. 18's.	.....	.....	.....		5,666 lb.	.706d.	16 13 4
Overseer,	1	12	24	5 0 0	1	24/	2 8 0
Assistant for do.	1	6	12	2 10 0			
Roller Coverer,	1	4	8	1 13 4	1	8/	0 16 0
			224	46 13 4			25 17 4

DRESSING AND WEAVING DEPARTMENT.

	AMERICA.				BRITAIN.		
	No. of Hanks or Quantity.	Rate.	Amount per fortnight — Dollars.	Amount per fortnight — Sterling. (£ s d)	No. of Hanks or Quantity.	Rate.	Amount per fortnight — Sterling. (£ s d)
Winding (the rate stated is per 1000 hks.)	181,440 hanks	.18	32.65	6 16 0½	101,520 hks.	7½d.	3 3 5
Warping, do. do. do.	181,440 do.	.16½	30	6 5 0	101,520 do.	7½d.	3 3 5
Dressing,	1,710 pieces	.4	68.40	14 5 0	1,408 pcs.	2¼d.	13 4 0
Drawing and Twisting,	216 beams	.20	43.30	9 0 0	140 bms.	4¼d.	2 9 7
Weaving,	1,710 pieces	.25	427.50	89 1 3	1,408 pcs.	10d.	58 13 4
Flour for Dressing,	1,710 do.	.2	34.20	7 2 6	1,408 do.	2d.	11 14 8
Overseers,	2 hands,	12	48	10 0 0	2 hands	26/.	5 4 0
Assistants for do.	2 do.	6	24	5 0 0			
Dressing Maker, and Brush-Washer,	1 do.	4.50	9	1 17 6	1 do.	13/.	1 6 0
Sweepers,	2 do.	2	8	1 13 4	1 do.	6/.	0 12 0
Cloth Pickers,	2 do.	2.50	10	2 1 8			
			734.95	153 2 3½			99 10 5

GENERAL CHARGES.

	AMERICA				BRITAIN		
	No. of Hands.	Rate per week.	Amount per fortnight — Dol.	Amount per fortnight — Sterling.	No. of Hands.	Rate per week.	Amount per fortnight — Sterling.
Calenderer or Packer, ...	1	4.50	9	1 17 6			
Hands for Measuring and Folding Cloth, &c.	2	2.25	9	1 17 6	3	7/.	2 2 0
Mechanics, ...	3	9	54	11 5 0	3	23/.	6 18 0
Porter, ...	1	5	10	2 1 8	1	16/.	1 12 0
Book-keeper, ...	1	9.50	19	3 19 8	1	23/.	2 6 0
Watchman, ...	1	5	10	2 1 8	1	12/.	1 4 0
Engine-keeper, ...	. . . .	. . . .	. . . .		1	21/.	2 2 0
Superintendent or Manager, . . .	1	25	50	10 8 4	1	50/.	5 0 0
			161	33 8 10			21 4 0

ANNUAL ON-COST.

	AMERICA				BRITAIN	
	Dollars.	Rate per Cent.	Amount — Dollars.	Amount — Sterling.	Rate per Cent.	Amount — Sterling.
Capital, with rate and amount of Insurance,	110,000	″	1,100	229 3 4	17/6	105 0 0
Tear and wear on Machinery, buildings, &c.	103,800	″	7,785	1,621 17 6	£7½	675 0 0
Coals, Oil, Tallow, and Gas, ...			2,250	468 15 0		368 0 0
Paper, Twine, Belting, &c.			750	156 5 0		120 0 0
Cloth and Skins for Rollers, ...			350	72 18 4		50 0 0
Materials for repairing Machinery, . . .			850	177 1 8		80 0 0
Carriage of Cotton and Cloth, ...			600	125 0 0		50 0 0
Feu duty and Water, ...						55 0 0
Incidental charges, ...			1,500	312 10 0		300 0 0
'Nett annual on-cost, ...			15,185	3,163 10 10		1,803 0 0
Nett on-cost for two weeks, ...			584	121 13 4		69 6 11

COMPARISON *of* PRODUCE *per fortnight in* BRITAIN *and* AMERICA.

SPINNING.

	No. of Yarn.	No. of Spindles.		Speed of Spindles.	Libs. produced	Hanks. produced.	Hanks per spindle.
		Throstle.	Mule.				
Throstle Warp Spinning in America,	18	2,880	. . . .	4,700	10,080	18,1440	63
Do. do. in Britain,	16	2,160	. . . .	4,400	6,345	10,1520	47
Do. Weft do. in America,	18	2,112	. . . .	4,700	7,744	13,9392	66
Mule do. do. in Britain,	18	. . . .	2,400	4,200	5,666	10,2000	42½

WEAVING.

	America.	Britain.
Pieces* produced from 128 Looms, ...	1,710	1,408
Yards do. do.	51,300	35,200
Speed of Looms per minute, ...	115	95
Effective Shots obtained, ...	92	77.42

*The Cloth assumed in the above calculations is, for America, a 9⁰⁰ three leaved tweel No. 18's warp, and weft three threads in the split, 2,400 threads, 30 inches broad, 30 yards long; "56 picks to the inch," weighing about 10 lbs. For Britain, a 10⁰⁰ shirting, No. 16's warp, No. 18's weft, 2000 threads 35 inches broad, 25 yards long, 63 picks to the inch, weighing about 8½ lbs.

ABSTRACT OF CHARGES FOR TWO WEEKS.

	AMERICA.				BRITAIN.		
	Dollars.	Sterling.			Sterling.		
Preparation Charges, ...	250.50	52	3	9	ˣ19	16	0ˣ
Spinning Charges, ...	224.	46	13	4	25	17	4
Dressing and Weaving Charges, ...	734.95	153	2	3½	99	10	5
General Charges, ...	161.	33	10	10	21	4	0
On-Cost for two weeks, ...	584.00	121	13	4	69	6	11
Nett amount of charges for two weeks, ...	1,954.45	407	3	6½	ʸ235	14	8ʸ
Nett charges per piece (see pieces produced as shewn in table of Comparative Produce,)	1.14		4	9	. . .	3	4
Nett charges per yard, (see yards produced as shewn in table of Comparative Produce,)	.3⅘			1.9	. . .		1.6
Difference in manufacturing charges per yard in favour of Britain equal to 19 per cent.	. . . .				. . .		.3

Comparative Cost of Manufacturing, including Raw Material.

Charges on Shipment to the British Manufacturer, . . .	4 per cent.	
Freight and Insurance, ...	12 ½ do.	
Importer's Profit, ...	5 do.	
Duty on Cotton Wool, ...	4 ½ do.	
Inland Carriage, (average,)	1 ½ do.	
Nett charges on Importation of Cotton to the British Manufacturer, ...	27½ percent.	
Average do. to the American Manufacturer, (*see pages* 145 *and* 146,)	11 do.	
1 Piece of 25 yards, 8½ lbs. = .34 lbs. per yard.		
Add ⅙th for Waste and Loss, .056		
.396 lbs. of cotton @ 7d. per lb.	2.772d.	
Add charges on Shipment, 27½ per cent.	.762d.	
Cost of Raw Material in Britain per yard, ...	3.534d.	
Charges of Manufacturing in do. do.	ʹ1.760d.ʹ	
Nett Cost of one yard of Cloth to the British Manufacturer, ...	. . . .	ᵃ5.294d.ᵃ
.396 lbs. of Cotton (as above) @ 7d. per lb.	2.772d.	
Add charges in America, 11 per cent.	.305d.	
Cost of Raw Material in America per yard, ...	3.077d.	
Charges of Manufacturing in do. do.	ᵇ2.150d.ᵇ	
Nett cost of one yard of Cloth to the American Manufacturer, ...	. . . .	ᶜ5.227d.ᶜ
Nett difference of Manufacturing in favour of America, equal to ᵈone per cent,ᵈ	. . . .	ᶜ .067dᶜ

It is not designed by the preceding estimates to point out the amount of profit realized by manufacturing cotton goods, as that will vary according to the state of the markets. The chief design is to show the real cost of manufacturing in the United States and in Great Britain: and the utmost accuracy has been studied in order to give a fair and impartial statement of the actual expense of manufacturing in both countries, so as to ascertain the real difference as nearly as possible.

The comparison has been made upon goods which require the same quality of cotton—the same expense in manufacturing,—and each yard contains the same weight of raw material; so that they may be supposed to sell at the same price. The various estimates have been submitted to the inspection and correction of experienced manufacturers in both countries, by whom they are considered accurate and impartial statements.

The amount of goods produced is much greater in America than in Great Britain; but the hours of labour are somewhat longer in the former than in the latter country.

The cost of the buildings, machinery, &c. is a great deal higher in America than in Britain, as well as the general rate of wages, particularly in the carding department.

After comparing the advantages and disadvantages of each, it appears that the British manufacturer can produce his goods, at least ᵈ22 percent.ᵈ cheaper than the American. This, however, is more than neutralised, by the cheaper rate at which the latter can purchase his cotton.

The circumstance of America being a cotton growing country, will always give to her manufacturers advantages of which the British cannot generally avail themselves. It is very common here for several manufacturers to join together, and appoint some person acquainted with the business, to go to the Southern States, and purchase cotton sufficient for a year's consumption. The person thus appointed goes to the first markets, and selects such cottons as he knows will suit those for whom he is purchasing—he buys it at the cheapest rate, and has it shipped to the nearest port to where it is to be manufactured. The whole charge for commission will not amount to one per cent. on the prime cost.

The ordinary way of purchasing, however, is through a cotton agent in the South; and in order to show the whole expense on transactions of this kind, I will here copy the amount of the charges from some invoices now before me.

Mobile, _____

Invoice of 300 square bales of cotton shipped by ᶜToulmin Hazard & Co.ᶜ per Brig Pioneer, Jordan, for S ____, Maine, consigned to P____ C____, Esq. Treasurer, for account and risk of the Y____ Manufacturing Co.[3]

			Dollars	Cents
300 square bales of cotton, 141,138 lbs. at $12\frac{1}{4}$ cents = $6\frac{1}{8}$d per lb.			17,289	40
CHARGES.	Dol.	Cts.		
Draggage to store, ...	$37\frac{1}{2}$			
Do. to vessel, ...	$37\frac{1}{2}$			
Wharfage, ...	30			
Storage, 25 cents per bale, ...	75			
Bills of lading, ...		50		
Postages till date, ...	2	25		
Brokerage, $\frac{1}{2}$ per cent, ...	86	45		
Commission, $2\frac{1}{2}$ per cent, ...	438	97		
Freight, 1 cent per pound, ...	1,411	38		
Total charges, ...			2,119	55
			19,408	95

The whole amount of charges on the above 300 bales is about $12\frac{1}{4}$ per cent. on the prime cost; and it is to be remarked that the whole were landed within 200 yards of the Factory where they were to be used.

New Orleans, _____

Invoice of 122 bales of cotton, shipped by Stetson Avery & Co. per Ship Ohio, C. Cutter, bound to Boston for account and risk of the Y__ Manufacturing Co. To the order of W____ B____ S____, Esq. and to him consigned.

			Dollars	Cents
122 Bales cotton, 53,913 lbs. at $10\frac{3}{4}$ cents = $5\frac{3}{8}$ d per lb.			5,795	64
CHARGES.	Dol.	Cts.		
Draggage, 12 dollars 20 cents: repairs on bagging, 1 dollar	13	20		
Brokerage, $\frac{1}{2}$ per cent = 28 dollars 97 cents. Commission, $2\frac{1}{2}$ per cent = 144 dollars 88 cents =	173	85		
Freight, $\frac{3}{4}$ cent. per lb.	404	35		
Total charges, ...			591	30
			6,386	94

The charges on the above 122 bales, amount to about $10\frac{1}{5}$ per cent. on the prime cost; and the storage, &c. in Boston, together with the carriage to the Factories, amounted to about $\frac{1}{2}$ per cent.; making the whole charges for having the cotton laid down at the Factories, something less than $10\frac{3}{4}$ per cent.

Savannah, Georgia, _____

Invoice of 100 bales of cotton, purchased and shipped by Woodbridge and May, on board Barque Richmond, Captain Andross, bound for Boston, consigned to W___ B___ S___, Esq. Presented by his order for account and risk of the Y___ Manufacturing Co.

100 bales gross wt. 36,810 lbs.
Tare, ____200
Nett, 36,610 lbs. at $10^{95}/_{100}$ cts. $= 5\frac{1}{2}$ pr. lb.

	Dol.	Cts.
	4,008	80

CHARGES.	Dol.	Cts.		
Draggage, 8 cents, wharfage, 5 cents, per bale, ...	13			
One week's storage, at $6\frac{1}{4}$ dollars,	6	25		
Mending bagging, twine, and labour, ...	4	75		
Commission on $4008^{80}/_{100}$ dollars at 2 per cent	80	17		
Freight, $\frac{3}{4}$ cent. per pound, ...	274	58		
Total charges, ...			378	75
			4,387	53

The charges on the above 100 bales, are nearly $9\frac{1}{2}$ per cent. on the prime cost, to which add $\frac{1}{2}$ per cent. for additional expenses of storage and carriage to the Factories, making the whole amount to about 10 per cent.

The above are fair specimens of ordinary transactions, from which it will be seen, that the whole expenses attending the purchase and carriage of cotton, until it is laid down at the Factories, seldom exceed 12 or 13 per cent. on the prime cost, and in many cases are much less. The Cotton Factories in this country are generally situated near the sea coast, so that the expense of inland carriage is very trifling, compared with that paid by the majority of Factories in Great Britain. The carriage from Boston to Lowell is two dollars per ton, while to many other Factories, the inland carriage is not above one dollar; there are other Factories, in different parts of the country, to which the carriage cannot be less than four dollars per ton.

The foregoing Tables bring out a view of the manufacture of the two countries, which is worthy of observation by manufacturers on either side of the Atlantic; viz. that in every description of goods in which the cost of the raw material exceeds the cost of production, the American manufacturers have a decided advantage over the British. And they have availed themselves of this advantage to improve the quality of their goods, as any person who has had an opportunity of comparing the domestics manufactured in the two countries, can have no hesitation in giving the preference to those manufactured in America; and the experience of every British manufacturer engaged in producing this description of goods has painfully convinced him, that the superior quality of the American goods is gradually driving him from every foreign market. On this subject Mr. William Gemmell of Glasgow states in his affidavit, (as given in Mr. Graham's pamphlet on "The impolicy of the tax on Cotton Wool,") that although he was for several years in the habit of supplying Chili with cotton domestics, he has latterly been obliged to abandon the trade, in consequence of being unable to compete with the manufacturers of the United States.[4]

Being well acquainted with the kind of domestics manufactured by Mr. Gemmell, and also with those of the same kind manufactured in various places in the United States, I do not think it is difficult to understand why Mr. Gemmell was obliged to abandon the trade. But if those kinds of domestics manufactured by Mr. John King of Glasgow,[5] could be sent to Chili on the same terms, they would be found to stand a competition with any of the kind that has yet been manufactured in any part of America. The coloured tweel stripes also, made on the principle invented by Mr. John M'Bride of Messrs. Sommerville and Sons, Glasgow,[6] would, from their beautiful texture and finish, successfully compete with any of the same description I have yet seen produced, either in this country or in Great Britain.

Hitherto the British have enjoyed a monopoly in the manufacture of fine goods, but the resources of the Americans will very soon enable them to compete successfully[f] in these. No people in the world are more enterprising, none more ready to[g] avail themselves of every improvement by which their interest is to be advanced; and there is no doubt, that, in a few years, they will adopt a more economical method of getting up their works, a more improved system of general management and conducting of the various processes, which will enable them, even in the finer goods, to compete successfully with the British.

It is worthy of remark, that, in printed goods, the French have the advantage over both countries; the French prints selling in this country 25 per cent. higher than either the British or American.

There are a few statements made by Dr. Ure, in his work entitled, "The Cotton Manufacture of Great Britain systematically investigated," which I think scarcely correct, and which I will here notice.

At page xxxix. he states: "capital required to carry on the manufacture in the best manner is considered to be at the rate of 100 dollars for each spindle; but, in general, not more than 60 dollars have been expended."[7] [h]Now[h] 40 dollars are amply sufficient for each spindle, allowing something above 20 for fixtures, and the remainder for business capital.

At page xli, he gives the general rate of wages paid in the United States, upon the authority of Mr. Kempton, a manufacturer from Philadelphia,[8] which statement may be correct as to the rate of wages paid in the Southern section of the United States, but differs very materially from those of New England, where very few hands are employed under fifteen or sixteen years of age, except in some Mills about Rhode Island. The Factories about Philadelphia cannot be admitted as a proper criterion of the general state of the cotton manufacture of America. The character and appearance of the manufacturing population in that section of the country, and the manner in which the business is generally conducted are very different from Massachusetts, which contains nearly one third of the Cotton Factories in the whole Union. In order to ascertain the actual state of the cotton manufacture of America, we must take our estimates of the cost of materials, rates of wages, &c. &c. from the latter State. The rate of wages given by Mr. Kempton, compared with that generally paid in the Factories of Massachusetts, is shown in the following Table.[9]

Upon the same authority as the preceding, is stated by Dr. Ure; "No. 16 water-twist, made entirely of good cotton, sells in the United States at 10½d. per pound; in England, No. 16 yarn, made from a mixture of waste and a small quantity of Uplands, sells at 11d. per pound."[10]

I have made inquiry regarding this statement at some old and

Rates of Wages according to Mr. KEMPTON's STATEMENT.		Rates of Wages paid in MASSACHUSETTS.
Card Tenters, ...	{ 10 Years of age.—3/per week, = 72 cents. { 12 do. do. —4/ do. = 96 do.	None employed. Do. do.
Attending Drawing Frames, ...	{ 14 do. do. —5/ do. = 120 do. { 16 do. do. —4/ do. = 144 do.	1 dollar 80 cents. 2 dollars to 2 dollars 20 cents.
Attending Roving Frames, ...	18 do. do. —8/ do. = 192 do.	[i]2 do. 12½ cents to 2 dollars 25 cents.[i]
Girls attending Throstle Frames, ...	5/to 8/do. = 120 to 192 do.	[j]2 do. 12½ do. to 2 do. 50 do.[j]
Machine-Makers, ...	5/per day = 120 do.	[k]1 do. do. to 1 do. 33 do.[k]
Overseers, ...	5/to 6/do. = 120 to 144 do.	[l]1 do. 75 do. to 2 do. do.[l]
Assistant Overseers, ...	3/to 4/do. = 72 to 96 do.	84 cents to 1 dollar 25 cents.

experienced manufacturers, none of whom can recollect the time
when No. 16 water-twist could be manufactured and sold in any
part of the United States at 10½d. equal to 21 cents per pound.
There is certainly some mistake in this statement; instead of 10½d. it
ought to have been 13d. equal to 26 cents; and even this is too
low.

It is stated at page xlvii. "The money prices of provisions have
been much higher in Great Britain than in the manufacturing
countries of the continent of Europe and America."[11] Now it may
be true, that the prices of provisions *have been* higher in Great
Britain than in America; but they are not so now. I can speak
from experience on this subject, and have no hesitation in asserting, that the price of living is higher in this country than in
Britain; I know of nothing that is cheaper here but spirits, tea,
and tobacco. I have no doubt but in the interior of the country,
potatoes, Indian corn, butter, milk, poultry, &c. may be[m] cheaper;
but in all the cities and manufacturing places, they are much
higher. It will be supposed that flour must be considerably cheaper
here than in[n] Britain; but it is not always so, these few years past,
there has been a vast quantity of wheat imported from[o] Britain
and the continent of Europe.

House rents are higher here than in Scotland, and fuel is at
least triple the price of what it is in Glasgow. All kinds of clothing
are higher, and particularly the making of clothes. The price of
making a coat in Boston is from eight to twelve dollars; as much
as would *purchase* one complete in Glasgow.

Dr. Ure proceeds to state, that "In the event of that more serious
struggle, which in the natural progress of competition is likely to
take place, the cheapness of the means of subsistence, by confer-
ring a higher condition upon the foreign workman, leaves more
room for a reduction of wages. Mr. Kirkman Finlay, a great authority in these matters, says, 'I think the difference would be this,
that if the amount of wages paid in Great Britain were absolutely
necessary for the comfortable subsistence of the workmen, it
would be quite clear that whatever pressure there might be, those
wages could not be permanently reduced; but if the money wages
paid in America are sufficient to get a great deal more than the
absolute necessaries and comforts of life, then, if there is a pressure upon its manufacturers, they can so reduce the wages as to
meet that difficulty, and by that means undersell the manufacturers here.'" (Britain).[p12]

Mr. Kirkman Finlay, in his letter to Lord Ashley in 1833, states, that the prices of spinning a given quantity of yarn from No. 10's to 20's, was 4/ in the United States, and 4/11 in Glasgow; and the prices of carding the same numbers, were 6/7½ in the former, and 7/1¼ per week in the latter.[13] I think there must be some misunderstanding in this. The prices of mule spinning in the New England States, was, previous to 1837, 10 cents per 100 hanks, or 100 cents = 4/2 per 1000 hanks, while, at the same time, the price of mule spinning in Glasgow, was 3/6½ for all numbers under 40's, being about[q] 15 per cent.[q] in favour of Glasgow.[r]

Throstle spinning is nearly as cheap in this country as in Britain, in consequence of the higher speed at which the spinning frames are driven, and the greater quantity of work produced in a given time. But the price of carding is fully double because men are generally employed to attend the cards, spreading, scutching machines, &c. while the same work is done by boys and girls in Britain. The lowest wages paid to any girl in the card room that I am aware of, is one dollar per week and her board; and taking her board at the lowest rate, viz.[s] 1 dollar 12½ cents[s] per week, her wages in all will amount to[t] 2 dollars 12½ cents[t], equal to 10/2 sterling per week. The average rate of wages for girls in the card room, may vary from[u] 9/ to 11/- per week.[u] Men's wages may vary from 13/ to 18/ per week. Thus, in every department, the rate of wages is higher in the United States than in Britain; nor do I think that they will, for many years, be so low in this country as in the latter country.[v]

[The Manufacturing population] are an entirely different class, and placed in very different circumstances from those of Great Britain, and[w] great changes must take place before the wages in the former can be so low as in the latter country: the manufacturers here can afford to pay higher wages than the British, they run their Factories longer hours, and drive their machinery at a higher speed, from which they produce a much greater quantity of work; they can also purchase their cotton at least one penny a pound cheaper, and their water power does not exceed one fourth of the cost of steam power in Great Britain.[x]

The British have, no doubt, attained to great perfection in the art of manufacturing cotton goods; but whether they will be able to maintain that high pre-eminence or have to yield to the increasing improvements of foreign nations, are questions of difficult solution. Their most powerful rivals are, doubtless, the Americans. The manufacturers of no other country can purchase their cotton so cheap, and it is presumed no country possesses so extensive water privileges; only a small portion of which has yet been occupied. If we add to these, the intelligence and enterprising spirit of the people, it will be obvious to every unprejudiced mind, that the American manufacturers are the most formidable competitors with which the British have to contend in foreign neutral markets.

The preceding brief details will be interesting chiefly to proprietors and practical manufacturers. What follows, being more of an historical and statistical nature, will, it is hoped, be found interesting both to them, and to the general reader.

HISTORICAL SKETCH OF THE RISE AND PROGRESS OF THE COTTON MANUFACTURE IN AMERICA.

[a]The introductory part of the following sketch has been compiled principally from White's "Memoir of Slater." The latter part has been collected by the Author from the most authentic sources upon which the utmost reliance may be placed.[a1]

As early as the year 1787, a Society was formed in Philadelphia, under the name of the "Pennsylvania Society for the Encouragement of Manufactures and the Useful Arts," which made some progress in the manufacturing of various kinds of goods, such as jeans, corduroys, fustians, plain and flowered cottons, flax linens, tow linens, &c. But the machinery employed in the manufacture seems to have been of the very rudest kind. A short time before the formation of this Society, an attempt to spin cotton yarn by machinery had been made at Bridgewater and Beverly, in the State of Massachusetts. Two mechanics from Scotland, Alexander and Robert Barr, brothers, were employed by a Mr. Orr, at East Bridgewater, to make carding, spinning, and roving machines, which they completed, and on the 16th November, 1786, the general court of Massachusetts made them a grant of £200, lawful money, for their encouragement, and afterwards added to the bounty, by giving them six tickets in the State land lottery, in which there were no blanks.

In March 1787, Thomas Somers, an English midshipman, constructed a machine, or model, under the direction of Mr. Orr; and by a resolution of the general court, £20 were placed in the hands of the latter, to encourage him in the enterprise.

The above remained in the possession of Mr. Orr for the inspection of all disposed to see them, and he was requested by the general court, to exhibit and give all information or explanation regarding them. It is believed that the above were the first machines made in the United States for the manufacture of cotton.

The Beverly Company commenced operations in 1787, and are supposed to be the first Company that made any progress in the manufacture of cotton goods; (that at Bridgewater had been on a very limited scale;) yet the difficulties under which they laboured —the extraordinary loss of materials in the instruction of their servants and workmen—the high prices of machines unknown to their mechanics, and both intricate and difficult in their construction, together with other incidents which usually attend a new business, were such, that the Company were put to the necessity of applying to the State legislature for assistance, to save them from being compelled to abandon the enterprise altogether.

In their petition to the Senate and House of Representatives of Massachusetts, presented June 2d, 1790, only three years after they had commenced operations, they state, "That their expenditure had already amounted to nearly £4000, whilst the value of their remaining stock was not equal to £2000, and a further very considerable advance was absolutely necessary, to obtain that degree of perfection in the manufacture, which alone could ensure success."

Accordingly a grant of £1000 was presented to them, to be appropriated in such a way as would most effectually promote the manufacturing of cotton piece goods in the Commonwealth.

The petition above referred to, and other collateral facts, sufficiently prove, that cotton spinning in this country, further than the hand card and one thread wheel, was carried through its first struggles by the Beverly Company in Massachusetts. And from this State the manufacture was carried to Rhode Island, though it must be acknowledged that both States were indebted to foreign emigrants for instruction and assistance in spinning and weaving, as well as in preparing the cotton.

Cotton spinning commenced in Rhode Island in 1788, in which year Daniel Anthony, Andrew Dexter, and Lewis Peck, all of Providence, entered into an agreement to make what was then called

"Home Spun Cloth." The idea at first was to make jeans of linen warp spun by hand; but hearing that Mr. Orr of Bridgewater, and the Beverly Company, had imported some models or draughts of machinery from England, they sent thither, and obtained drawings of them, according to which they constructed machinery of their own. The first they made was a carding machine, which was something similar to those now used for carding wool, the cotton being taken off the machine in rolls, and afterwards roped by hand. The next was a spinning frame, something similar to the water frame, or rather the common jenny, but a very imperfect machine. It consisted of eight heads of four spindles each, being thirty-two spindles in all, and was wrought by means of a crank turned by the hand; this, after being tried for some time in Providence, was taken to Pawtucket, and attached to a wheel propelled by water: the work of turning the machine was too labourious to be done by the hand, and the machine itself was too imperfect to be turned by water. Soon after, these machines were sold to Moses Brown of Providence; but as all the carding and roving was done by hand, it was very imperfect, and but very little could be done in this way. Such were the rude machines used for spinning cotton previous to 1790; and the wonder is, not that the manufacturers failed in their undertakings, but rather that they were able to persevere. And we can now perceive that from these small beginnings, the present brightened prospects received their foundation.

Previous to 1790, the common jenny and stock card had been in operation in various parts of the United States: and mixed goods of linen and cotton, were woven principally by Scotch and Irish weavers. Mr. Moses Brown of Providence, had several jennies employed in 1789, and some weavers at work on linen warps. The jennies were used for making weft, and operated by hand in the cellars of dwelling houses. During 1790, Almy and Brown of Providence, manufactured 326 pieces, containing 7823 yards, of various kinds of goods. There were also several other Companies and individuals in different parts of the Union, who manufactured goods from linen warps and cotton weft. But notwithstanding these most laudable and persevering efforts, every attempt failed of success, and they saw their hopes and prospects entirely prostrated. There was no deficiency of enterprise or exertion; no want of funds, or of men ready and willing to engage in the business; and no lack of patronage from the government, they having learned from the privations to which the country was subjected during the revolutionary war, the absolute necessity of promoting and encouraging domestic manufactures. The great cause of these failures is to be found in the fact, that during all these incipient struggles to establish the cotton manufacture in America, Great Britain had in full operation a series of superior machinery, which the manufacturers in this country had in vain endeavoured to obtain.

It is to be remembered that Sir Richard Arkwright took his first patent for an entirely new method of spinning cotton yarn for warps in 1769, at which period his first Mill was put in operation at Nottingham in England, and his second Mill, which was much larger, was erected at Cromford, Derbyshire, in 1771. After which his mode of spinning by water frames extended rapidly all over the kingdom; so that during the period when the most persevering exertions were being made, by various enterprising individuals, in different parts of the United States, to improve and perfect this most important manufacture, England was enjoying all the benefit of Arkwright's patents, by means of which cotton yarn was produced at much less expense, and of a superior quality to any that had ever been made by machinery before that period: and, at the same time, the British government were using every means in their power, to prevent models or drawings of these machines from being carried out of the country. Every effort to erect or import this machinery into the United States had hitherto proved abortive. Much interest had been excited in Philadelphia, New York, Rhode Island, and Massachusetts, but they found it impossible to compete with the superior machinery of England. The difficulties under which these incipient measures, towards the establishment of the business, were pursued, can hardly be conceived at the present day, even by a practical machinist or manufacturer. Besides the difficulties experienced in consequence of imperfect machinery, the period at which the business commenced in this country, was also most unfavourable, as from the peculiar state of the manufacture in England at that time, and other causes, many in that country became bankrupts, their goods were sold at auction, and shipped to the United States in large quantities, where they were again sold at reduced prices. Agents were also sent from England to the various manufacturing towns with goods, which were sold at low prices and long credit given, extending in some instances to eighteen months. It is likewise said, that British manufacturers formed themselves into societies, for the purpose of sending goods to this country, to be sold on commission, when they could not be disposed of to advantage at home.

Such was the state of the cotton manufacture in the United States in 1790: every endeavour to introduce a proper system of spinning had been fruitless; and nothing but the introduction of the water frame spinning, which had superseded the jennies in England, could have laid a foundation for the successful prosecution of the business in America; and that was happily accomplished by one who was personally and practically acquainted with the business in all its details. The individual here referred to was Mr. Samuel Slater, who has justly been called the FATHER OF THE COTTON MANUFACTURE OF AMERICA.

Mr. Slater was born in the town of Belper, Derbyshire, England, on 9th June, 1768; and when about fourteen years of age, he was bound an apprentice, at Milford, near Belper, to Jedediah Strutt, Esq., (the inventor of the Derby ribbed stocking frame, and for several years a partner with Sir Richard Arkwright in the cotton spinning business.) At that time Mr. Strutt was erecting a large Factory at Milford, where Slater continued to serve him for some time in the capacity of clerk, but during the last four or five years of his apprenticeship, his time was solely devoted to the Factory, as general overseer, both as respected the making of the machinery and in the manufacturing department. After having completed the full term of his engagement, viz. six and a half years, he continued for some time longer with Mr. Strutt, for the purpose of superintending some new works that were then erecting; his design in doing so, was to perfect his knowledge of the business in every department, as previous to this time his thoughts had been directed to America, by various rumours which had reached Derbyshire, of the anxiety of the governments of the different States in that country to introduce and encourage manufactures. A newspaper account of a liberal bounty of £100 having been granted to a person who succeeded in constructing a very imperfect carding machine, for making rolls for jennies, and the knowledge that a society to promote manufactures had been authorised by the same legislature, finally determined him to try his fortune in the western hemisphere.

Mr. Slater had a perfect knowledge of the Arkwright mode of spinning; and from the confidential situation he occupied under Mr. Strutt, few enjoyed the same opportunities of acquiring a complete knowledge of all the minutiae of the business; and being a person of retentive memory, close observation, and attentive to his engagements, it can easily be supposed that he must have been eminently qualified to introduce cotton manufacture into America upon the same improved scale in which it was then in operation in England, especially as his mind had been for some time directed to that object. For, having once determined to leave his native country, and give to the land of his adoption all the benefits of his practical knowledge and enlarged experience, there is every reason to suppose that he would embrace every opportunity of preparing himself for the great object he had in view. He knew that it was impossible to take any patterns or drawings along with him, as the government restrictions were very severe, and the customhouse officers scrupulously searched every passenger for America. It was therefore necessary that he should be fully qualified to superintend the building and arrangement of the Mills, the construction of the machinery, and to direct the details of the manufacture, without the aid of a single individual: as the whole business was new to the people of this country, he could not expect any one to assist him except by his own directions. He, accordingly, stopped with Mr. Strutt, until he considered himself qualified to engage in this important enterprise.

He embarked at London for New York, on the 13th September, 1789, and landed at the latter on the 17th November, after a passage of sixty-six days. He was immediately after his arrival introduced to the New York Manufacturing Company; but finding that the state of their works did not suit his views, he left that place in the January following for Providence, Rhode Island, and there made arrangements with Messrs. Almy and Brown, to commence preparations for spinning cotton entirely upon his own plan: on the 18th of the same month, the venerable Moses Brown took him out to Pawtucket, where he commenced making the machinery, principally with his own hands; and on the 20th December, 1790, he started three cards, drawing, and roving, together with seventy-two spindles, entirely upon the Arkwright principle, being the first of the kind ever operated in this country. These were worked by the water wheel of an old fulling mill in a clothier's building, in which place they continued spinning about twenty months; at the expiration of which time several thousand pounds of yarn were on hand, notwithstanding every exertion had been used to weave it up and sell it.

Early in 1793, Almy, Brown, and Slater, built a small Mill in the village of Pawtucket—known to this day by the name of the

Old Factory*ᵇ—in which they put in operation seventy-two spindles, with the necessary preparation, and to these they gradually and slowly added more and more as the prospects became more encouraging. After a short time, besides building another Factory, they considerably enlarged the first.

Such then were the circumstances under which the Arkwright mode of spinning was introduced into this country, and such was the individual to whom belongs the entire merit of its introduction. Previous to 1790, the year in which Mr. Slater arrived at Providence, and which is justly denominated the era of the American cotton manufacture, there had been introduced at various places, particularly at New York, Providence, and Massachusetts, jennies, billies, and cards, for spinning cotton weft, to be woven into velverets, jeans, fustians, &c. with linen warp; but the history of those times shows the imperfection of the above-named machines to have been such, as to preclude the manufacture of cotton cloth, or cotton yarn for warps—that they were defective in their operations—deficient and expensive in their application—and that, under such difficulties and perplexities, it was entirely beyond the power of American manufacturers to compete with foreign goods introduced by British agents and American merchants, even though assisted by legislative aid, as was done at Beverly.

The citizens of Massachusetts, perplexed and involved in their incipient and imperfect attempts to manufacture cotton goods, and fully aware of the importance of introducing a better system of machinery, which they knew to be in successful operation in Great Britain, exerted themselves to obtain a model of the Arkwright patent spinning frame; but finding no person able to construct that series of machinery, and unable to obtain one from England, in consequence of the severe penalties imposed by the government on the exportation of machinery, they entirely failed in all their efforts. In this gloomy period of the American manufacture, Mr. Samuel Slater, as already stated, then in the employ of Strutt and Arkwright, having seen a premium offered by the Pennsylvania Society for a certain machine to spin cotton, was induced to leave the land of his fathers, where he had every prospect of succeeding in business, and embark for America. After

---

*ᵇ Two of Mr. Slater's old water spinning frames, are still kept in this old factory. They are venerable relics of the olden time, which ought never to be destroyed. They are certainly worthy of being deposited in some public museum, both as a memorial of the respected individual, who first introduced them; as well as to exemplify the rapid improvements that have been made, in the arts and manufactures since [that] day.[2]

his arrival, being informed that Moses Brown of Providence had made some attempts at water spinning, he repaired thither; but on seeing Mr. Brown's machinery, he pronounced it entirely worthless, and advised him to lay it aside. At this period, without the aid of a single individual skilled in making machinery, Mr. Slater constructed the whole series of spinning machines upon the Arkwright principle, and put them in operation so perfectly, as to supply all the establishments with cotton warp superior to linen; and in fourteen months Mr. Brown informed the Secretary of the Treasury, that machinery and Mills could be erected in one year, to supply the whole United States with yarn, and thus render its importation unnecessary.[3] Such is the amount of evidence regarding the introduction of the Arkwright machinery into the United States; and if the manufacturing establishments are in reality a blessing, as has been well observed, the name of Slater must ever be held in grateful remembrance by the American people.

Mr. Slater laboured under every disadvantage in the construction of his machinery; for although he had perfect confidence in his own remembrance of every part and pattern, and in his ability to perfect the work according to his agreement, yet he found it difficult to get mechanics who could make anything like his models. But, perhaps, one of his greatest difficulties was to get card sheets made to suit his machines, as the card-makers in this country were entirely unacquainted with the operations of his machinery; indeed, the first carding machine he put in operation, had almost turned out an entire failure, in consequence of the defective manner in which the card teeth were set. But he persevered until he overcame this, as well as all his other difficulties; and his case furnishes one other bright example of the never-failing success which always attends patience and perseverance in the pursuit of any laudable object.

In 1798 Mr. Slater entered into partnership with Oziel Wilkinson, Timothy Green, and William Wilkinson; the two latter, as well as himself, having married daughters of Oziel Wilkinson. He built his second Mill on the East Side of Pawtucket river; the firm was Samuel Slater & Co., as he owned one half of the stock. A short time afterwards the hands in this Mill revolted, or struck work for higher wages; five or six of them went to Cumberland, and erected a small Mill, owned by Elisha Waters and others: from these men and their connections, several Factories were commenced in various parts of the country, and, in fact, most of the establishments erected from 1790 to 1809, were built by men who had directly or indirectly, drawn the knowledge of the business from Pawtucket, the cradle of the American Cotton Manufacture. Some of his servants stole his patterns and models, and by that means his improvements were soon extended over the country; so that the business has, from that to the present time, been rapidly extending over the United States.

Mr. Slater's business was so prosperous, that about the year 1806, he invited his brother, Mr. John Slater, to come to this country.[c] The[d] village of Slatersville was then projected in which that gentleman embarked as a partner and in June of the same year, removed to Smithfield as superintendent of the concern. In the spring of 1807, the works were sufficiently advanced for spinning; and up to the present time, [e]they have continued in full operation.[e] This fine estate was owned in equal shares by four partners, but now wholly belongs to John Slater and the heirs of his brother.[fg]

STATISTICAL NOTICES OF VARIOUS MANUFACTURING DISTRICTS IN THE UNITED STATES.

The preceding historical sketch details the introduction of the cotton manufacture into the United States, and the names of the several gentlemen through whose enterprising exertions it was first established. But in order to know its success, or the extent to which it has arrived, it is necessary to give some account of the various manufacturing districts. As the Cotton Factories of America, however, are widely scattered over a great extent of country, it is impossible here to take notice of them all. Some observations on a few of the principal districts is all that will be attempted.

It has already been stated in a former part of this work, that Massachusetts is the principal manufacturing State in this country. An Act was passed by the Senate and House of Representatives of that State in 1837, for the purpose of obtaining "Statistical information in relation to certain branches of Industry within the Commonwealth." The following Table is copied from the report of the Secretary of the Commonwealth, which he prepared from the returns of the assessors in the various towns and cities in the State.

Statement of the Cotton Manufacture of the State of Massachusetts in 1837. Compiled from the Returns of the Assessors in each Town and County, by John P. Bigelow, Secretary of the Commonwealth.[1]

Counties.	Number of Mills.	Number of Spindles.	Pounds of Cotton consumed yearly.	Yards of Cloth manufactured yearly.	Value of Cotton Goods manufactured yearly.	Males employed.	Females employed.	Capital invested in the Cotton Manufacture.
					Dollars.			Dollars.
Suffolk, ...								
Essex, ...	7	13,300	804,222	2,301,520	372,972	115	402	337,500
Middlesex, ...	34	165,868	17,696,245	52,860,194	5,971,172	1,054	6,435	6,909,000
Worcester, ...	74	124,720	5,292,018	20,280,312	1,991,024	1,384	1,998	2,015,100
Hampshire, ...	6	8,312	563,000	1,574,000	176,060	72	233	216,000
Hampden, ...	20	66,552	4,727,302	15,107,583	1,504,896	626	1,886	1,698,500
Franklin, ...	4	5,924	135,045	1,081,140	76,125	48	140	90,000
Berkshire, ...	31	35,260	1,390,162	7,530,667	575,087	339	766	633,725
Norfolk, ...	32	25,782	1,365,953	4,953,816	509,383	280	583	609,500
Bristol, ...	57	104,507	4,814,238	18,382,828	1,678,226	987	2,015	1,622,778
Plymouth, ...	15	13,298	480,884	2,052,061	182,474	85	279	230,616
Barnstaple, ...	2	1,508	6,848	195,100	19,240	7	20	7,000
Dukes County,								
Nantucket,								
Total, ...	282	565,031	37,275,917	126,319,221	13,056,659	4,997	14,757	14,369,719

161

The total population of the State of Massachusetts at this period, was 701,331. The total number of hands employed in all the different branches of industry, was 117,352. This number, (with the exception of those engaged in the rearing of sheep and the fisheries,) does not include any of those employed in the various branches of agriculture and commerce; neither does it embrace store-keepers, clergymen, physicians, lawyers, bankers, hotel-keepers, labourers, stage coach drivers, nor those employed on rail roads, in steam vessels, &c. It only includes those employed in the various manufacturing and mechanical arts, from the ship builder down to the manufacturer of snuff and cigars, together with wool growers, and those employed in the fisheries.

Out of the 117,352 engaged in the various branches of industry, there were employed in manufacturing cotton goods,   19,754 hands.
Cotton batting, thread, warp, and candle wicks,   151
Calico printing, ...   1,660
Total employed in the Cotton Manufacture, ... 21,565 hands,

being fully 3 per cent of the whole population, and upwards of 18 per cent of all those employed in the different manufacturing and mechanical arts.

The annual value of the produce of all these arts and manu-
factures was estimated at .... 86,282,616 dollars.
Value of cotton goods manufactured, ... 13,056,659 dollars
Cotton batting, thread, warp, and wicking,   169,221 do.
Calico printing, ...   4,183,121 do.
Total annual value of the Cotton Manufac-
ture, ... 17,409,001 dol.
being about 20 per cent. of the value of all the manufactures of the State.

The amount of capital invested in the various branches of industry was estimated at .... 54,851,643 dollars.
In the manufacture of cotton goods, ... 14,369,719 dollars
Cotton batting, thread, warps, and
  wicking, ...   78,000
Calico printing, ...   1,539,000
Total capital invested in the Cotton Manufacture, ... 15,986,719 dol.
being a little over 29 per cent. of the capital invested in all the different branches of industry.[a2]

The valuation of property in the State of New York, from the Comptroller's Report of January 1835, was as follows:

Real Estate, ... 350,346,043 dollars
Personal Estate 108,331,941
Total, ... 458,677,984 dollars.

The capital invested in the cotton manufacture being 3,669,500 dollars, is nearly one per cent. on the valuation of the whole property of the State.[3]

Statement of the Cotton Manufactures in 12 of the States in 1831.[4]

States.	Capital.	Number of spindles.	Yards of Cloth produced yearly.	Pounds of Cloth produced yearly.	Pounds of Cotton consumed yearly.
	Dollars.				
Maine, ...	765,000	6,500	1,750,000	525,000	588,500
New Hampshire,	5,300,000	113,776	29,060,500	7,255,060	7,845,000
Vermont, ...	295,500	12,392	2,238,400	574,500	760,000
Massachusetts, ...	12,891,000	339,777	79,231,000	21,301,062	24,871,981
Rhode Island, ...	6,262,340	235,753	37,121,681	9,271,481	10,414,578
Connecticut, ...	2,825,000	115,528	20,055,500	5,612,000	6,777,209
New York, ...	3,669,500	157,316	21,010,920	5,297,713	7,661,670
New Jersey, ...	2,027,644	62,979	5,133,776	1,877,418	5,832,204
Pennsylvania, ...	3,758,500	120,810	21,332,467	4,207,192	7,111,174
Delaware, ...	384,500	24,806	5,203,746	1,201,500	1,435,000
Maryland, ...	2,144,000	47,222	7,649,000	2,224,000	3,008,000
Virginia, ...	290,000	9,844	675,000	168,000	1,152,000
Total, ...	40,612,984	1,246,703	230,461,990	59,514,926	77,457,316

In the State of Pennsylvania there were 500,000 dollars, and in Delaware 162,000 dollars invested in hand looms, both of which sums are included in the amount specified in the preceding Table, as the capital invested in the cotton manufacture.

The preceding Table shows the extent of the cotton manufacture of the United States in 1831; since that time there has been a considerable increase. The amount of capital invested in manufactures in the State of Massachusetts was then 12,891,000 dollars; in 1836, it had increased to 14,369,719 dollars, being nearly 12 per cent. in the space of only five years; but allowing the ratio of increase since 1831 to be 10 per cent. all over the Union, the amount of capital now invested in the cotton manufacture cannot be less than forty-five millions of dollars, equal to £9,375,000 Sterling, being about a fourth part of the capital invested in the cotton manufacture of Great Britain.

The following Table contains the number of Mills, rate of weekly wages, and the number of hands employed in the Factories in 1831.[5]

States.	Mills.	Looms.	Males employ'd	Average Wages of Males weekly.		Females employ'd	Average Wages of Females weekly.		Children under 12 employ'd	Average Wages of children	
				dols.	cts.		dols.	cts.		dols.	cts.
Maine, ...	8	91	84	5	50	205	2	33			
New Hampshire,	40	3,530	875	6	25	4,090	2	60	60	2	0
Vermont, ...	17	352	102	5	0	363	1	84	19	1	40
Massachusetts, . . .	256	8,981	2,665	7	0	10,678	2	25	3,472	1	50
Rhode Island, . . .	116	5,773	1,731	4	25	3,297	2	20	439	1	50
Connecticut, . . .	94	2,609	1,399	4	50	2,477	2	20	484	1	40
New York, ...	112	3,653	1,374	6	0	3,652	1	90	217	1	40
New Jersey, ...	51	815	2,151	6	0	3,070	1	90			
Pennsylvania, . . .	67	6,301	6,545	6	0	8,351	2	0			
Delaware, ...	10	235	697	5	0	676	2	0			
Maryland, ...	23	1,002	824	3	87	1,793	1	91			
Virginia, ...	7	91	143	2	73	275	1	58			
Total, ...	801	33,433	18,590			38,927			4,691		

[b]The following table shows the number of Factories and printing establishments, the persons employed and capital invested in 1840.

Names of the States wherein the Factories are situated	Number of Cotton Factories	Number of Spindles	Printing & Dying Establishments	Value of Manufactured Articles	Number of Persons Employed	Capital invested in dollars
Maine	6	29,736	3	970,397	1,414	1,395,000
New Hampshire	58	195,173	4	4,142,304	6,991	5,323,202
Massachusetts	278	665,095	22	16,553,423	20,928	17,414,099
Rhode Island	209	518,817	17	7,116,792	12,086	7,326,000
Connecticut	116	181,319	6	2,715,964	5,153	3,152,000
Vermont	7	7,254		113,000	262	118,100
New York	117	211,659	12	3,640,237	7,407	4,901,772
New Jersey	43	63,744	13	2,086,104	2,408	1,722,810
Pennsylvania	106	146,494	40	5,013,007	5,522	3,325,400
Delaware	11	24,492		332,272	566	330,500
Maryland	21	41,182	3	1,150,580	2,284	1,304,400
Virginia	22	42,262	1	446,063	1,816	1,299,020
North Carolina	25	47,934		438,900	1,219	995,300
South Carolina	15	16,355		359,000	570	617,450
Georgia	19	42,589	2	304,342	779	573,835
Alabama	14	1,502		17,547	82	35,515
Mississippi	53	318		1,744	81	6,420
Louisiana	2	706		18,900	23	22,000
Tennessee	38	16,813		325,719	1,542	463,240
Kentucky	58	12,358	5	329,380	523	316,113
Ohio	8	13,754		139,378	246	113,500
Indiana	12	4,985	1	135,400	210	142,500
Arkansas	2	90			7	2,125
Total	1,240	2,284,631	129	46,350,453	72,119	51,102,359

The above is copied verbatim from the statistical tables published by Congress in connection with the census for 1840 and though not strictly correct it is the only public document from which such information can be obtained. It will be observed that the number of power looms are not specified. It is however estimated that there are upwards of 40,000 in operation in the United States.[b6]

The preceding tables show the particular distribution of the cotton manufacture in the United States, from which it will be seen that the greatest number of Factories and spindles employed are in the State of Massachusetts, next to it are Rhode Island and New York; but Rhode Island is a very small State compared with either of the other two, and in proportion to its extent may be said to contain more than three times the number of Cotton Factories in New York State. The cotton manufacture commenced in Massachusetts and Rhode Island, and ever since these two have continued the principal manufacturing States in the Union.

LOWELL, (MASSACHUSETTS.)[7]

THE principal manufacturing town in the United States is that of Lowell, which may justly be denominated the Manchester of America, as regards the amount of capital invested for manufacturing purposes, the extent of the business, and the spirited manner in which it is conducted. And here, too, the Factory system is perhaps in more perfect operation than in any other part of the United States. Here are the largest establishments, the most perfect arrangement, and the richest corporations. And it may, without fear of contradiction, be asserted, that the Factories at Lowell produce a *greater quantity of yarn and cloth from each spindle and loom (in a given time,) than is produced in any other Factories without exception in the world* [Fig. 26].

[c]The[cI] territory of Lowell extends over a space of about four square miles and according to the census of 1840 contains 20,981 inhabitants.[cII] [cIII]Twenty-two[cIII] years ago the whole of this was owned by a few honest farmers who supported themselves and families by the cultivation of this comparatively barren spot which was then called Chelmsford Neck as being situated at the confluence of the Merrimack and Concord rivers. By the Indians it was originally called *Wamaset*.

[cIV]The Merrimack is rendered impassable at this place in consequence of a fall in the river—upwards of twenty feet in height and extending over half a mile in length—called the Pawtucket falls. To form a passage for rafts, logs, &c round these falls a canal was cut by a company who obtained an act of incorporation as early as 1792. This canal which is nearly two miles in length receives the waters from the Merrimack above the falls and empties it into the Concord just above their junction.[cIV]

The project for employing the waters of this canal for manufacturing purposes was first conceived in 1821 as stated in a preceding part of this work.[cV] Mr. Patrick T. Jackson of Boston having been fully satisfied of its utility and advantages is said to have employed Mr. Kirk Boot as his agent to purchase as much of the

Fig. 28. Merrimack Manufacturing Company's five mills alongside the Merrimack Canal (to right) at Lowell with a line of boarding houses receding to the left. Nearer and to the north of the canal flows the Merrimack River. Detail of a view drawn by E. A. Farrar in 1834. Courtesy of the Museum of American Textile History.

canal stock as could be obtained. And that the latter [in order not
to excite suspicion] [deleted] visited the place in the habit of a
hunter and purchased a large majority of the stock from the
original owners before the project was made public. As soon as
Mr. Jackson had secured the greater proportion of the stock together with a large tract of land adjoining the canal he immediately
formed a new company who (under the original corporate name
of "Proprietors of Locks and Canals on Merrimack River") in 1822
commenced enlarging and deepening the canal and putting it in
a fit condition to become the feeder of such manufactories as
might subsequently be erected. A rude dam was thrown across
the Merrimack river which turned the waters into the canal. This
dam has been made more effective from time to time as additional
mills required a larger supply of water. The width of the large
canal is sixty feet and carries eight feet in depth of water. Lateral
branches are cut which carry the water to the several manufactories from which it is discharged into the Merrimack and Concord
rivers.[cV]

The Proprietors of Locks and Canals Co., being the owners of
all the water power with the adjoining land, they sell out privileges
to manufacturing companies, dig the canals, erect the mills, and
build the machinery all ready for being put in operation. They
do all this cheaper than any other company will do it and these
are the only terms on which they sell water privileges.

Lowell was incorporated in [cVI]1826 into a town distinct from
Chelmsford and in 1836 it became a chartered city. It now contains
thirty-two factories[cVI] besides print works, bleacheries, &c and it
is believed that the water power might yet be increased so far as
to supply water for at least [cVII]ten more manufactories.[cVII] A part
of the companies' lands have been sold out to individuals for house
lots, &c at an enormous advance on the original price. Some which
cost them only twenty or thirty dollars per acre have been sold
again for one dollar per square foot. Kirk Boot, Esq.* acted as

---

* Mr. Kirk Boot was a native of Boston but received his early education in England. After
having spent one or two years at Harvard College he joined the British army under the
Duke of Wellington and served some time in the Peninsular War, on his return from
which, he spent some years at the Military school at Woolwich, England—where he
acquired considerable proficiency as a draftsman and engineer. Some time after his
return to Boston he was appointed superintendent of the works then projected at Lowell.
In cutting the canals, building factories, planning and laying out the whole city his talents
and acquirements rendered his services invaluable. During his agency a large thriving
city was raised from a comparative wilderness. He died April 11th 1837 deeply regretted
by a numerous circle of friends. He was a man of a generous public spirit possessed of

the companies' agent from the commencement of operations in this place until his death in 1837.

Lowell is connected with Boston by a railroad of two tracks, the distance being twenty-six miles. Passenger cars run four times a day (Sabbath excepted) fare one dollar. Transportation of goods: two dollars per ton for common merchandize; one and a half dollars for compressed goods—one dollar for coals besides 25 cents for loading and disloading.[cVIII]

The Lowell railroad connects with the Nashua and Andover branches about ten miles below the city, and the latter branch unites Lowell with Andover, Haverhill, Exeter, and Dover in New Hampshire; from which it again unites with the Eastern railroad; that connects Boston with Salem, Newburyport, Portsmouth, Saco, and Portland, so that Lowell by means of railroads, stage coaches, &c has a ready and quick means of communication with all parts of the country.[c]

The total amount of capital invested for manufacturing purposes at Lowell is now[d] (1843) 10,700,000 dollars equal to £2,229,666. 1¾ sterling. There are eleven incorporated companies.[d9]

1st, Locks and Canals Co.—capital 600,000 dollars, = £125,000. This company originally owned the whole water power, which they sell out to the different manufacturing companies at the following rate. A Mill power is estimated at 3,584 throstle spindles, with the necessary machinery for preparing the cotton and manufacturing No. 14 yarn into coarse heavy cloth and is sold for four dollars per spindle, = 14,336 dollars for the whole Mill power, together with about four acres of land surrounding the site of the Mill, for the Mill court and other necessary buildings. This company has a large machine shop for making machinery for the cotton and woollen manufactures, railroad cars, engines &c. They employ in general upwards of 500 hands; when building Mills they employ directly and indirectly from 1,000 to 1,200.

2d, The Merrimack Co.—capital 2,000,000 dollars, = £416,666.13.4. This company have five large Cotton Mills, besides print works. They run[e] 38,304 spindles[e], 1,300 looms, and give employment to[f] 1,250 females, and 550 males[f], they make upon an average[g] 250,000 yards[g] of cloth per week, and use about[h]

---

energy and enterprise sufficient for accomplishing great undertakings. While his gentlemanly manners and urbanity endeared him to his friends, his disinterested generosity and kindness gained for him the respect and esteem of the public.[8]

56,000 lbs.[h] of cotton in the same time. They generally spin No. 22's and 40's yarn for making printed goods and sheetings.

3*d*, The Hamilton Co. — capital 1,000,000 dollars, = £208,333 .6 .8. This company have a large printing establishment, and three Cotton Mills. They run[i] 21,248 throstle spindles, 590 looms[i], and give employment to[j] 550 females and 200 males[j]: they make about 100,000 yards of cloth, and use about[k] 42,000 lbs.[k] of cotton weekly: they generally spin No. 14's to 40's yarn for making drilling (three leaf tweel) printed and coloured goods.

4*th*, The Appleton Co.—capital 500,000 dollars, = £104,166 .13 .4. This company have two Cotton Mills, and run about 11,776 throstle spindles,[l] 400 looms, and give employment to 340 females[l] and 65 males. They make about 100,000 yards of cloth, and use[m] 36,000 lbs.[m] of cotton weekly; they generally spin No. 14's yarn for making shirtings and sheetings.

5*th*, The Lowell Co.—capital 500,000 dollars, = £104,166 .13 .4. This company have one Cotton and one Carpet Factory contained in one building, but divided in the middle. They run[n] 6,000[n] throstle spindles besides those used in the woollen manufacture;[o] 174 cotton and 74 carpet looms[o]; and give employment to 400 females, and 200 males. They make about 2,500 yards of carpeting, 150 rugs, and[p] 85,000 yards[p] of cotton cloth per week: they spin[q] No. 10's[q] yarn for making coarse negro cloth.

6*th*, The Suffolk Co.—capital 600,000 dollars, = £125,000. This company have two Cotton Mills, and run[r] 11,776 throstle spindles[r], 352 looms, and give employment to 460 females, and 70 males. They make about 90,000 yards of cloth, and use about 32,000 lbs. of cotton weekly; they spin No. 14's yarn for drillings.

7*th*, The Tremont Co.—capital 600,000 dollars, = £125,000. This company have two Mills, and run about 11,520 throstle spindles, [s]409 looms[s], and give employment to [t]360 females[t], and 70 males. They make about [u]115,000 yards[u] of cloth and use [v]30,000 lbs.[v] of cotton weekly: they generally spin No. 14's yarn, for making shirtings and sheetings.

8*th*, The Lawrence Co.—capital 1,500,000 dollars, = £312,500. This company have five extensive Factories and a Bleachery; they run [w]32,640 throstle spindles, 950 looms[w], and give employment to [x]900 females, and 170 males.[x] They make about [y]193,000 yards of cloth, and use about 62,000 lbs.[y] of cotton weekly: they spin No. 14's to 30's yarn for making printed cloth, shirtings and sheetings.

9*th*, The Middlesex Co.—capital 500,000 dollars, = £250,000.

This company manufacture broad cloths, cassimeres, &c. They have two Mills and a dye-house, and give employment to [z]500 females, and 250 males[z]: they run about[a] 6,120 spindles, 37 broad-cloth and 122 cassimere looms.[a] They make about[b] 9,000 yards of cassimere and 1,800 yards[b] broad cloth weekly. They use[c] 1,000,000 lbs.[c] of wool, and 3,000,000 teasels yearly.

10*th*, The Boot Cotton Mills Co.—capital 1,200,000 dollars, = £250,000. This company have four large elegant Factories in operation, containing [d]30,373 throstle spindles, and 858 looms[d]. They employ upwards of 950 females, and 120 males, and produce upwards of [e]180,000 yards[e] of cloth, and consume [f]59,000 lbs.[f] of cotton weekly. They spin from No. 14's to [g]40's[g] yarn for making drillings, shirtings, and cloth for printing.[h]

[i]11*th*, The Massachusetts Co.—capital 1,200,000 dollars = £250,000. This company have four large Mills containing 24,576 throstle spindles, 782 looms giving employment to 665 females and 150 males. And producing 225,000 yards of cloth per week consuming 77,000 lbs of cotton. The goods Manufactured are shirtings, sheetings and drillings from yarn Numbering 13's and 14's.[i]

To the above-named principal establishments may be added, the extensive Powder Mills of O.M. Whipple, Esq.[10]; the Lowell Bleachery; Flannell Mills; Card and Whip Factory; Planing Machine; Reed Machine; Flour, Grist and Saw Mills, together employing above 300 hands, and a capital of 300,000 dollars, = £62,500; and in the immediate vicinity of Lowell, there are Glass Works, and a Foundry supplying every description of castings.

The Locks and Canals Co.'s Machine Shop can furnish complete machinery for a Mill containing 5,000 throstle spindles with weaving in proportion, in four months, having lumber and materials always at command, to build or re-build a Mill in that time if required.

[j]The following table compiled from the most authentic sources, and published yearly, gives a more condensed view of the Lowell Manufactories. It may be remarked however, that the water power at this place is nearly exhausted, and consequently, no additional Mills are likely to be erected, unless some means are adopted for obtaining a larger supply of water, or economising that already used.[j][11]

Cotton consumed at Lowell per annum, (say one half Uplands, and one half New Orleans and Alabama, ... 18,059,600 lbs.

Statistics of Lowell Manufactures,[k] January 1, 1843. Compiled from Authentic Sources.

Corporations,	Locks & Canals.	Merrimack.	Hamilton.	Appleton.	Lowell.	Middlesex.	Suffolk.	Tremont.	Lawrence.	Boott.	Massachusetts.	Total.
Incorporated,	1792	1822	1825	1828	1828	1830	1830	1830	1830	1835	1839	
Commenced operations,	1822	1823	1825	1828	1828	1830	1832	1832	1833–34	1836	1840	
Capital Stock,	600,000	2,000,000	1,200,000	600,000	600,000	600,000	600,000	600,000	1,500,000	1,200,000	1,200,000	10,700,000
Number of Mills,	2 Shops, Smithy, Furnace.	5 & Print Wks	3 & Print Wks	2	Cotton & carpet Mill in 1 building.	2 and 2 Dye-houses	2	2	5	4	4	32, exclusive of print-wks, &c.
Spindles,		38,304	21,248	11,776	6,000 Cotton, beside Wool.	6,120	11,776	11,520	32,640	30,373	24,576	194,333
Looms,		1,300	590	400	174 Cotton, 74 Carpet.	37 Br'de-cloth, 122 Cassim.	352	409	950	858	782	6,048
Females employed,		1,250	550	340	400	500	460	360	900	950	665	6,375
Males employed,	500	550	200	65	200	250	70	70	170	120	150	2,345
Yards made per week,	1225 tons wr't and cast iron per annum.	250,000	100,000	100,000	2,500 Car. 150 Rugs, 85,000	9,000 Cassim. 1,800 Br'del.	90,000	115,000	193,000	180,000	225,000	1,351,450
Bales of Cotton used in do.		130	100	90	110	1,000,000 lbs.	90	75	175	137	188	1,095
Pounds of Cotton wro't in do.	Machinery, R R Cars & Engines.	56,000	42,000	36,000	40,000	wool per ann. and 3,000,000 teasels.	32,000	30,000	62,000	59,000	77,000	434,000

	1	2	3	4	5	6	7	8	9	10	11	Total
Yards dyed and printed do.		210,000	63,000									273,000
Kind of Goods made, ...	15,000 bush. char-coal, 200 chal. smith's coal	Pr'ts & Sheetings, No. 22 to 40.	Prts, Flan-nels Sheet-ings &c No. 14 to 40.	Sheetings & Shirt-ings, 14.	Carpets, Rugs & Negro Cloth	Broadcloth & Cassi-mere.	Drillings, 14.	Sheetings & Shirt-ings, 14.	Print'g Cloths Sheet & Shirt. No. 14 to 30.	Drillings, 14. Shirt-ings, 40. Pr'g Cloth, 40	Sheetings, 13. Shirt-ings, 14. Drilling, 14.	
Tons Anthracite Coal per ann.	200 tons hard coal.	5,000	3,000	300	500	600	300	250	650	750	750	12,300
Cords of Wood per annum, ...	200	200	500		500	1,300	70	60	120	70	70	3,090
Gallons of Oil per annum, ...	2,300	13,000	6,500	3,440	Olive, 4,000 Sperm, 4,000.	Lard, 12,000 Sperm, 5,000	3,840	3,692	8,217	7,100	7,100	80,189
Diameter of Water-wheels,	13 ft.	30 ft.	13 ft.	13 ft.	13 ft.	17 & 21 ft.	13 ft.	13 ft.	17 ft.	17 ft.	17 ft.	
Length of do. for each mill, ...	14 ft.	24 ft.	42 ft.	42 ft.	60 ft.	23 & 21 ft.	42 ft.	42 ft.	60 ft.	60 ft.	60 ft.	
How warmed, ...	Hot Air Furn.	Steam.	Steam & H.A.	Steam & H.A.	Hot Air Furn.	Fur. & Steam.	Steam.	Steam.	Steam.	Steam & H.A.	Steam.	

Yards of Cloth per annum, ...70,275,400
Pounds of Cotton consumed, ...22,568,000
Assuming half to be Upland, and half New Orleans and Alabama, the consumption in bales,
averaging 361 lbs. each, is 56,940
  A pound of Cotton averages 3 1-5 yards.
100 lbs. Cotton will produce 89 lbs. Cloth.
  As regards the health of persons employed, great numbers have been interrogated, and the result shows that six
of the females out of ten enjoy better health than before
being employed in the mills—of males, one-half derive the
same advantage.
  As regards their moral condition and character, they are
not inferior to any portion of the community.

Average wages of Females, clear of board, per week, $1 75
  "   "   Males,   "   per day,   70
Medium produce of a Loom No. 14 yarn, yds. pr day,   44 to 45
  "   "   "   No. 30   "   "   "   30
Average per spindle, ...   11–10
Average amount of wages paid per month, ... $150,000
  A very considerable portion of the wages are deposited
in the Lowell Institution for Savings.
Consumption of Starch per annum, (lbs.) 800,000
  Flour for Starch in Mills, Print Works,
    and Bleachery, bbls. per ann. 4,000
  Charcoal. bushels per ann. 600,000

The Locks & Canals Machine Shop, included among the 32 Mills, can furnish machinery complete for a Mill of 5000 Spindles in four months; and lumber and materials are always at command, with which to build or rebuild a Mill in that time, if required. When building Mills, the Locks & Canals Company employ directly and indirectly from 10 to 1200 hands.

To the above-named principal establishments may be added, the Lowell Water-Proofing, connected with the Middlesex Manufacturing Company; the extensive Powder Mills of O. M. Whipple, Esq.; the Lowell Bleachery, with a capital of $50,000; Flannel Mill; Blanket Mill; Batting Mill; Paper Mill; Card and Whip Factory; Planing Machine; Reed Machine; Foundry; Grist and Saw Mills—together employing about 500 hands, and a capital of $500,000.[k]

STATISTICS *of* LOWELL MANUFACTURES, 1*st January*, 1839.—*Compiled from authentic sources.*

Corporations.	Locks and Canals.	Merrimack.	Hamilton.	Appleton.	Lowell.	Suffolk.	Tremont.	Lawrence.	Middlesex.	Boot Cotton Mills.	Total.
Capital Stock,	Dollars. 600,000	Dollars 2,000,000	Dollars 1,000,000	Dollars. 500,000	Dollars. 500,000	Dollars. 600,000	Dollars 600,000	Dollars. 1,500,000	Dollars. 500,000	Dollars. 1,200,000	Dollars. 9,000,000
Num. of Factories,	2 shops and 1 smithy.	5 and print works.	3 and print works.	2	1 cotton & carpet	2	2	5	2 and dye-house.	4	29 exclusive of print works.
Spindles,		37,984	20,992	11,776	5,000 cotton, besides woollen.	11,264	11,520	31,000	4,620	29,248	163,404
Looms,		1,300	564	380	154 cotton. 70 carpet.	352	404	910	38 broad cloth 92 cassimere.	830	5,094
Females employed,		1,300	830	470	400	460	460	1,250	350	950	6,470
Males employed,	500	437	230	65	200	70	70	200	185	120	2,077
Yds. made weekly,		220,000	100,000	100,000	60,000 cot. 2,500 carpet 150 rugs.	90,000	125,800	200,000	6,300 cassimere. 1,500 broad cloth.	155,000	1,061,250
Bales of Cotton used per week,	....	120	100	100	80	90	90	180	....	130	890
Pounds of Cotton used per week,	....	50,000	40,000	40,000	34,000	32,000	34,000	64,000	600,000 lbs. wool. 3,000,000 teasels per an.	53,300	347,300

Yards dyed and printed per week,	255,000	......	......	......	......	......	......	......	70,000	185,000	......
Tons wrought and cast iron used per an.	......	......	......	......	......	......	......	......	......	......	1,225
Kinds of Goods made,	......	Printing cloth, Drillings and Shirtings, No. 14 to 50.	Broad cloth, & Cassimeres.	Printing cloth, Sheetings & Shirtings, No. 14 to 30	Sheetings, & Shirtings, No. 14.	Drillings, No. 14.	Carpets, Rugs & Negro cloths.	Sheetings & shirtings, No. 14.	Flannels, Prints and Drilling, No. 14 to 40.	Prints and sheeting, No. 22 to 40.	Machinery, engines, cars, &c. for railroads.
Tons Anthracite coals used per year,	11,360	750	500	650	330	330	400	400	2,800	5,200	......
Chaldrons smiths' coal,	200	......	......	......	......	......	......	......	......	......	200
Tons hard coal,	200	......	......	......	......	......	......	......	......	......	200
Cords of wood used per year,	3,880	70	1,000	60	60	70	500	......	1,250	570	300
Gallons of oil used per year,	65,289	7,100	Olive, 11,000. Sperm, 2,500	8,217	3,692	3,840	Olive, 4,000. Sperm, 4,000	3,440	6,500	8,700	2,300
Diameter of water wheels,		17	17 & 12	17	13	13	13	13	13	30	13
Length of water wheels,		60	46 & 21	60	42	42	60	14	42	24	14
When incorporated		1835	1830	1830	1830	1830	1830	1828	1825	1822	1792
Commenced operations,		1836	1830	1833—4	1832	1832	1828	1828	1825	1823	1822
How warmed,		Steam & Hot air.	Wakefield furnace and Steam.	Steam.	Hot air furnace.	Steam & Hot air.	Hot air furnace.	Hot air furnace.	Steam & Hot air.	Steam & Hot air.	Hot air furnace.

Cloth manufactured at Lowell per annum, ... 55,185,000 yds. being rather more than 3 yards from each pound of Cotton.
One hundred pounds of Cotton will produce eighty-nine pounds of Cloth.
Average wages of females at Lowell, 2 dollars per week, besides their board.
Average wages of males, 80 cents. or 3/4 per day, besides their board.

As regards the health of persons employed, great numbers have been interrogated, and the result shows that six females out of every ten enjoy better health than before being employed in the Mills,—of males one half derive the same advantage: as regards their moral condition, they are not inferior to any portion of the community.

Medium produce of each loom at Lowell, on No. 14's, 44 to 55 yards per day.
Medium produce of each loom at Lowell, on No. 30's, 30 yards per day.
Average produce per spindle, ... $1\frac{1}{10}$ yds of cloth per day.
Average amount of wages paid per month, 145,000 dollars, = £22,083 .6 .8.
Consumption of Starch per annum, ...600,000 lbs.
.. .. .. of Flour for Starch in the Mills, Print Works and Bleachery per annum, ... 3,800 bar.
Consumption of Charcoal per annum, ...500,000 bu.

[1]Lowell being the largest, and most important manufacturing town in the United States; I have carefully collected the most authentic statistical information regarding it, believing that such information would be interesting to manufacturers generally. The rates of wages—produce of the factories, cost of materials, and other regulations established here, will always exert a considerable influence on other manufactories, in this part of the country. But the most remarkable circumstance connected with Lowell is the rapidity with which it has sprung into existence. The population in 1820 was less than 200 souls, while the valuation of the property did not exceed 100,000 dollars. In 1840, the census shows a population of 20,981, and the assessors' valuation of property makes it equal to 12,400,000 dollars.
From the above facts, an inhabitant of Britain may readily suppose the population of Lowell will form a heterogeneous mass of

people, collected from all quarters, and that the moral character of such a community must be of the lowest kind. This however is by no means the case. The proprietors and agents of the different establishments have used *every laudable endeavour* to guard against such an occurrence and have always maintained the strictest supervision over the character of those they employ. Nor does this remark apply to Lowell only, but also to the manufacturing establishments in the eastern states. For moral elevation, or intellectual improvement, the manufacturing population of these districts can bear an honourable comparison with the middle classes of society in any part of the country.

Dr. Bartlett in his "Vindication of the character and condition of the females, employed in the Lowell Mills, against the charges contained in the Boston Times, and Boston Quarterly Review;" gives a very gratifying account of the moral and physical condition of this interesting class of females. He says that "Hundreds and hundreds of these girls in leaving the neighbourhood of their nativity, and coming here, (Lowell) are removed from influences that are either negatively or positively bad, to those of the most active and excellent character. The observations and enquiry of every successive year, have confirmed in me (Dr. B.) more and more strongly the opinion, that the aggregate change which is wrought in the moral character and condition of the young females who come here from the country is eminently happy and beneficial. The great preponderance of influence is enlightening, elevating and improving, not darkening, debasing, and deteriorating. Their manners are cultivated, their minds are enlarged, and their moral and religious principles developed and fortified. Hundreds and hundreds of these girls will live to refer the commencement of their best and highest happiness to their residence in the city of Lowell."[12]

There are eighteen religious societies in Lowell, which maintain as many regular clergymen; being one to every 1,200 of the whole population. Fourteen of these societies worship in elegant churches. The other four meet in convenient halls. Ten of the religious societies constitute a sabbath school union. And according to their last annual report for 1841, the number of scholars connected with their schools was 4,936, the number of teachers 433, making an aggregate of 5,369. If to this we add the teachers and scholars belonging to the other schools not connected with the Union, the whole number will be upwards of "*six thousand, nearly one third of the entire population.*"

There are in the city of Lowell, "*thirty-one public free schools*; viz.

one high school, seven grammar, and twenty-three primary schools," *being one entire school to every 700* of the whole population. "These schools employ twenty-one male, and thirty-five female teachers, in all fifty-six." The grammar schools receive the pupils from the primaries. The High school is divided into a male and female department, with a male principal in the former, and a female principal in the latter. In this school young men are fitted for any of the colleges; and all the pupils, whether male or female, have every facility for acquiring a complete Mathematical, English, and Classical Education.

The average number of pupils generally attending the schools is six thousand. The schools now established and in operation are sustained by the city at an expense of 20,500 dollars = £4,270 16s. 8d. per year.

Lowell is well supplied with public institutions which have always met with the greatest encouragement, and been liberally supported by the different manufacturing corporations: only a few of these will be noticed in this place.

The Middlesex Mechanics Association was incorporated as early as 1825. Some years afterwards the Locks and Canals Company gave them a lot of land worth 5,000 dollars, on which a costly brick edifice was erected, called "Mechanics Hall," for the building of which the several manufacturing companies contributed the sum of 17,970 dollars. This association procures one or more courses of lectures to be delivered annually in their spacious hall. They have also an extensive Reading Room, supplied with the best newspapers, and other periodicals; besides a library containing upwards of 2,000 volumes, and over 4,000 mineralogical specimens, with rooms appropriated for chemical and other scientific and philosophical purposes.

The *Lowell Lyceum* and *Lowell Institute* are voluntary literary associations. They furnish lectures on moral, literary, and scientific subjects once a week during six months of the year. Besides these regular ones, there are occasional lectures and discussions on various questions during the winter.

There is also a well furnished museum, in a central part of the city, which is a place of considerable resort by a portion of the citizens.

There are two weekly and two tri-weekly newspapers published in Lowell, besides five other periodicals devoted to religion and literature, one of which deserves particular notice viz. "The Lowell Offering," *every article in it being original, and written entirely by*

Females employed in the Factories. It is published monthly, under the direction of two respectable clergymen. For purity and dignity of style and correct moral sentiment, it is alike honourable to its fair contributors and worthy Editors, while its literary merits may stand a fair comparison with "The Ladies' Book," or the average of the other literary journals of the day.[113]

Hours of labour at the Lowell Factories.

From the first of September to the first of May, work is commenced in the morning as soon as the hands can see advantage, and stopped regularly during these eight months, at half-past seven o'clock in the evening.

During four of these eight months, viz. from the first of November to the first of March, the hands take breakfast before sunrise, that they may be ready to begin work as soon as they can see: but from the first of April till the first of October, 30 minutes are allowed for breakfast at seven o'clock, and during the months of March and October at half-past seven.

During the four summer months, or from the first of May to the first of September, work is commenced at five o'clock in the morning, and stopped at seven in the evening.

The dinner hour is at half-past twelve o'clock throughout the year; the time allowed is 45 minutes during the four summer months, and 30 minutes during the other eight.

The following Table of the average hours of labour, has been furnished me by an experienced manufacturer, and is deemed as correct an average as could be given.—The time given is for the first of each month.

Average hours of work per day throughout the year.

	Ho.	Min.		Ho.	Min.
January, ...	11	24	July, ...	12	45
February, ...	12		August, ...	12	45
March, ...	11	52*	September, ...	12	23
April, ...	13	31	October, ...	12	10
May, ...	12	45	November, ...	11	56
June, ...	12	45	December, ...	11	24

*The hours of labour on the first of March are less than in February, even though the days are a little longer, because 30 minutes are allowed for breakfast from the first of March to the first of September.

Taking one day for each month, the whole number of working hours in the year, according to the preceding Table, are 146 hours

44 minutes, which, divided by twelve for the number of months, gives a result of 12 hours 13 minutes as the average time for each day, or 73 hours 18 minutes per week; therefore about 73½ hours per week may be regarded as the average hours of labour in the Cotton Factories at Lowell, and generally throughout the whole of the Eastern District. In many, perhaps in the majority of the Cotton Factories in the Middle and Southern Districts,* the hours of labour in summer are from sunrise to sunset; or from half-past four o'clock in the morning, till half-past seven in the evening; being about 13¾ hours per day, equal to 82½ hours per week. In these Factories the average hours of labour throughout the year will be about 75½ per week.

As the days in this country are shorter in summer, and longer in winter, than in Britain, the following Table is given to show the time of the sun's rising and setting on the first and fifteenth days of each month in the year. It is compiled from the American Almanack, and adapted to the latitude of Boston, viz. 42°20′23″ North.[14]

	Sun rises.		Sun sets.		Length of day.	
	H.	M.	H.	M.	H.	M.
January 1...................	7	30	4	38	9	8
Do. 15...................	7	28	4	51	9	23
February 1...................	7	14	5	14	10	″
Do. 15...................	6	57	5	32	10	35
March 1...................	6	35	5	50	11	15
Do. 15...................	6	12	6	7	11	55
April 1...................	5	43	6	26	12	43
Do. 15...................	5	19	6	40	13	21
May 1...................	4	54	6	59	14	5
Do. 15...................	4	39	7	18	14	39
June 1...................	4	25	7	29	15	4
Do. 15...................	4	22	7	38	15	16
July 1...................	4	26	7	40	15	14
Do. 15...................	4	36	7	35	14	59
August 1...................	4	52	7	20	14	28
Do. 15...................	5	6	7	2	13	56
Septem. 1...................	5	26	6	35	13	9
Do. 15...................	5	39	6	12	12	33
October 1...................	5	56	5	43	11	47
Do. 15...................	6	12	5	20	11	8
Novem. 1...................	6	33	4	55	10	22
Do. 15...................	6	51	4	39	9	48
Decem. 1...................	7	10	4	29	9	19
Do. 15...................	7	23	4	28	9	5

[m]*The hours of labour in Paterson New Jersey, and at Baltimore are the same as in Britain, viz. 69 hours per week.[n]

As much vague and contradictory information has been circulated in Great Britain regarding the hours of labour in the American Factories, I have endeavoured to give as accurately as possible, both the average hours per week for the year, and the length of working hours at the different seasons. From the preceding statements, it will be seen that the average hours per week throughout the year, are 73½ in the Eastern, and part of the Middle and Southern Districts, and 75½ in a considerable number, probably four-fifths of the Factories in the Middle and Southern Districts.

In Great Britain the hours of labour per week are limited by Act of Parliament to 69, or 11½ hours per day, but the general regulation in all the Factories is 9 hours on Saturday, and 12 hours on each of the other five days.[15] It is also enacted, that there shall be six holidays in the course of the year. In the United States, there are only three holidays·in the year. The first is called a general fast, and is entirely devoted to religious exercises. It is generally kept about the middle, or the 20th of April, being about the time the battle of Lexington was fought, near Boston, when the first blood was shed by the Americans in the cause of Independence. The second holiday is the 4th of July, called Independence Day, and is wholly devoted to public rejoicing, being an anniversary commemorative of the memorable Declaration of Independence, published at Washington on the 4th of July 1776, by the representatives of the then thirteen confederated colonies. The third holiday is generally about the" end of November" and denominated Thanksgiving-day. It is partly devoted to religious exercises in the morning, and social intercourse in the afternoon. Upon this day, the scattered members of each family endeavour to meet at what may be called their home, for the purpose of enjoying the social company of each other, and gratifying their filial attachments. °Christmas is also kept as a holy day in some few of the States.°[16]

In this country the time for breakfast is seven o'clock, and for dinner half-past twelve: supper at seven, or half-past seven in the evening. In Scotland the Mills begin work at six o'clock in the morning throughout the year, and stop at nine o'clock for breakfast, and at two o'clock for dinner, 45 minutes being allowed for each. In England the Mills begin work in the morning at six o'clock, and stop at eight o'clock for breakfast, and Phalf-pastP twelve for dinner, and four o'clock in the afternoon for tea. Those that do not stop for tea, allow the workers to have it carried into the Mill at four o'clock. Thirty minutes are allowed for breakfast, and an hour for dinner. Thus each country has its own peculiar regulations, and each will no doubt prefer their own; but certainly for three meals a day, the Scotch proportion their time better than the others.

RHODE ISLAND.

THERE are no manufacturing towns in this State equal to Lowell, yet there are a great number of manufacturing villages, and in proportion to its extent, this State contains a much greater number of Factories than that of Massachusetts.

The following notices of the various manufacturing districts in this State are chiefly compiled from White's Memoir of Slater. The first which claims attention, is that of

NORTH PROVIDENCE.[17]

THIS place was incorporated in 1767, and is now distinguished for its manufactures. There are ten Cotton Mills, one of which is the first that was built in America; and in Pawtucket, S. Slater erected the first water-frame spinning machinery. The extent of this business having concentrated a large capital, and an immense aggregate of industry has, within the last thirty years, given rise to this large flourishing village, which is situated on the North-East section of the town, four miles North-East of Providence, on the border of the Seekonk river, which affords numerous natural sites for manufacturing establishments of almost every description.[18] The rapid march of manufacturing and mechanical industry which the short annals of this place disclose, has few examples in this country. The village is built on both sides of the river, which, at this place, divides Rhode Island from Massachusetts. The part of it which is in Rhode Island, is principally built on four streets, qcontaining a number of elegant dwelling houses and mercantile stores.q There are six shops engaged in the manufacture of machinery, having the advantage of water power; and various other mechanical establishments, affording extensive employment, and supporting a dense population. Upon the Massachusetts side of the river, the village is of nearly equal extent.

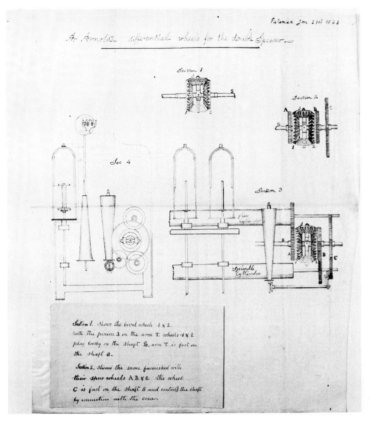

Fig. 29.   Patent drawing of differential gear patented for the roving frame by Aza Arnold of North Providence, Rhode Island, on 21 January 1823. Drawing in Zachariah Allen Papers, Rhode Island Historical Society. Courtesy of Rhode Island Historical Society.

Besides the Cotton Factories, there are in the town two Furnaces for casting, one Slitting Mill, two Anchor Shops, Screw Manufactories, Nail Factory, Flour Mills, and Machine Manufactories: here some very superior machinery is made, and sent to various parts of the country.

SMITHFIELD.[19]

AT this place there are nine Cotton Factories, all of which contain more than 11,000 spindles, one half of which belong to Slaters & Co., whose establishment is situated on a branch of the Pawtucket river, about one mile and a half from its junction. At this place there is a large and flourishing village, called Slaterville, which is but of recent date, having grown up with the cotton manufacturing business.

There is also a large establishment, containing 8,000 spindles owned by Butler, Wheaton & Co. of Providence.

There are twenty public schools in Smithfield, and several private ones, three incorporated academies, four social libraries, and four places of worship.

WOONSOCKET FALLS.[20]

'The village of Woonsocket is situated on the Blackstone river about 14 miles from Providence, where there is a fall of water upwards of 20 feet; not perpendicular but over a precipice of rocks for some distance. '¹A canal has been cut from the top of the falls, which supplied water power sufficient for driving 14 cotton and 2 satinet factories, a large furnace, machine shops, sash factory &c. And on the opposite side of the river there are two large factories belonging to the Woonsocket Manufacturing Co. which are decidedly the best in the whole district. They are fitted up in good style, containing many of the best improvements of the day, and deriving a profitable business: but with the exception of these two, and the mills at Lonsdale, the whole of the cotton factories throughout the district, from Blackstone to Pawtucket, are of an inferior grade. Much of the machinery is old and dirty, while the work produced is deficient in both quantity and quality.[r121]

At Woonsocket, which may be denominated the capital of Smithfield, there is a great amount of business done in machinery, and manufacturing. It is said there are upwards of 50,000 spindles in operation, besides other machinery. The mill sites too are very valuable, and could not be purchased for half a million dollars. The village and most of its dependencies belong to capitalists in Providence.'

DOUGLAS.[22]—At this town, on the Munford river, the Douglas Manufacturing Co. have two Mills, both five stories in height. They have in operation 4,000 spindles, 119 looms, and employ about 200 hands: the cloth manufactured is generally for printed goods.

SLATERVILLE embraces a part of Douglas and a part of Dudley. This place derived its name from Samuel Slater, who generally resided there and had several Mills, (said to be seven in all,) part of which derive their power from French River, and part from a large pond called Slater's Lake, about four miles long, which is a never-failing source of supply.

Besides the above-named villages, there are many others of equal, and some of greater extent and importance, in various parts of the State, such as Cranston, Warwick, Scituate, New-Port, Lonsdale, Coventry, Cumberland, Johnston, &c. At New-Port, there are four Cotton Factories all moved by Steam. There is also a large Steam Mill at Providence, owned by the heirs of Samuel Slater. At Lonsdale, there are three elegant Factories, and here, too, Nankeens are manufactured from Cotton of a very deep dirty cream colour, a small quantity of which is raised annually in Georgia of an inferior quality.[23]

Passing from Rhode Island into the State of Connecticut, the traveller is greatly delighted with the sight of a number of beautiful and handsome manufacturing villages. Indeed, this State is famed for its neat villages and beautiful landscapes.
"The pleasant village of Cabotsville," says Mr. White, "has grown up with astonishing rapidity, and bids fair to become, at no very distant day, a second Lowell."[24]
"'There is a neatness, too, and good taste, in the location of the streets, and the arrangement of the buildings, which is not common in manufacturing villages. The Cotton Factories are extensive, and in appearance resembling those at Lowell."[25]
'The village is situated on the Chickopee river, about four miles from Springfield and contains about 4,000 inhabitants. There are at present six cotton factories in operation containing in all about 36,000 throstle spindles and 1,400 looms besides a machine shop, a sword and cannon factory.
The mills are elegant and substantial, the general arrangement excellent. The style of the machinery is similar to that at Lowell but in some things superior. They indeed possess every facility for equalling the best mills in the country. The whole property is valuable and must ultimately stand high in the stock market.
Chickopee is about two miles farther up the river or about six miles from Springfield. At this place there are four factories besides machine shops. The style of the mills and machinery is per-

haps older than the former but the goods are equal to the average of other mills in the country.[26]

Another flourishing village in this State, called Williamantic, is situated in Windham county, on the Williamantic river, near its confluence with another small river called the Natchang. It extends about a mile along the former stream. Twelve years ago there were less than a dozen houses, and those very indifferent ones, on the site of the present village. Now there are four manufacturing establishments, containing 12,800 spindles, and making annually about 2,915,000 yards of cotton cloth. There is also a Paper Mill and a small Satinet Factory. The village, as well as the surrounding country, seems to be prospering, and advancing in moral and intellectual improvement. It contains three places of public worship, two free, and three private schools, and a public library.[27]

Greeneville is another beautiful village, situated on the West bank of Shetucket river, a little below its junction with the Quinebaug, and five hundred rods above steam and packet navigation. This village has had almost as rapid a growth as the villages of the West. The General Assembly of Connecticut granted a charter in 1828, to a Company under the name of the "Norwich Water Power Company," the object of which was the construction of works to bring into use the immense water power at this place, then wholly unoccupied. The capital of the Company was 40,000 dollars; and having purchased a large tract of land lying on both sides of the river, they proceeded to erect a dam, and dig a canal, through which the water of the river necessary for manufacturing purposes might flow. The dam is built of stone, in length 280 feet, and is both solid and substantial, so that there is little reason to apprehend that it will be carried away, although the river is subject to great annual freshets. The abutments are very handsome and durable specimens of stone masonry. The canal is about a mile in length, forty-six feet wide at the surface, and ten feet deep. These works were completed in 1830.

There are several manufacturing establishments at this place, the largest of which is that of the Thames Company, for the manufacture of cotton cloth. This is one of the finest edifices of the kind in New England, being built of brick, five stories high, 138 feet in length by 44 in width. There are employed in it about 180 persons of different ages and sexes, and about 42,000 lbs. of cotton are consumed monthly, while, in the same time, about 132,000 yards of cloth are produced.

The Shetucket Factory for the manufacture of bed ticking, contains 1,650 spindles, employs about 70 persons, and consumes about 14,000 lbs. of cotton, producing about 28,000 yards of cloth per month.

The Greeneville Manufacturing Company employ about 50 persons, and produce about 12,000 yards of flannel monthly, using for that purpose about 4,800 lbs. of wool.

There are also two Carpet Factories, a Paper Manufactory, a Machine Shop, and other small Factories at this place, which is still increasing. A number of very eligible sites for Manufactories are yet unoccupied, and a large amount of water power is unemployed.

The village is situated in a delightful tract of country. The dwellings are very neat and attractive, being all painted white; they have a uniform and handsome appearance, and seem to be the abode of industry and contentment.[28]

There are various other manufacturing villages about this part of the Union, of which I have not been able to obtain correct information. A vast number of small Factories and manufacturing villages are scattered over the State of New York, which I have not yet had the opportunity of visiting, neither have I been able to obtain any written account of them.

PATERSON, NEW JERSEY.[29]

THIS town, next to Lowell, is one of the greatest manufacturing towns in America. The Mills are not so large and splendid, nor is the business conducted with the same "spirit as at Lowell; yet there are some" superior goods manufactured, and machinery made, at this place, "upon a very extensive scale."

The cotton manufacture commenced at Paterson at a very early period. A society was formed in the early part of 1791, denominated "The Society for the Establishment of Useful Manufactures;" the immediate object of which was the manufacture of cotton cloths. At this period no cotton had been spun by machinery in America, except at Pawtucket in Rhode Island. The number of shares originally subscribed was 5000 at 100 dollars a share, but 2267 shares only were fully paid up. It was ascertained by this Society, that the Great Falls of the Passaic river in New Jersey, had an elevation of 104 feet above tide water, and were calculated

by their elevation and volume of water, to furnish power sufficient
for driving 247 undershot water wheels. And at Little Falls, four
miles higher up, a thirty-six feet Fall was deemed capable of driving 78 water wheels. Becoming thus satisfied, from various sources
of information, regarding the superiority of this situation, they
selected the Passaic as the principal site of their proposed operations, and gave to the town the name of Paterson after the Gov-
ernor of New Jersey, who signed their charter, vesting them with
power over, and possession of, the waters of the Passaic river at
this place. The Society, soon after the grant of its charter and
purchase of the ground, proceeded to establish their first Cotton
Factory and printing house; which were attended however with
considerable loss. They invited and encouraged skill by leasing
privileges and aiding manufacturers with capital. This was well
calculated to induce numbers to come and share in the advantages
of the vast water power. Experienced Mill owners from various
parts came hither, bringing wealth even from England; artisans
of various descriptions from Britain, were invited and encouraged
to settle in the place. A raceway and canal were commenced by
the direction of the Company, designed to unite the Upper Passaic
with the Lower, at the head of tide, near the present village of
Acquackanonck; the engineer to whom the execution of this work
was committed, spent vast sums of money to little purpose: latterly,
Mr. Colt from Connecticut,[30] was appointed superintendent of
the affairs of the Company, with full powers to manage its concerns: he completed the race-way, but the canal[x] was abandoned.
The first Factory, 90 feet long by 40 wide, and four stories high,
was finished in 1794, when cotton yarn was spun in the Mill; but
yarn had been spun the preceding year by machinery moved with
oxen. Calico shawls and other cotton goods, were printed the same
year; the bleached and unbleached muslins being sold in New
York. Although the concerns of the Company became involved
in difficulty, they persevered in their enterprise, and in the years
1795 and 1796, much yarn of various sizes was spun, and several
pieces of cotton goods manufactured. At length, however, they
were forced to abandon the manufacture, and discharge their
workmen. This result was produced by a combination of causes.
Nearly 50,000 dollars had been lost by the failure of parties, to
certain bills of exchange purchased by the Company, to buy in
England plain cloths for printing: large sums had been wasted by
the engineer and the machinists, and the manufacturers brought
from other countries, were presumptuous and ignorant of many

branches of the business they engaged to conduct; the Company was inexperienced in the business, and the country unprepared for manufactures. The Cotton Mill belonging to the Company was subsequently leased to individuals, who continued to spin candlewick and coarse yarn, until 1807, when it was accidentally burned, and never rebuilt.

The water power of the Company, however, was not wholly unemployed. In 1801, a Mill site was leased to Messrs. Kinsey & Co.[31]; in 1807, a second; and in 1811, a third to other persons; between 1812 and 1814, several others were sold or leased. In 1814, Mr. Colt purchased, at a depreciated price, a large proportion of the shares, and re-animated the association. From this date the growth of Paterson has been steady, except during the three or four years that followed the peace of 1815. The advantages derivable from the great fall in the river, have been improved; a dam of four and a half feet high, strongly framed and bolted to the rock, in the bed of the river above the falls, turns the stream through a canal excavated in the trap rock of the bank, into a basin, whence, through strong guard-gates, it supplies in succession, three canals on separate elevations giving to the Mills on each head, a fall of about 22 feet. By means of the guard-gate the volume of water is regulated at pleasure, and a uniform height preserved, avoiding the inconvenience of back water. 40,000 dollars have been expended to complete this privilege.

'When the company first commenced operations there were not more than ten houses at Paterson. In 1838 there were 975 dwelling houses and a population of 9,048 whites [and] 239 blacks [making the total number 9,287] [deleted].

There are 11 houses of public worship, 23 pay schools containing upwards of 900 scholars, and one free school with about 160. There is also an infant school supported by ladies in which are gratuitously taught on an average 200 scholars. There are 8 sabbath schools in which about 1,700 scholars are instructed.

Paterson has few public institutions compared with manufacturing towns in the New England states and the few she has are not in a very flourishing condition. There is only one incorporated library containing about 500 volumes; a museum of natural and artificial curiosities; a public dispensary; and a *young men's society*, whose meetings are held two evenings in the week for discussion, lectures, essays, &c. The society has a library and reading room.

There are 34 public works of various kinds in Paterson consisting of 16 cotton factories, one woollen, and one satinet factory besides flax mills, print works, bleachery and dyeing, four large machine shops, millwrights, card making and bobbin factories. There is also a patent arms manufactory employing about 90 hands and expending 4,000 dollars per month, 3,000 of which is for wages alone.

Number of spindles in operation at
   Paterson   46,067
   do. looms do.   481

Cotton consumed annually	3,880,840	lbs.
Flax do. do.	600,000	lbs.
Wool do. do.	238,000	"
Total consumption	4,718,840	lbs.

Quantity of pig iron used in machine and millwrights' shops

chine and millwrights' shops	1,456	tons
Bar iron	460	"
Boiler iron	60	"
Copper	56	"
Steel	65	"
Total	2,097	tons

Employed in the different machine and [illegible] factories

Males over 16 years of age	1,104
do. under 16 do.	265
Females over 16 do.	564
do. under 16	364
Total	2,297

The hours of labour at Paterson are the same as in Britain, viz. nine hours on Saturday and twelve hours on each of the other five days equal to 69 hours per week.

Water power is sold at Paterson by the square foot, under a head of 2 feet 6 inches, and 22 feet fall; which admits a water wheel of 18½ feet in diameter. This is considered equal to 20 horses' power and for which an annual rent of 500 dollars is charged, equal to 25 dollars per horse power.[32] There are yet (1842) 20 mill privileges for sale of 20 horse power each.

The cost of transportation between New York and Paterson is 13 cents per 100 lbs., or three dollars per ton, each way.

From the preceding statements it appears that Paterson, next to Lowell, is one of the largest manufacturing towns in America. The business commenced at this place at a very early period. Yet

however strange it may appear, it has scarcely advanced beyond
the first or second stages of improvement. There is no other place
in the United States where the cotton manufacture is upon a
[larger?] scale than at Paterson, New Jersey. There is here a total
want of system in the style and arrangement of the mills and
machinery. [The whole town has an air of slovenliness, a want of
taste and public spirit, an entire absence of the enterprising spirit
of the New England manufacturers. As compared with Lowell it
is certainly an age behind]. [deleted] Had the same master minds
that built Waltham and Lowell been applied to Paterson; or had
it been situated in the immediate vicinity of Boston, within the
influence of the capital and enterprise of the Boston merchants,
instead of a third rate, it would long ere this have been the first
manufacturing town in the country. But as it is, there is nothing
about it to admire, except the machine shops and excellent water
privileges, and the sublimely beautiful falls on the Passaic river,
a sight of which is well worth the expense of some hundreds of
miles travel.[y]

The cotton manufacture is carried on to some extent near to
Philadelphia, but I have not been able to obtain correct information regarding the number or extent of the Factories, either
at this place or Baltimore. Cotton Factories are also extending
rapidly in various parts of the Southern States, in Virginia, North
Carolina, Tennessee, and towns along the Ohio. "At a fine water
privilege in Athens, Georgia, there was established a Cotton Mill,
with machinery from England, by Dearing & Co.; it is still in
operation, and one also in Columbus."[33] The time, indeed, seems
to be fast approaching, when Cotton Factories will be established
at the South, the North, East and far West of America. And there
can be no doubt but this country is destined, at no very remote
period, to be the great emporium of the cotton manufacture of
the world, as it possesses all the necessary requisites for that purpose, viz. extensive available water power, an intelligent and enterprising population, and having within itself an abundant supply
of the raw material. Those Factories established in the South, must
possess decided advantages over all others, as the manufacturers
there will also be the cultivators of their own cotton, which may
be brought from the fields where it is raised, to store-houses connected with the Factory in which it is to be cleaned and spun into
yarn, and thereby all the expenses of baling and transporting
saved. And if the experiment of slave labour succeed in the Factories, as is confidently expected, the cost of manufacturing the

cotton into cloth, will be much less there than anywhere else; so
that it will not be surprising if, in the course of a few years, those
Southern Factories should manufacture coarse cotton goods, and
sell them in the public markets at one half the price at which they
can be manufactured in England. There are several Cotton Factories in Tennessee operated entirely by slave labour, there not
being a white man in the Mill but the Superintendent; and according to a letter lately received from the Superintendent of one
of these Factories, it appears that the blacks do their work in every
respect as well as could be expected from the whites.

"Cotton Factories are rapidly springing up in North Carolina;
but with two or three exceptions they are chiefly employed only
in spinning cotton yarn. The two oldest Factories in this State are
one at the Falls of Tar River in Edgecombe county, established
in 1818; and another near Lincolnton, in 1822. Factories have
since been established at Mocksville, Greensborough, Fayetteville,
Lexington, Salem, Milton, and in the counties of Orange and
Randolph; there are eleven Factories in all. Arrangements are in
progress for establishing similar works in various other parts of
the State."

^zThe Saluda cotton factory situated on the Saluda river about
two miles distance from Columbia bridge in Lexington district,
South Carolina has been in operation for some years. The building
is four and a half stories high and constructed of beautiful gray
granite. It is said to have cost about 150,000 dollars and employs
nearly 200 persons. The machinery is excellent and the goods
manufactured are of a very good quality.[234]

"Cincinnati," says Mr. White, "is the great commercial emporium of Ohio, and, next to New Orleans, the largest city in the
valley of the Mississippi. The value of its manufactures is about
2,500,000 dollars annually. There are ten foundries, including a
brass and bell foundry, and one for casting types. There are four
Cotton Factories, fifteen Rolling Mills, Steam Engine Factories
and Shops, three Breweries, Button Factory, Steam Coopering
Establishment, five or six Saw Mills, two Flour Mills, and one
Chemical Laboratory. There are not less than forty different Manufacturing Establishments driven by steam power."[35]

A writer quoted by Mr. White, visited the Cotton Works of one
company situated at Matoaca, on the North bank of the Appomattox, about four miles from Petersburgh, and "was no less grati-
fied by the beauty and substantial appearance of the buildings,
than surprised at the expedition with which they have been

erected." The works here referred to consist of two Cotton Mills, three stories high; a Machine Ship and Sizing House, built of granite of a superior quality, obtained from a quarry on the Company's land. The principal Mill is 118 feet long by 44 wide; the other 90 feet long by 40 wide. They contain about 4000 spindles, and 170 looms.[36]

"The manufactures of Virginia, like her coal mines, are just beginning to rise into importance. But recently the attention of her citizens has been directed to the subject, and few out of the State are aware how far she has already advanced, and how rapidly she continues to advance in this branch of industry."[37]

In Richmond and Manchester, Virginia, there are in full operation two Cotton Factories, and three Iron Foundries, to one of which a Steam Engine manufactory is attached. There are also a number of other Establishments, and few places can boast of such large or superior Flour Mills. The Galego Mill, which is, perhaps, the largest in the world, runs 22 pairs of stones, and makes 500 barrels of flour daily. Haxal's Mill is but little inferior to this, and Rutherford's and Clark's though less than the others, are considerable Mills.[38]

"The water power at and near Richmond, is immense, and easily available; it is the entire James River, which is nearly half a mile wide, and falls more than a hundred feet in a few miles. The advantages of its position are many and great; situated at the head of good navigation,—open nearly all the year,—adjacent to a rich coal field,—connected with the interior, as it soon will be, by a canal leading through a fine iron district,—with a healthy and pleasant climate, surrounded by a good soil, nothing can prevent its becoming one of the greatest manufacturing cities in the Union.[39]

"Next to Richmond in importance, and in some respects in advance of it, is Petersburgh, at the head of the tide water of the Appomattox. Here Cotton Factories grow up and flourish as if by magic; there are now five or six all in full operation, and all of them extensive establishments. One of them, a short distance from Petersburgh, is inferior to few, if any, in the Northern States, and with the houses built for the workmen, forms quite a village. All these Manufactories employ white labourers. The experiment, however, of negro or slave labour, has been made in one of the Factories at Richmond, and has proved fully successful. Other Manufactories are about to be erected near Petersburgh, in some of which it is expected that negro labour will be introduced generally, if not exclusively. Indeed, there is every reason to believe that it is better adapted to the manufactory than to the field, and that the negro character is susceptible of a high degree of manufacturing cultivation. Should this kind of labour be found to succeed, of which I think from some years' acquaintance with it, there can be no doubt, it will give a decided advantage to the Southern over the Northern or European manufacturer. This kind of labour will be much cheaper, and far more certain and controllable. The manufacturer will have nothing to do with strikes, or other interruptions that frequently produce serious delay and loss to the employer. Before the present year, [1835,] the average expense for a good negro man per year, might be estimated at 100 dollars, = £20 . 16 . 8 for field labour. Some superior hands, well acquainted with tobacco manufacturing, or good mechanics would, perhaps, go to 150 dollars; these prices include hire, food, clothing, &c. They are now, in consequence of the great demand for labourers on the railroads, at least 20 dollars higher; that is, about 170 dollars, = £35 . 8 . 4 for a good negro man for a whole year; of course, females and young men will be much cheaper.[40]

"The water power of Petersburgh, though inferior to that of Richmond, is yet very considerable. It is also without the advantages of an immediate connection with the coal and iron regions; nor has it so good navigation as the latter, as vessels only of six feet draught of water can come to it, while those drawing eleven may go to Richmond. Yet Petersburgh is as well, if not better situated for the cotton manufacture, than Richmond. A railroad of sixty miles in length connects it with the Roanoke, and brings to it daily large quantities of cotton, from which it can have the first and best selections. This together with the cheapness of water power, building materials, and all other articles that enter into the consumption of those who labour, give to it great advantages.

"Besides these two prominent places, many others might be found in Eastern Virginia, Georgia, Carolina, Tennessee, &c. equally favourably situated for Manufactories. At Fredricksburgh, on the Rappahannock, is a considerable water power, and on nearly all the rivers that empty into the Chesapeake, there are more or less sites. On the James River, between Richmond and the mountains, they are almost innumerable; and when the State improvements are completed, they will be in good location.

"Manufacturing is carried on at Wheeling on the Ohio, but Western Virginia is identified with the great valley of the Mississippi; the future greatness or prosperity of which, no imagination can reach; it is a world in itself, and the world beyond it cannot change its destinies."[41]

[a]The following from the *New Orleans Bulletin of March 9th, 1839* shows that the extension of cotton factories in the South is not confined to the Middle districts but reaching to the very extremities.

"A cotton factory has been established in our city and is now in successful operation. Our enterprising fellow citizen, Mr. Whitney, has the honour of being the first to make the experiment in New Orleans, and from present indications, there seems to be no doubt of ultimate success.[42] At present the manufacture is limited to the coarser fabrics, and such stuffs as are suitable for negro clothing. In this character of cloths, the advantages of a location in New Orleans, has enabled Mr. Whitney to commence a successful competition with the Northern factories. His cottons are much heavier and stouter than those of the same class manufactured at Lowell. The cheapness of the raw material at this place, and the many opportunities for buying the inferior staples at a low price, puts it in the power of the manufacturer to give more weight and substance to the texture at a more reasonable rate. For this reason factories established at New Orleans will always be able to furnish coarse fabrics better and cheaper than it can be done elsewhere. All the cost of carriage is saved, and other expenses incident to transportation to foreign ports. The planters in the neighbourhood of New Orleans have already discovered the superiority of Mr. Whitney's cloths, and hesitate not to give them the preference over Northern and foreign manufactures. We have no doubt that his enterprise will meet with ample encouragement. There is a growing disposition at the South to foster domestic manufactures, and if a liberal patronage be extended to this infant establishment, we may expect to see many others springing up around us, and a system of mechanical industry put in motion upon an extensive scale, adequate to the wants of this section of the South."[a]

From the above extracts, it appears that the Southern States possess many facilities for extending the cotton manufacture, such as cheap labour, materials, &c. with an abundant water power, land and water carriage, &c. but though the manufacturers in the Northern States cannot provide the raw material so cheap as those in the South, they have the advantage of the manufacture being already established, with mechanics and artisans of every descrip-

tion intimately acquainted with the business. The standing which
the cotton manufacture has acquired in the North, and the un-
congenial nature of the Southern climate, render it a matter of
doubt to many, whether the latter will ever outrival the former
in this business.

At present the Northern States are still progressing in the manufacture of cotton goods; the greater part of which, has hitherto
been made for home consumption. A considerable quantity, however, has been exported to South America, India[b] & China[b] in
both of which markets the American manufacturers have been
able to compete successfully with the British. At present, the attention of manufacturers in this country, is directed to exporting
cotton yarn in the bundle; and should this business succeed, I
have no doubt but the British will find them rather formidable
rivals, as whatever yarns may be thus exported will, (like the cloth,)
maintain its character in any market where it may be sold, and
this, I know, is not always the case with much that is exported
from Great Britain.

There is yet much unoccupied water privilege in the Northern
States. A dam has lately been completed across the Kennebec
River, in the State of Maine, by means of which the whole water
of that river may be directed into canals, and thereby furnish
valuable sites for, it is said, nearly one hundred Cotton Factories.[43]
At Brunswick, in the same State, there is abundance of unemployed water power, where one Factory has lately been put in
operation, and others are expected soon. At Saco there are [c]four[c]
Mills in operation, and sufficient water power for [d]sixteen[d] more.[44]
At Amoskeag, about 40 miles from Lowell, a canal [e]has been[e] cut,
which, it is said, may take in the whole of the Merrimack River,
and supply water power for upwards of fifty Factories.[45] But instead of availing themselves of these water privileges, the attention
of manufacturers has been for some time directed to the advantages of steam, as a means of propelling machinery; the advantages of a good location being considered equal to the extra expense of steam power. Mills propelled by steam may be situated
in seaport towns, where there will be no expense for land carriage,
and being in a thickly populated neighbourhood, are likely to
have a good supply of helps; whereas those driven by water, are
subject to interruptions in winter, in consequence of the canals
freezing up, and a deficiency of water in the drought of summer,

besides the expense of land carriage; although the latter, no doubt, depends upon the location of the Mills.

Three large Mills driven by steam are in operation at Newbury-port in Massachusetts, and another is soon to be erected[46]; it is likewise probable that so soon as coals become a little cheaper, there will be a number of Mills with steam power erected in various parts of the country.

There are several manufacturing towns and villages to the East-ward of Massachusetts which are worthy of a passing notice before leaving this part of the subject. In New Hampshire there are New Market, Dover, and Great Falls, at each of which there are exten-sive Manufactories. At New Market there are three large Cotton Factories, in which goods of a very superior quality are manufac-tured, chiefly shirtings and sheetings. These works bear a very high character, and it is said, they have always been well con-ducted, and have paid the proprietors a very handsome profit, at least until the depression of 1837.[47]

At Dover there are three large cotton factories containing 28,666 throstle spindles and a calico printing establishment which are so situated as to form a solid square one side of which being the print works and the other three the cotton mills. 'The number of persons employed are upon an average about 1,160 and the value of the goods produced annually are about 590,000 dollars, the amount of capital invested is 1,056,000 dollars.'[48]

Dover in New Hampshire with Lowell and Fall River in the State of Massachusetts; Providence, in Rhode Island; and Hudson in New York, are the principal printing establishments in the United States[49]*: at all of these places printed goods are produced equal to any in England.

---

* Average amount of cotton goods printed in the United States annually with the number of printing establishments, number of yards, average and total value.[50]

States	Factories	Yds. per ann.	Av. value	Total value
New Hampshire	2	5,546,667	13 cents	$ 721,066
Massachusetts	10	38,162,667	"	4,831,146
Rhode Island	9	26,624,000	"	3,461,220
New York	7	12,202,667	9	1,098,240
New Jersey	2	6,101,334	"	549,120
Pennsylvania	4	8,874,667	"	798,720
Maryland	2	2,600,000	8 cents	208,000
	36	100,112,002		$11,667,512[g]

At Great Falls about four miles from Dover there are [h]five[h] large elegant Cotton Factories, containing in all about 39,840 spindles, (16,000 of which are mules) and 1132 looms. This is one of the pleasantest and most beautiful manufacturing villages I have ever seen. There are [i]five[i] large Mills, but two of them form one connected building. These are situated[j] at the distance of about one hundred yards from each other; the canal which supplies them with water runs in front of the Mills, leaving a level space between them about 30 or 40 yards broad; on each side of this there is a row of young trees planted so as to form a delightful promenade in front of the Mills; on the opposite side of the canal from the Factories, there is a large open space of ground rising with a gentle acclivity, about 100 yards broad, on the outer verge of which, is the main street of the village: [k]a branch of the eastern railroad passes through the midst of[k] this open ground between the canal and the Main Street,[l] and bordering the outside of Main Street are the boarding or dwelling-houses for the Mill workers; these are neat brick buildings, three stories in height, and each building contains four tenements: there are seven of these boarding houses, set at equal distances from each other, which gives to the whole an appearance of neatness and uniformity. The Main Street, the Canal, and the Mills, all running in parallel lines with a large open area between them, have a most delightful effect upon the mind of the stranger when he first enters the village. The whole plan of the village displays good taste, and its general appearance is delightful and beautiful in the highest degree. A great sum of money has been sunk in this place; and it is said that the proprietors have never realized the interest of the money advanced. [m]But the whole establishment is now undergoing a complete renovation. The general management having lately fallen into the hands of Gentlemen well qualified to render it one of the best manufacturing establishments in the United States.[m][51]

SACO.—The last of the manufacturing villages which we shall notice, though not the least in importance, is that of Saco, in the State of Maine.[52] This village takes its name from the river *Saco*, which runs through it, and is one of the largest rivers in New England: it rises in the White Mountains of New Hampshire, the summits of which are so elevated as to be covered with snow throughout the greater part of the year; the melting of which in summer is a never-failing source of supply to the river, so that in the greatest drought there is water sufficient for driving [n]upwards of twenty[n] large Factories. The river is about 160 miles in length; but being much broken by falls, it is not navigable for more than four or five miles from the sea. There are four principal Falls; first, Great Falls at Hiram, the height of which is about 72 feet; second, Steep Falls at Lymington, about 20 feet; third, Salmon Falls at Hollis and Buxton, about 30 feet; and fourth, Saco Falls, about 34 feet: the latter are about four miles from the mouth of the river, which is navigable for vessels of nearly 200 tons, till within one hundred yards of the Falls. From the mouth of the river a most beautiful beach of smooth level sand, stretches along ten or twelve miles to the Eastward: this is a place of great resort for pleasure parties in the summer season, and it is said this beach is unequalled by any on the American coast. It is only about four miles from the village of Saco, and as the population of the place increases, its close vicinity to the sea cannot fail to add to its value and importance.

The river at this place may be about one hundred yards broad, and as it approaches the village, it divides into two branches, and forms three separate Islands, the largest of which is called Factory Island, and contains about thirty acres of ground; on each side of this Island there are large waterfalls of about thirty-four feet; that on the North side is perpendicular, but on the South, the water rolls over several ledges of huge rocks; and in the spring, when the river is swollen by the melted snow, as it tumbles over these Falls it presents a scene of the most terrific grandeur. This river, however, is not like the rivers in Britain, which swell with every heavy rain, and again subside as soon as the rain is over. The Saco swells a little in the fall of the year in consequence of the continued rains at that season, but it gradually subsides during the frost in winter; and again, when the snow melts in the spring, it rises, perhaps, two or three feet, and afterwards gradually and slowly subsides during the summer; but a heavy rain continued for a whole day, will scarcely affect it.

Factory Island belonged to a family of the name of Cutts, some of whose ancestors purchased it from the Indians, who, it is said, resorted to this place in considerable numbers, for the purpose of fishing at the Falls, and to obtain a secure retreat from their enemies. In 1825 it was purchased by a Company, principally from Boston, for the purpose of erecting a Cotton Factory. The whole cost to the Company was 110,000 dollars; they, at the same time, bought a considerable part of the privileges on the opposite side of the river for 10,000 dollars. A Mill was erected in 1826, and a

canal cut from the head of the Falls to the Mill site. The length
of the Mill was 210 feet, the °breadth 42,° and seven stories in
height. This was the largest Factory ever attempted in America;
it was calculated to operate about 12,000 spindles and 300 looms.
In 1829 there were about 500 workers employed about the establishment, the greater part of whom occupied the Company's
tenements, which were erected upon the Island. The whole machinery, on which the sum of 200,000 dollars had been expended,
was completed in the early part of 1830. This Mill, however, was
no sooner completed, than it was burned down to the foundation,
and the Company lost all their stock. The wreck of the Mill, with
all the other property, was sold at a very low price to another
Boston Company, who obtained a charter under the name of the
"York Manufacturing Company of Saco." This Company immediately commenced building their first Factory on the foundation
of the one which had been burned, the length and breadth of
which is the same as the former, but the height is only four stories
with an attic. This Mill was completed in 1833, the canal was
afterwards lengthened for the purpose of conveying water to the
other two Mills, one of which was completed in 1835, the other
in 1837 (*See View of York Factories facing title page.*)[p]

[q]A fourth mill of five stories has just been completed.[ql] In the
four mills there are 23,944 throstle spindles, 734 looms, and the
amount of cotton consumed is over 500,000 lbs. producing upwards of 136,000 yards of cloth per week,[ql] giving employment
to 1,200 females and 270 males. The usual pay roll amounts to
over 3,000 dollars per week or 156,000 dollars per year. The mills
are entirely heated by steam which consumes about 800 tons of
anthracite coals per year at the average cost of seven dollars per
ton. The capital stock of the company is one million of dollars.[qII]
The goods manufactured are fine jeans, bed-tickings, and a variety
of coloured and striped goods all of superior quality from yarns
averaging from No. 12 to No. 24.[qII]

In 1838 another company was formed at this place under the
name of the "Saco Water Power Company" which purchased the
whole property and water privileges owned by the York Company,
with the exception of one half of Factory Island, and sufficient
water power to operate four factories.[53] This company in 1841
built a large machine shop, 146 feet by 45, and five stories high,
where machinery of the best description and most improved style
is now manufactured at as reasonable prices as any in the United
States. They have also cut a canal 40 feet wide by 10 feet in depth

and are now prepared to sell mill privileges upon as advantageous terms as can be found any other where in the country. And few places possess equal facilities for manufacturing purposes. Situated at the head of tide water which flows up to the falls, cotton may be imported from any part of the world and landed within a few yards of the mills and the manufactured goods shipped at the same place, the cost of transportation to and from Boston each way being only one dollar per ton.^{qIII}* ^{qIV}Besides, the eastern rail road passes through the village which greatly facilitates the intercourse between this and all other parts of the country.[54]

The population of Saco is upwards of 4,000 souls and Biddeford on the opposite side of the river contains above 3,000. There are in the two villages ^{qV}eight^{qV} places of public worship, ^{qVI}six^{qVI} free schools besides private ones, and an academy where the languages and other branches of useful education are taught. There are also a Circulating Library, Public Reading-room, three large ^{qVII}Temperance^{qVII} Hotels, two Banks, and ^{qVIII}a weekly newspaper regularly published.^{qVIII} The place is still increasing in wealth and extent as every new factory that is put in operation adds about 500 to the population.^{q}

MISCELLANIES.

DYNAMOMETER.

A MACHINE has been used at Lowell, called a Dynamometer, for the purpose of ascertaining the power required to move any of the different machines used in Cotton Factories, but having never seen it, I cannot give any description either of its construction, or of the mode of applying it. It is said, however, to be rather complex, and that the results given by it cannot be entirely relied on. My design at present is merely to describe a very simple one lately constructed by Samuel Batchelder, Esq., Agent of the York Manufacturing Co. Saco [Fig. 28].[1]

Plate VIII [Fig. 29]. *figures* 1*st* and 2*d*. gives two views of this machine, which is constructed on the principle of what is called the "differential box," and consists of the two pairs of belt pullies A A, B B, mounted on the shaft C C: one of these pullies on each side is loose, while the other is fast. The fast pulley on the side A A, and the bevel wheel D, are both fastened to the shaft C C. The

PLATE VIII.

DYNAMOMETER.

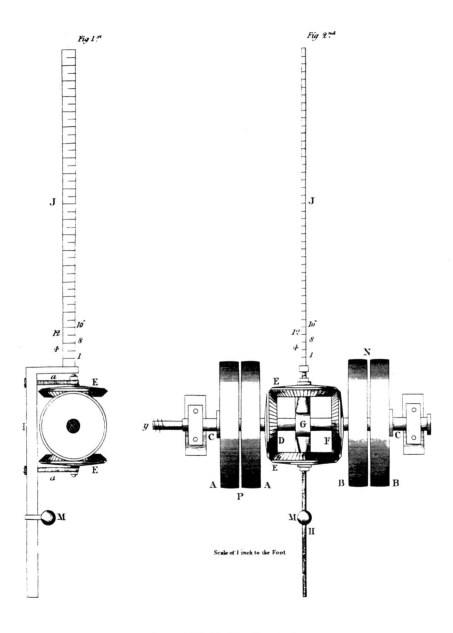

Scale of 1 inch to the Foot

Glasgow Published by John Niven Junr

Fig. 30. Dynamometer incorporating differential gear, plate VIII from *CM1*.

bevel wheel F is fastened[a] to the arms of the pulley B by means of two bolts.[a] The wheels E E, are connected by G, which is constructed so as to revolve round the shaft C C. To apply this simple machine, a belt from a drum on the main driving shaft is brought to the pullies A A, whilst another belt is carried from the pullies B B, to the machine or machinery, the weight of which is to be ascertained. And it is plain, that if the pullies A A, and the wheel D, are once put in motion, the wheels E E will also revolve on their axis, and at the same time the connection shaft G will revolve round the shaft C C, thus leaving the wheel F and the pullies B B, standing still; but, if the wheels E E are kept in their present horizontal position, and prevented from revolving round the shaft C C, it is equally obvious, that the wheel F and the pullies B B will then be moved at the same speed as the wheel D and the pullies A A; hence the weight required to keep the wheels E E in their present position, is equal to the weight required to move the pullies B B. The weight thus required, is found by means of the lever H J. The arm H is attached to the centre of the wheels E E, by the straps *a a, fig.* 1*st*. The arm J is divided upon the principle of the Roman steel yard. The weight M is merely intended to balance the arm J, and being fastened with a set screw, can easily be shifted on the arm H, as may be found requisite. Therefore, when the wheels E E are kept in their present position by means of the lever J H, it is evident that a weight of 20 lbs. acting upon the pullies A A at P, will balance another of the same weight at N, of the pullies B B. Now the distance from the centre of the shaft C C to the division on the lever J marked 1, is equal to the radius of the belt pullies; hence a weight of 20 lbs. at 1 will counterbalance the same weight at P, that is making no allowance for friction, the amount of which is ascertained by the additional weight required to balance the given weight at P; and having once ascertained the proper allowance for friction, the machine is put in motion by shifting both belts on to the fast pullies, and moving the balance weight along the lever J from 1 to 4, 8, 12, or to whatever number will balance the wheels E E, and the weight thus indicated on J, is the weight required to move the machine or machines, from which deduct the allowance for friction. A worm at *y* on the end of the shaft C C, working into a wheel with an index and pointer indicates the speed at which the machine is driven and also serves to determine the weight at different speeds. From the above description it is presumed that the principle upon which this dynamometer is constructed, as well as the mode of

Fig. 31.   Differential gear designed for a dynamometer and described in James White, *A* ➤
New Century of Inventions, Being Designs and Descriptions of One Hundred Machines Relating to
Arts, Manufactures and Domestic Life (Manchester: The author, 1822), plates 2 and 3.

applying it, will be easily understood.[b] It is properly a compound
lever, of which P is the power, N the fulcrum, and J the weight.
But in practice the power acts at K instead of P.[b]

PRICES OF MACHINERY AND VARIOUS OTHER ARTICLES USED IN COTTON FACTORIES, BOTH IN GREAT BRITAIN AND AMERICA.

Prices of Machinery, &c. at Lowell, Massachusetts.[2]

	Dols.	Cts.		£	s.	d.
Conical Willow,	110	0	=	22	18	4
Whipper used, instead of a Willow, for beating Cotton,	95	0	=	19	15	10

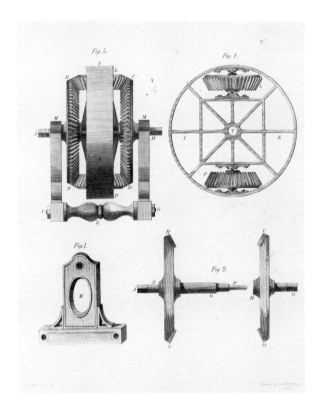

	Dols.	Cts.		£	s.	d.
Picker or Spreading Machine (two beaters)	550	0	=	114	11	8
Carding Engine, 37 inches broad, and 36 inches diameter, clothed,	260	0	=	54	3	4
Lap Winder or Lapping Machine,	240	0	=	50	0	0
Drawing Frame of three heads,	165	0	=	34	7	6
Speeder containing 24 spindles,	850	0	=	177	1	8
Stretcher or Extenser, 36 do.	1,000	0	=	208	6	8
Spinning Frame, 128 do. (dead spindle)	575	0	=	119	15	10
Warping Machine, complete,	135	0	=	28	2	6
Dressing Machine, do.	500	0	=	104	3	4
Loom, do.	75	0	=	15	12	6
Belt Leather, per pound,	0	26	=	0	1	1

	Dols.	Cts.		£	s.	d.
Lacing or thong Leather for sewing belts, per side, about	1	0	=	0	4	2
Sheep skins for covering rollers, cut into strips to suit the rollers, per dozen,	8	0	=	1	13	4
Calf skins for covering rollers, per dozen,	22	0	=	4	11	8
Picker Leather, cut into strips for shuttle cords, per side,	6	0	=	1	5	0
Shuttle Drivers, made of Buffalo hide, per dozen.	0	67	=	0	2	9½
Sperm Oil, per gallon,	0	90	=	0	3	9
Tallow, per lb.	0	7	=	0	0	3½
Dust or hand Brushes, per dozen,	4	0	=	0	16	8
Floor Brushes, per dozen,	10	0	=	2	1	8
Reeds, (Steel) per beer, or 20 dents,	0	4	=	0	0	2
Do. (Brass) do. do.	0	5	=	0	0	2½
Reeds, (Cane) per beer, or 20 dents,	0	3¼	=	0	0	1⅝
Card Sheets and Fillets, per square foot,	1	12½	=	0	4	8¼
Shuttles, (Apple tree) per dozen,	5	0	=	1	0	10
Bobbins for Speeder, 6 inches by 4, each	0	6	=	0	0	3
Do. for Spinning Frame, (Birch) do.	0	1	=	0	0	0½
Do. for Do. (Apple tree) per 1,000	12	50	=	2	12	1
Potato Starch for dressing, per lb.	0	5	=	0	0	2½
Brass Casting, per lb.	0	30	=	0	1	3
Iron do. do. 5 cents to ..	0	6	=	0	0	3

'At Pawtucket, Rhode Island[3]

	Dol.	cts.		£	s.	d.
Spreading machine (2 beaters)	420		=	87	10	
Carding engines (30 inch broad, 36 in dia)	140		=	29	3	4

	Dols.	Cts.		£	s.	d.
Drawing frame per head	90		=	18	15	
Speeder (Waltham) per spindle	42		=	8	15	
Extenser do. do.	35		=	7	5	10
Spinning Frame (dead spindle) do.	5		=	1	0	10
Hand mule do.	2	37½	=		9	10³⁄₈
Self-acting do. (Sharpe, Roberts & Co) do.	3		=		12	6
Loom	75		=	15	12	6
Dressing frame	450		=	93	15	

In addition to the above we might have given statements of the prices of machinery at Saco, Taunton, Paterson, and Matteawan, but as they all differ so very little from the preceding it was not deemed necessary. Those already given present very fair specimens of the average prices of machinery in the United States. It is proper to remark however that since these notes were taken some reduction has taken place in the rates of wages and prices of several materials so that at the present time (April, 1843) the above machinery can be furnished in various places for from 5 to 10 per cent cheaper than the prices herewith given.

Large orders for machinery have been executed in this country for Mexico and Russia, especially the former.[4] But since the laws prohibiting the exportation of machinery from Great Britain have been repealed it is known that extensive orders are in the course of execution in Manchester for both those markets and in consequence it is believed that this trade will be in a great measure lost to the United States.[5] A small quantity of machinery has just been imported from England into Massachusetts. It is not supposed however that this can be a very extensive trade as the duty of 20 per cent on all machinery imported into this country with the expense of transportation will amount almost if not entirely to a prohibition.

In addition to the duty on imported machinery the United States charge a duty on the boxes in which it is packed. One of the Lowell companies imported a spreading machine from Manchester for which they had to pay for the boxes alone 150 dollars duty included, which is equal to £31 - 5/ sterling. Under such circumstances it is not likely that the importation of machinery shall be an extensive business.[c]

The Providence Machine Company[6]

	Dols.	Cts.		£	s.	d.
Make very superior Mules for per spindle,	2	75	=	0	11	5½

At Andover, in Massachusetts[7]

	Dols.	Cts.		£	s.	d.
Speeders containing 18 spindles, ...	666	0	=	138	15	0
Stretchers or Extensers, containing 36 spindles, ...	1,080	0	=	225	0	0
Spinning Frames, per spindle, ..	4	0	=	0	16	8

According to the statements of some experienced machine makers, a Mill containing 4,000 spindles, with all the necessary machinery for weaving, &c. might be filled with machinery ready for operation, for about 55,000 dollars, equal to 13 dollars 75 cents, or £2 . 17 . 3½ per spindle; and including buildings, gearing, &c. for 20 dollars, or £4 . 3 . 4 per spindle.

Prices of Machinery, &c. in Glasgow, Scotland.

	£.	s.	d.		Dols.	Cts.
Cylindrical Willow,	20	0	0	=	96	0
Scutching Machine, with two beaters,	58	0	0	=	278	40
Spreading Machine for 24 inch cards,	52	0	0	=	249	60
Do. do. 36 do.	58	0	0	=	278	40
Carding Engine, 24 inches broad, and 36 in diameter, clothed and ready for operation,	£28 to £30	0	0	=	144	0
Carding Engine, 36 inches broad, and 36 in diameter, clothed, &c.	44	0	0	=	211	20
Lapping Machine or Lap Winder, 24 inches, with 4 calender rollers,	25	0	0	=	120	0
Fly Frames, per spindle,	1	18	0	=	9	12
Drawing Frames, per head,	9	0	0	=	43	20
Mules, per spindle,	0	6	0	=	1	44

	£.	s.	d.		Dols.	Cts.
Spinning Frames, (live spindle) per spindle,	0	10	6	=	2	52
Looms,	9	0	0	=	43	20
Dressing Machine,	40	0	0	=	192	0
Warping Machine,	17	0	0	=	81	60
Winding Machine,	10	0	0	=	48	0
Belt Leather, per lb.	0	1	6	=	0	36
Lacing or Thong Leather, per skin,	0	1	6	=	0	36
Sheep Skins for covering rollers, per dozen,*	1	2	0	=	5	28
Shuttles, (Boxwood) with wheels, per doz.	1	14	0	=	8	16
Do. do. with iron slides do.	1	10	0	=	7	20
Shuttle Drivers, (Birch) do.	0	0	7	=	0	14
Reeds, (Brass) per hundred splits,	0	0	$3\frac{1}{2}$	=	0	7
Do. (Steel) do.	0	0	3	=	0	6
Dust brushes, per dozen,	0	11	0	=	2	64
Floor do. do.	1	10	0	=	7	20
Sperm Oil, per gallon,	0	7	0	=	1	68
Tallow, per lb.	0	0	$5\frac{1}{2}$	=	0	11
Bobbins for Fly Frames, 6 inches by 3, per dozen,	0	1	$9\frac{1}{2}$	=	0	43
Bobbins for Spinning Frames, per dozen,	0	0	8	=	0	16
Flour for Dressing, per barrel, (1839)	1	17	0	=	8	88
Brass Castings, per lb.	0	1	2	=	0	28
Iron do. do.	0	0	$1\frac{1}{2}$	=	0	3
Card Sheets, (Breakers) 24 inches by $3\frac{1}{2}$ per sheet,	0	3	3	=	0	78
Card Sheets, (Finishers) 24 inches by $3\frac{1}{2}$ per sheet,	0	3	$5\frac{1}{2}$	=	0	83
Filleting, (Breaker) $1\frac{1}{2}$ inches broad, per foot,	0	0	10	=	0	20
Filleting, (Finisher) $1\frac{1}{2}$ inches broad, per foot,			$10\frac{1}{2}$	=	0	21

*The sheep skins used in Britain are double the size of those used in America and equal in quality to the American calf skins.

*Prices of various Machines in Manchester, England, from Dr. Ure's
Work on the Cotton Manufacture of Great Britain.*[8]

	£.	s.	d.		Dols.	Cts.
Conical Willow,	70	0	0	=	336	0
Scutching Machine, (two beaters)	70	0	0	=	336	0
Spreading Machine, do.	70	0	0	=	336	0
Carding Engine, 37 inches broad, and 42 inches diameter, un-clothed,	42	0	0	=	201	60
Clothing Furniture of do.	24	0	0	=	115	20
Drawing Frame of six heads,	37	10	0	=	180	0
Fly Frame, (coarse) per spindle,	2	6	0	=	11	4
Do. (fine) do.	1	11	10	=	7	64
Mules, per spindle,	0	4	9	=	1	14
Self-acting Mule, per spindle,	0	8	0	=	1	92
Spinning Frame, (live spindle) per spindle,	0	10	6	=	2	52

From the preceding statements it will be seen, that the prices of machines used in the cotton manufacture, are much higher in this country than in Great Britain: besides, the British machinery generally contains a greater number of improvements, or those little contrivances by which they can be more easily adjusted, and operated to the best advantage.

Mr. White, in his Memoir of Slater, gives the following Extract from Baines's History of the Cotton Manufacture in England.[9]

	Prices of machinery in England, 1834.		Prices of machinery in the United States, 1834.		Actual prices sold in U.S.
		D.C. D.C.		D.C. D.C.	D.C. D.C.
Carding Engines,	£30 to £40	144 to 192	£40 to £50	192 to 240	100 to 250
Throstles, per spindle,	8s. to 9s.	1.92 to 2.16	24s. to 26s.	5.76 to 6.24	4.25 to 6
Mules, per do.	4s. 6d. to 5s.	1.8 to 1.20	13s. to 14s.	3.12 to 3.36	2.12 to 2.25
Dressing Machines,	£30 to £35	144 to 168	£80 to £90	384 to 432	400
Power Looms,	£7½ to £8½	36 to 40.80	£12 to £16	57.60 to 76.80	50 to 75

On the above Extract Mr. White has the following note: -

"The fact respecting the higher prices of American machinery, arises from their ornamental work, which the English think un-necessary; as they regard only the utility and durability of the machine. This circumstance may be worthy the attention of our machinists; whether it is best to spend so much for polishing the appearance of the works."[10]

In the preceding statement, Mr. White is greatly mistaken, as the English machinery in general, is more highly polished than any I have yet seen in America. A great part of the framing of machinery in this country is made of wood, painted green, [d] which can never appear so rich as bright polished iron, with which all the machinery made in Glasgow and Manchester is mounted. Let any one who supposes that the British do not expend labour and expense in polishing or ornamenting their machinery, only visit Mr. Orrell's Mill at Stockport, England, and he will find there a Factory fitted up like a palace.[11] Indeed, all the Factories lately built in England and Scotland have a splendid and elegant appearance, at least much more so than those of this country.

COST OF STEAM POWER IN THE UNITED STATES

The following estimate of the cost of steam power is taken from a Mill in ᵉNewburyportᵉ Massachusetts containing[12]

One Willow and Spreading Machine, with three beaters.
26 Carding Engines, 18 inches broad, and 36 inches diameter.
One Drawing Frame of three double heads.
Six Mules containing 264 spindles each = 1,584 spindles.
Spinning Frames, (dead spindle).   = 2,116   do.
                                           3,700 spindles

Three Warping Machines.
Three Dressing Machines dressing 35 pieces of 30 yards, = 1050 yards each, or 3,150 yards in all, per day.
100 Looms, making 112 picks per minute, yarn No. 30, for printing Cloth.

The Mill is driven by two large belts, from a high pressure steam engine of 40 horses' power—length of stroke, 4 feet—diameter of cylinder, 1 foot—makes 20 double, or 40 single strokes per minute. There are four round boilers, each 15 feet long by $2\frac{1}{2}$ in diameter—requiring 300 gallons of water, and consuming $1\frac{1}{4}$ chaldrons of bituminous coal per day—pressure of steam, 68 lbs. to the square inch.

	Dols.	Cts.	
1¼ chaldrons of bituminous coal @ 8 dol. per chald.	10	0	per day
Wages of Engineer, ...	1	33	do.
Do. Fireman, ...	0	87	do.
	12	20	

Thus the daily cost of the above is 12 dollars 20 cents = ʼ30½ cents per horse power or 94¼ dollars per horse power per year of 309 days = £19-12-8 being more than one half cent on the yard.ᶠ

ᵍThere are two mills in New Port, Rhode Island driven by very superior steam engines of nearly equal power and which (in the opinion of those on whose judgement we can rely) furnish a fair and just criterion of the average cost of steam power in the United States.[13]

The first mill contains the following machinery
One Mason's whipper; 2 spreading machines; 44 cards 24 inches broad; 3 drawing frames 3 heads each;

speeder spindles		144
Mule stretching frame spindles		660
Mule	do.	7,728

207 Looms; 5 warpers; 5 dressing frames; spinning No. 35 yarn; producing 6,000 yards of cloth per day, 30 inches wide, 5 yards per lb., 72 picks of weft per inch. The whole of the above are driven by a high pressure engine rated at 60 horse power; cylinder 16 inches in diameter, 6 feet stroke, 28 strokes per minute; 5 steam boilers 30 feet by 2½; pressure of steam 112 lbs. per square inch; consuming about 2 tons of coals per day which at the average price of 6 dollars per ton laid down at the mill makes 12 dollars per day, to which add 20 cents for tallow and 10 cents for oil, engineer and fireman's wages viz. 2 dollars, making in all 14 dollars and 30 cents = £2–18–4 pr. day, being something over one fourth of a c[ent?] on the yard of cloth or 24 cents per horse power per day besides the expense of repairs, cleaning boilers, flues &c. The cost per year of 309 days is 74¹⁶/₁₀₀ dollars per horse power = £15–9–(9?), not including repairs, cleaning &c.

The second mill referred to contains the following machinery: One Mason's whipper; 2 spreaders; 44 cards, 24 inches; 2 drawing frames, 3 heads each

Speeder	spindles	112
Extenser	do.	336
Mule	do.	4,032
Mason's cylindrical flyer do.		3,456

200 Looms; 5 warpers; 6 dressing frames; spinning No. 44 yarn, producing 2,500 yards of cloth per day, 36 inches wide, 5 yards per lb.

The engine by which the above are operated is estimated as working up to 75 horse power, cylinder 6 feet by 20 inches, 6 feet stroke, 26 strokes per minute; 6 steam boilers, 30 feet by 2½; pressure of steam 110 lbs per square inch; consumes 2½ tons of coals per day at the average price of 6 dollars laid down at the mill. Expense per day estimated as follows

	Dol. cents
2½ tons coals at 6 dollars	15.00
Oil, tallow &c	30
Engineer and foreman's wages	2.00
	17.30 = £3-12-1

Daily expense, repairs, cleaning flues &c. not included
Expense per horse power per day $23^{06}/_{100}$ [dollars, or?][paper torn away] $-71^{27}/_{100}$ per year.[g]

CALCULATIONS OF THE COST OF WATER POWER[h] IN THE UNITED STATES.[h]

These calculations are entirely for breast wheels, all those at Lowell being of that description.

At Lowell 3,584* throstle spindles, with all the necessary machinery for preparing cotton, and manufacturing the cloth, are

---

[i]* The entire water power at Lowell as already stated was purchased from the original owners by the proprietors of the Waltham Factories. And when they came to sell out privileges to manufacturing companies, having no given data by which to fix the price, they took the Waltham mills as their standard, each of which contained 3,584 spindles with all the necessary machinery for making coarse heavy cloth from No. 14 yarn. This they called a mill power; to operate which required 60 cubic feet of water per second on a fall of 13 feet, and was sold to manufacturing companies at 4 dollars per spindle. But in consequence of the great improvements that have been made upon machinery since that time the power then required to operate 3,584 spindles with the subordinate machinery will now operate a great deal more so that though a mill power is sold by [the . . . .] Locks and Canals Company at the nominal price of 4 dollars per spindle,

considered a Mill power, [14] and sold at 4 dollars per spindle, or 14,336 dollars in all, with about 4 acres of land. In conveying the Property, the Company give a right on the fall of 13 feet, to draw 60 cubic feet of water per second for 24 hours; and 24 cubic feet on the fall of 30 feet. On the fall of 13 feet, one foot is deducted, leaving 12 feet for effective fall.

Cubic feet.   Feet.   lbs.

$60 \times 12 \times 62\frac{1}{2}$ (weight of one cubic foot) = 45,000 lbs. per second.

45,000 lbs. $\times 60'' = 2,700,000$ lbs. of power expended per minute; from which deduct $\frac{1}{3}$ for the difference between the power expended and the effect produced; 2,700,000 lbs. - $\frac{1}{3}$ = 1,800,000 lbs. being the effective force of 60 cubic feet per second, on a fall of 13 feet; which, if divided by 33,000 lbs. (Watt's estimate of one horse power,) will give the number of horses' power on the above fall.

$1,800,000 \div 33,000 = 54.54$ horses' power.

Therefore $54\frac{1}{2}$ horses' power, (together with four acres of land,) are sold for 14,336 dollars, and are estimated as sufficient to propel 3,584 spindles, with all the other machinery connected.

The whole cost divided by the number of horses' power gives the cost of each.

Dols.   Horses' power   Dols.

$14,336 \div 54.54 =$ 262.85. The interest on this at 6 per cent., 15 dollars 77 cents = £3 . 5 . $8\frac{1}{2}$ Sterling, being the annual cost of each horse power.

If we suppose the land worth 857 dollars per acre, being a little over 17 cents to the square yard, (and it is known that much of the Company's ground brings as much for the square foot,) then four acres will cost 3,428 dollars, leaving 10,908 dollars for the water power, *i.e.* 200 dollars for each horse power; the interest

---

it is actually purchased by the manufacturing companies at a much lower rate especially those that make fine goods, As for example in a mill where fine goods are manufactured from No. 40 yarn, if the original cost of the power required was calculated by the running spindle it would be found not even to exceed two dollars per spindle. And though the nominal allowance of water to the mill power be 60 cubic feet on the fall of 13 feet, and 24 on the fall of 30 feet, yet the actual supply is so very liberal as to reduce the cost considerably. Water power at Cabotsville is understood to be sold upon more liberal terms than at Lowell but being a more inland situation it is subjected to a heavier cost for transportation of cotton and goods.

The Saco Water Power Company have some valuable mill sites and water privileges which they are prepared to sell on as reasonable terms as any in the United States with all the additional advantages of being situated on a navigable river.[i]

of which is 12 dollars, [j] = £2 . 10 being the [j] annual cost of each horse power to the different manufacturing corporations.

At Manayunk, near Philadelphia, the water is leased or sold by the square inch of the gates, under a head of 3 feet. The whole fall is 23 feet, from which 3 feet is deducted; leaving 20 feet for effective fall.

"The square root of the head or depth of water multiplied by 5.4, gives the velocity of feet per second: this multiplied by the area of the orifice in feet, gives the number of cubic feet which flows out in one second." *See Brunton's Compendium of Mechanics, seventh edition, page* 147.[15]

$\sqrt{3}$ = 1.732 × 5.4 = 9.3528, velocity per second, or the number of cubic feet per second, running through an orifice of one square foot.

9.3528 × 20 feet, height of effective fall = 187.056 feet × 62½ lbs, weight of cubic foot = 11,691 lbs. per second × 60 seconds = 701,460 lbs. power expended per minute, from which deduct ⅓ for the difference between the power expended and the effect produced.

701,460 lbs. − ⅓ = 467,640 lbs. effect produced ÷ 33,000 lbs. = 14.17 horses' power for each square foot of the gates on the above fall of 23 feet, and under a head of 3 feet.

The above water power is sold at 100 dollars per square inch of gate.

Square inch. Inches in a foot. Dollars. Dollars.
1 :   144 : :   100 : 14,400 cost of square foot.
14,400 dollars ÷ 14.17 horses' power = 1,016 dollars for each horse power. The interest of which at 6 per cent is 60 dollars 96 cents = £12 .14s. Sterling, the annual cost of each horse power; being about equal to 6 dollars for each square inch of gates on the above fall and head of water.

[k]Water power at Paterson, New Jersey, is sold by the square foot of gate under a head of 2½ feet and 22 feet fall from which 2 feet is deducted leaving 20 feet for effective fall, which calculated by the preceding rule will stand as follows:

($\sqrt{2.50}$ = 1.581) × 5.4 = 8.5374 velocity per second
8.5374 × 20 × 62½ × 60 = 640.305 lbs. power expended per minute from which deduct ⅓ and divide by 33.0
640.305 − ⅓ ÷ 33.0 = 12.93 horses' power
The above is sold at 500 dollars per year

$$500 \text{ dol.} \div 12.93 \text{ h. power} = 38\,^{67}/_{100} \text{ dollars per}$$

horse power per annum

$$= £8-1-1 \text{ sterling}$$

There are yet 20 mill privileges to be disposed of at Paterson.

The results derived from the preceding estimates of the cost of steam and water [expenses]?] in the United States are the following:

	Daily	Yearly
Cost of Steam Power		
Steam mill in Newburyport		
Expense per horse power	$30\frac{1}{2}$ cents	$94^{24}/_{100}$ dollars
Steam mills at New Port		
Expense per horse power (1st mill)	24 cents	$74^{16}/_{100}$ dollars
do. do. (2nd mill)	$23^{2}/_{15}$ "	$71^{27}/_{100}$ do.

	Daily	Yearly
Cost of Water Power		
At Lowell and Saco		
Cost per horse power	.03 cents	$15^{77}/_{100}$ dollars
At Manayunk	$19\frac{3}{4}$ "	$60^{96}/_{100}$ "
At Paterson	$12\frac{1}{2}$ "	$38^{67}/_{100}$ "

These are simply the cost of power but in addition to the water privileges, the purchaser receives land sufficient for the mill site, stores, and other offices required for a manufacturing establishment. The price of this land should be deducted from the above estimated cost of water power. Without this deduction we find the average yearly cost of steam power $72^{71}/_{100}$ dollars per horse power   do. water power $38^{46}/_{100}$ do.

Difference $34^{25}/_{100}$ do.

Being over 48 per cent in favour of water power. And if we compare the average cost of steam power with the nominal cost of water power at Lowell and Saco, we find the latter to be a little over one sixth of the former.[k]

[l]Interest tear and wear $7\frac{1}{2}$ per cent
on original cost of engine viz £3,000   225 — . — .
Nett expense of steam power for one year   £923 — 15 — .

The above engine is now working full up to 70 horse power

according to the indicator although originally rated at something less which makes the yearly expense £13—3/11 per horse power. But as part of the oil and tallow were used for other purposes, and the coals were also used to generate heat for seven flats of a large mill 150 feet by 37 which were generally heated during the night as well as holy day(s), the annual cost is estimated to average £12—10/per horse power.

It is very common in Glasgow and other manufacturing cities in Britain for small companies to hire one or two flats of a mill for the purpose of carrying on some particular branch of the manufacturing business. In such cases the proprietor supplies the *power,* the main shafting, and the necessary offices, for which he charges a yearly rent of from £20 to £25 per horse power according to the location of the factory.

The writer of this is aware of only one place in Britain where water power is sold, that is by "The Shaws Water Company" at Greenock.[16] Mostly all the other water privileges were purchased at a very early period of the cotton manufacture and purchased not as water privileges but as real estate generally at a very low price.[17]

The conditions on which The Shaws Water Company sell their privileges are £3—10/ per horse power yearly for those nearest the town of Greenock. And as they recede from the town the water privileges decline in price. The lowest is £2—10/ per horse power yearly. Besides the water power the company supply sufficient land for mill sites and other necessary appendages to a manufacturing establishment.[1]

[m]RESULT OF AN EXPERIMENT TO ASCERTAIN THE WEIGHT OF ANTHRACITE COALS CONSUMED IN HEATING A COTTON FACTORY FOR ONE WEEK, IN WINTER, BY JOHN M. BATCHELDER, SACO[18]

The Mill in which the following experiment was made, consists of four stories, with an attic, and is 42 feet in width by 142 in length; the average height of each story about 10 feet. The attic is about 29 feet in width, and between 8 and 9 in height.

The basement story contains the water wheels; second, the carding; third, the spinning; fourth and attic, the weaving and dressing. The Mill is warmed by means of two upright steam boilers, one at each end of the Factory. One of these boilers is 6 feet in height, by 3 feet in diameter; the other is of the same height, and 2½ feet in diameter. The temperature of each room, as well as of the external air, was observed each hour of the 24 on the first day; during the remainder of the week it was observed four times each day, at such times as would give a correct average.

The water supplied to the boilers was of the temperature of 32°.

The average temperature of

The wheel room or basement story, was	44.24°
—— external air,	14.05
Temperature of wheel room was raised	30.19°

Average temperature of carding room, was	64.72°	
Do. do. spinning room,	65.94	
Do. do. weaving room,	66.57	
Do. do. attic,	71.04	
Average temperature of the four rooms,	268.27° ÷ 4 = 67.07°	
Average temperature in the Mill	67.07	
Do. do. of external air,	14.05	
Temperature of the four rooms was raised	53.02°	

The temperature was taken night and day, therefore the above was the average of the whole 24 hours of each day during the week.

The lowest temperature of the external air observed during the week, was 8 degrees below, and the highest range 38 degrees above zero; yet the temperature of the several rooms was quite uniform, not varying at any time above four degrees from the average given.[m]

The quantity of fuel consumed in the six days, was 7,484 lbs. of anthracite coal.

7,484 ÷ 6 = 1,247 lbs. of coal per day.

Cost of 7,484 lbs. anthracite coals at 7 dollars per ton.

Lbs.   Dols.   Lbs.   Dols.   Cts.
2,240 :   7   : : 7,484 :   23 . 39

Thus, the cost of coals to heat a Mill of the above extent for one week or six days, during the coldest month in winter, was 23 dollars 39 cents = £4 . 17 . 5½, Sterling.

In order to ascertain the annual cost, we may fairly estimate ten weeks in the coldest part of winter equal to the above, and twenty weeks in spring and autumn at one half the above cost. The annual cost would therefore stand as follows: -

Weeks. Dols. Cts. Dols. Cts. Weeks. Dols. Cts. Dols. Cts.

$$10 \times 23 . 39 = 233 . 90 + 20 \times 11 . 69\frac{1}{2} = 467 . 80$$

The above may be regarded as the fair average annual cost of heating a large Factory in this country by steam. I have not had an opportunity of ascertaining the cost of heating by hot air, as those that heat by the latter use a considerable quantity of wood for fuel.

TEXTUAL FOOTNOTES

Title page

"with comparative estimates of the cost of manufacturing in both countries":

Preface

[a-a] v,3: strongly urged
[b] v,8: more
[c] v, 18: in this country
[d-d] vi, 1–13: *aims* to give simple facts *without distortion*
[e-e] vi, 21–27: *promises estimates of costs of buildings*

Contents

[a-a] ix, 15–22: *longer title and subtitles:* comparative estimates *of costs*
[b-b] x, 21–25: *specific subtitles mentioning* Lowell *and* Manayunk

Plan and arrangement of the mills

[a-a] 13, 22–24: Springfield and Three Rivers, Massachusetts *deleted*; Newburyport and Amoskeag *added*
[b-b] 14, 12
[c-c] 14, 28: its manufactories
[d] 15, 5: considerably
[e] 15, 20: the plan generally adopted
[f-f] 16, 1: seven to eight dollars % six to seven dollars %
[g-g] 16, 1–17, 2
    [gI] 16, 8: considerably
    [gII-gII] 16, 16
    [gIII-gIII] 16, 18
    [gIV] 16, 22: nearly
    [gV-gV] 16, 22
    [gVI] 16, 31: like guard houses
[h] 17, 3: It is said that Cotton Mills in this country are very liable to take fire, for which I cannot assign any particular cause, at least for such as are heated by steam; those heated with hot air may be more liable to such accidents, especially when wood is used for fuel.

ⁱ⁻ⁱ 17, 22–18, 12
    ^{il-il} 17, 25: with ladders extending right over the top of the building
    ^{ill} 18, 3: as, for example, at the end of every half hour
^j 20, 24: in consequence of the multiplied friction
^{k-k} 20, 27: very generally adopted
^{l-l} 21, 10: 5,000
^m 21, 18: Some of the most recently built Mills at Lowell have only one broad belt from F to D, instead of two; which seems to perform its operations equally well.
ⁿ 24, 18: The writer has seen some of these which have sunk down four and five inches in the course of four % a few % years.
^o 24, 30–25, 5: *Two adjacent mills at Fall River seem to settle the question: both receive the same water power but the one equipped with belting drives more machinery than the one equipped with shafting.*

Notices of the various machines

^{a-a} 25, 15: used in this country are
^b 26, 1–31: (This page was deliberately covered over with the annotated leaf for page 27, i.e. JM meant to dispose of it permanently.) There is another machine, called a Picker, used in a number of Factories in this country. It consists of a small cylinder, about 14 inches diameter, set full of short spikes: besides which there is a scutcher, or beater, combined in the same machine. The cotton is led into the picker by a pair of fluted rollers, having been previously spread upon a revolving cloth, or apron; and after passing the picker and scutcher, it is forced up through a funnel, by a pair of fanners, to a chamber above. This machine is very injurious to the cotton, and likely to be laid aside.

A machine called a whipper, is also used in some Factories, and very highly spoken of. It is merely a substitute for the old mode of beating the cotton with switches, and consists of a flake table, or an oblong frame, the top or cover of which is composed of elastic cords, with two parallel shafts, fitted up with arms extending across above the cords, one shaft on each side of the frame, and moved by cranks, so as to make the arms strike alternately and rapidly upon the cords; one end of the frame being higher than the other, the cotton is thrown in upon the cords at the higher end; and by the operation of the arms, or reiterated strokes of the beaters, gradually passes down until it drops out at the lower: during this process of beating, the cotton is perfectly opened, and the seeds and dust drop down between the cords. The whole machine is covered with a kind of wicker work, to prevent the cotton from being thrown out by the beaters.

Another modification of the whipper has been lately introduced: the writer has had one of them under his charge, and regards it as the best and simplest as well as the cheapest machine of the kind he has yet known in either Great Britain or America.
^{c-c} 27, 31: the greatest
^{d-d} 28, 12
^{e-e} 28, 26–29, 29
    ^{el} 29: but there is one great error which seems to pervade all the Cotton Factories in America, that is, to have too little room in their picking houses: few of those that I have seen have more than barely room for two or three bales of cotton, besides a willow and lap spreader.
^f 30, 10: *Whitin's patented scutcher is in principle the same as the newest made in Manchester and Glasgow*
^{g-g} 30, 13–31, 7: *the scutching process is not carried out to the same extent in American factories as in English ones; exceptions are the mills at Three Rivers and Thorndike, Massachusetts. There cotton was worked by four scutching machines, in fact a picker and three scutchers. Consequently these factories made perhaps the best and finest goods in America. Using short staple cotton such as New Orleans and Upland, from which British mills spun Nos. 70–80, these mills made Nos. 50–60 spun on mules from double rovings.* % Mr. Brown, the Agent of the above named factories seems to be the first in this country who thoroughly understood the true theory

of cleaning cotton by scutchers. He says "that a beater of 12 inches diameter, should not revolve less than 2,400 times per minute; and that it should be set as close to the feeding rollers as possible without touching; that the under edge of the blade, which strikes the cotton, should be very thin, but not sharp, as by that means the seeds &c. will be better beat out of the cotton without injuring the staple." If successful practice is any proof of correct theory, Mr. Brown's is certainly the true one; as few factories in this country have paid handsomer profits, or sustained their character better, or even equal to those he superintends. %

[h] 31, 27: The average speed of the main cylinders is about 100 revolutions per minute
[i-i] 32, 5–17
    [iI-iI] 32, 6: 24 to 30 inches *in the Southern district and some with cast iron cylinders, covered with sheets instead of fillets. Main cylinders in the Middle and Southern districts are 36 inches in diameter.*
[j-j] 34, 6–20
    [jI-jI] 34, 8
    [jII] 34, 9: as the flat tops are here placed in the most convenient position (for being frequently stripped)
[k] 34, 32: to be set
[l-l] 35, 10: which I have yet seen
[m-m] 36, 4: 24 inches
[n-n] 36, 22–37, 5: *the first part of this revision is a re-phrasing of CM1 36, 22–33; the rest is amplification.*
[o] 37, 22: Great
[p-p] 38, 9: neither is it so well carded
[q] 38, 21: (what is technically denominated)
[r] 41, 32: Great
[s-s] 43, 33–44, 19
    [sI-sI] 44, 1–2: at least the Eastern district; *about Southern factories; JM, not having inspected them, repeats the report that they follow British practice.*
    [sII-sII] 44, 17: three or four years
[t-t] 44, 24: both considerably
[u-u] 45, 19
[v-v] 46, 27–47, 17
    [vI] 46, 28–32: *by the saving of one or two hands, whose wages are generally the lowest paid in the whole Factory, and superseding the use of a number of card cans*
    [vII] 47, 6–10: But the same objection will equally apply to the English system of making the ends from several carding engines wind on to a large bobbin for the purpose of being set up at the first heads of the drawing frame.
[w-w] 47, 21
[x-x] 48, 11: number
[y-y] 48, 14
[z-z] 49, 20: for the purpose of being sold to country people, or others who may want it, to be
[a] 50, 30: called a latch
[b] 51, 21: each frame containing only three single heads
[c-c] 53, 7
[d] 53, 8
[e] 53, 9: with which I am acquainted
[f] 53, 12: and I know it is in a great many
[g-g] 53, 12: I am not surprised
[h] 53, 17: I have
[i-i] 53, 25–54, 22: *longer description of effects of static electricity set up by the friction of belting: cotton flying in the air and mild shocks for those touching the belt. JM recalls seeing twelve drawing frames in one card room rendered idle as a result of static electricity.*
    [iI] THE LAST LINE OF THIS INSERTION HAS BEEN MOSTLY CUT AWAY AT THIS POINT

[j-j] 55, 5–13: *most American drawing frames have three heads and the cotton is drawn, or passed through the drawing frame, three times*
[k] 55, 13: Great
[l-l] 56, 21: manager of factories
[m-m] 56, 23–33: *two girls in Britain operate drawing frames twice the size of those in America (six single heads compared to three single heads) driven at the same speed in both countries*
[n-n] 57, 10: great numbers
[o] 57, 17: Great
[p] 59, 6: entirely
[q] 60, 7: entirely
[r] 60, 26: in this country
[s] 60, 26: entirely
[t-t] 61, 3: are altogether the best that I have yet seen
[u-u] 62, 28–32: The Plate Speeder, whilst it possesses all the merits, and obviates many of the defects of the tube frame, is not considered equal to the eclipse speeder, either in the quantity of work produced in a given time, or in simplicity of construction.
[v] 63, 6: Great
[w] 63, 10: I have yet seen
[x-x] 63, 15: require double the power to move it
[y-y] 65, 4–7: therefore I consider this method of bevelling the beam rollers an improvement of much importance, which has not yet been generally adopted in Great Britain
[z-z] 66, 2: Messrs. Cocker and Higgins
[a-a] 66, 9–24
  [aI] 66, 20–24: *British fly frames cost 35–36 shillings per spindle compared to American double speeders at 35–36 dollars per spindle, four times more than fly frame spindles.*
[b-b] 67, 18: No. 18
[c-c] 67, 23: 73 3/5 lb
[d-d] 68, 2: 48 cents each per day = 2/ Sterling
[e-e] 68, 3: 60 cents per day = 2/6 Sterling
[f-f] 68, 6: 60 cents per day = 2/6 Sterling
[g-g] 68, 8: 50 cents per day* = 2/1 Sterling
  [*] 68, 30–32: These rates of wages were furnished to the author by a gentleman of Lowell, who had ample means of ascertaining the different rates of wages paid in the various Factories in that place.
[h] 69, 7–9: for as the one side comes into contact with its opposite, the other is just going out
[i] 69, 22: recently built
[j-j] 69, 23: No. 50
[k] 69, 29–70, 2: with the exception of those at Baltimore, which follow the Lowell style of machinery
[l-l] 70, 11–18
  [lI] 70, 11: Great
  [lII] 70, 14: so far as I can learn
  [lIII-lIII] 70, 15: has not been so successful
  [lIV] 70, 15: Great
[m] 70, 27: greatly
[n] 70, 29; much
[o] 71, 13: Great
[p] 71, 19: Great
[q-q] 72, 18: a higher speed
[r-r] 74, 23–31: *Gore's spindle is said to have been tried in the fly frame in Manchester before being taken to America*
[s-s] 75, 3–76, 15
  [sI] 75, 5–11: They are altogether upon the plan of the mules made in Scotland . . . their only deficiency is the want of the taking in apparatus for returning the carriage home to the beam-rollers.
  [sII-sII] 75, 13: will average eight to ten cents per 100 hanks = 4d to 5d sterling

[sIII-sIII] 75, 14–32: *JM recalls visiting Newport, Rhode Island, and meeting spinners from Glasgow; working from 4.30 a.m. to 7.30 p.m., they were paid seven cents (or 3½ d sterling) for spinning 10 hanks of No. 30. When trade was good the rate rose to eight or nine cents. Some fine spinning mills paid mule spinners eleven cents for No. 40 powerloom warps prior to the trade depression (of 1837). Since then these, and other Rhode Island factories have* reduced their rate of paying about 40 per cent.

[sIV] 75, 33–76, 15: The Self-acting Mule, invented by Mr. Smith of Deanston, has been introduced into America, and put into operation in the State of New York, but there are several circumstances which militate against its rapid or extensive use in this country: *girls are not employed as mule piecers; boys of sufficient age are scarce and highly paid, as are overseers; few mechanics are qualified to maintain self-actors. Yet JM, mindful of the* well known energy and ability of the talented inventor of these mules *and their advantages, proved in Britain, expects them in all likelihood to be extensively adopted in the USA.*

[t-t] 77, 19: wheels

[u-u] 79: footnote

[v-v] 81, 18: ropening

[w] 81, 29: now

[x] 82, 9: very

[y] 82, 17: I consider

[z] 82, 18: Great

[a] 82, 20: in every respect

[b-b] 82, 24: so far as I am informed, has not yet made a beginning

[c-c] 83, 4: I consider much

[d] 84, 31: liable to be

[e-e] 85, 6–14

[eI-eI] 85, 7: All those I have seen

[eII-eII] 85, 12–14: the origin of which is ascribed to the inventive genius of Mr. Perkins, inventor of the celebrated steam gun

[f] 89, 30: instantly

[g-g] 90, 23: 27 cents per beam, being 3 dollars 51 cents = 14/ 7¼

[h-h] 90, 28: 33 cents per beam, being two dollars 97 cents, = 12/ 4¼

[i-i] 91, 4: any I have seen either in England or Scotland

[j-j] 91, 5–9: *cost of American dresser made in Britain put at half cost of British ones*

[k-k] 96, 31: 14 pieces

[l-l] 96, 33–97, 1: 20, in others 30, 40, 50, and 60 pieces per day

[m] 97, 6: great

[n-n] 97, 15: the greatest

[o-o] 97, 31

[p-p] 97, 23: 4½ cents = 2¼ sterling

[q-q] 98, 14–28: 2½ Gallons of yeast, and 2 quarts of vinegar, to be well mixed with about 9 gallons of water, which has been previously heated to 120°, or as hot as the hand will bear to work in it. To these are added 125 lb. of potato starch. The whole is then allowed to stand in a warm place about 10 or 12 days, or until it is perfectly fermented, then 3¼ lb. of common clean tallow is dissolved in 75 gallons of water, heated to 160°, to which are added 75 lb. of the fermented starch. The whole is well stirred, until all the ingredients are perfectly incorporated. The size is then to be used immediately before, or after it is perfectly cooled down. To the above some add about 2½ lb. of the sulphate of copper, to prevent mould.

[r] 100, 29–32: *JM, unable to obtain a drawing of the machine, believes that a description would not be intelligible to readers unfamiliar with it.*

[s-s] 100, 32

[t-t] 101, 15–102, 6: *different phrasing except for*

[tI-tI] 102, 2–6: but the principle upon which the newest crank looms are constructed, is the same in both countries; the only difference lies in some of the minor modifications, which do not affect the general merits of the machine.

[u-u] 102, 7: I have seen in this country

ᵛ⁻ᵛ 102, 11: I am only surprised

ʷ 102, 13: Great

ˣ 102, 21: Great

ʸ 102, 29: Great

ᶻ 102, 33: to some of the looms about Lowell that

ᵃ⁻ᵃ 103, 23–25: *rephrased except*

  ᵃᴵ 103, 23: in all the power looms I have seen in this country

ᵇ⁻ᵇ 104, 6

ᶜ⁻ᶜ 104, 27–105, 5: 1 Gallon Linseed Oil,   ¹/₂ Lib. Umber

    1 Lib. Lithorage   ¹/₂ "   Gum Shalac

    1   "   Red Lead   ¹/₄ "   Sugar of Lead

All these, except the shalac, are first well boiled over a moderate fire, until the strength is out of the lead; the shalac is then added, but only a little at a time, while the whole is still boiling, and it requires to be well stirred all the time. When the shalac is entirely dissolved, the whole is then cooled down to blood heat, then %when% a sufficient quantity of the spirit of turpentine is added, to make it fit for use. Such articles as require it, are to be pulverized.

ᵈ⁻ᵈ 106, 23–107, 34

  ᵈᴵ⁻ᵈᴵ 106, 23: one half cent, equal to one farthing per beer

  ᵈᴵᴵ⁻ᵈᴵᴵ 106, 25: 120 picks per minute

  ᵈᴵᴵᴵ 106, 25–107, 15: *JM describes the work of a weaving room, possibly one of his own at Saco in view of his familiarity with it, where twilled jean is woven on 48 three shaft (harness) powerlooms. The jean is 38 inches broad with 82 warp threads per inch or 3,120 in the warp and 56–60 picks per inch. Over the four weeks prior to JM's report the whole room averaged 51 pieces per day (each piece being 30 yards long) or nearly 32 yards per loom daily.*

    The girls' wages are 27 % 22 % cents per piece, equal to 1/1¹/₂ sterling % 11d % A smart girl can weave 14 pieces per week on a pair of looms, for which she will be paid $3.78 % $3.08 % equal to 15/9 sterling % 12/10%; her hours would be 12¹/₄ daily or 73¹/₂ per week.

  ᵈᴵⱽ⁻ᵈᴵⱽ 107, 16–21: *Sometimes one girl attend three looms, but this is rare, unless labour is scarce; in this case it is better to give three looms to the best weavers than to leave the machines idle.*

  ᵈⱽ⁻ᵈⱽ 107, 25: *shuttles are $5 a dozen*

  ᵈⱽᴵ 107, 27–31: *Persimmon wood, also used for shuttles, is much superior to apple, but dearer; it grows chiefly in Maryland and Southern states.*

  ᵈⱽᴵᴵ⁻ᵈⱽᴵᴵ 107, 33: about eight dollars per gross = £1 13s 4d sterling

ᵉ⁻ᵉ 109: *rephrased; but JM adds that his foregoing technical descriptions noted where American machines* differed from those which I was accustomed to in Scotland.

ᶠ⁻ᶠ 109, 20–110, 30: nor do I think that, in this respect, they are at all equal to the British: and, indeed, most of those machines that have been taken from this country, are in a much higher state of perfection in Britain, than any of the same kind I have seen here.

A great proportion of the machinery about this part of the country, is fitted up as if never to be altered. The machine makers seem to proceed upon the supposition that their machinery is already perfect. The machines are calculated for one kind of goods, and only one system of working. Hence, in making alterations in the adjustment of any machine, so as to introduce a different system of working, or to suit a different quality of goods, it is frequently necessary to make some new patterns, in order to accomplish the alterations required. Now the machinery used in Great Britain is, in this respect, greatly superior to the American. There, every machine is so constructed, that all its parts can be adjusted with the greatest accuracy, to suit the various qualities of cotton, or whatever kind of goods may be wanted: and, consequently, the manufacturers can easily arrange and alter their machinery, so as to make just such goods as will for the time being suit the market. But here they can make only one kind of goods, whether they suit the present demand or not, as they have not the same facilities for changing the style of their goods, so as to take advantage of the variations in the markets.

The framing of a great part of the machinery in this country is made of wood painted green; indeed, this is the prevailing colour for painting the wooden parts of the machinery, such as that of cards, speeders, spinning frames, spooling, warping, and dressing machines, looms, &c. There are also a great many machines made with cast iron framing. Of these, some very superior machinery is made at Pawtucket and Providence, State of Rhode Island; and at Matteawan, in the State of New York. At this latter place, the drawing rollers for throstles and mules are covered with fine velvet cork-wood, for which a patent has been taken out, and are said to suit the purpose much better than those covered in the ordinary way with cloth and leather.[55]

Comparative estimates

[a-a] 114, 13: £960
[b-b] 114, 14: 25 horses' power
[c-c] 114, 17: £2, 480
[d-d] 115; 10: 2 Drawing Frames
[e-e] 115, 10: 12 hds
[f-f] 115, 10: £108
[g-g] 115, 13: 9 Fly-frames
[h-h] 115, 13: 432 Sps
[i-i] 115, 13: £820 16s.
[j-j] 115, 17: £3,000 12s.
[k-k] 117, 5: 6 warpers @ $150: $900 or £187 10s.
[l-l] 117, 10: $14,620 or £3,045 16s. 8d.
[m-m] 118, 9: Do. at Drawing Frames
[n-n] 118, 9: 4 hands @ 7/-: £2 16s.
[o-o] 118, 12: Do. at Fly Frames, (3 @ 7/6, and 3 @ 4/6,)
[p-p] 118, 12: 6 hands: £3 12s.
[q-q] 118, 17: £18 4s.
[r-r] 122, 4, 5, 13, 14: *JM circled these figures and lines, indicating perhaps intentions of further corrections.*
[s-s] 123, 11: 120
[t-t] 123, 12: 104
[u-u] 123, 14: 62 picks to the inch
[v-v] 124, 4: £18 4s.
[w-w] 124, 9: £234 2s. 8d.
[x-x] 125, 14: 1.6d.
[y-y] 125, 15: 5.134d.
[z-z] 125, 19: 1.9d.
[a-a] 125, 20: 4.977d.
[b-b] 125, 21: 3 per cent
[c-c] 125, 21: .157d.
[d-d] 126, 30: 19 per cent
[e-e] 127, 20: J. Hazard & Co.
[f] 131, 11: even
[g] 131, 12: pick up and
[h-h] 131, 33: I conceive
[i-i] 133, 9: $2.20 to $2.50
[j-j] 133, 10: $2.50 to $2.75
[k-k] 133, 11: $1.20 to $1.50
[l-l] 133, 12: $1.75 to $2.25
[m] 134, 27: much
[n] 134, 30: Great
[o] 134, 33: Great *JM circled the latter part of this sentence, suggesting that he contemplated changes.*
[p] 135, 27–136, 10: That the general rate of wages is higher in the United States than in Britain is admitted, particularly the wages of females employed in the Factories. The greater part of these are farmers' daughters, who go into the Factories only for a short time until they make a little money, and then "clear out," as it is called; so that there is

a continual changing amongst them, and in all the places I have visited, they are generally scarce; on that account the manufacturers are under the necessity of paying high wages, as an inducement for girls to prefer working in the Factories to house-work: and while this state of things continues, it is not to be expected that wages in this country will be so low as in Great Britain; and although they have undergone a considerable reduction, during the late depression, still they are higher than in any part of Britain.

q-q 136, 22: 17 per cent

r 136, 23–29: But during 1837, the price of mule spinning was greatly reduced in the New England Factories; so far as I could learn, the average rate was about 7 cents per 100 hanks = 2/11 per 1,000, which, supposing the rate still paid in Glasgow to be 3/6½ would be 21 per cent in favour of America.

s-s 137, 8: 1 dollar 20 cents

t-t 137, 9: 2 dollars 20 cents

u-u 137, 11: 11/ - to 12/6 per week

v 137, 18–33: The price of living here is higher, and the hours of labour longer; besides, the greater part of the Factory workers being connected with farming, whenever wages become reduced so low, as to cease to operate as an inducement to prefer Factory labour above any other to which they can turn their attention, then a great many Factories will have to shut up. During a stagnation of trade, it is common for the manufacturers here to stop a part, or the whole of their Factories, and then the workers retire to their farms: such was the case in 1837, when a vast number of Factories were entirely shut up. Yet it seemed not to affect the workers very materially; indeed, many of the girls who had been some time in a Factory, seemed to rejoice and regard it as a time of recreation.

w 138, 3: very

x 138, 13–27: But though wages cannot be reduced much lower than they are at present, there are other means by which manufacturers might abridge their expenditure. Their establishments might be erected at much less expense—a more improved arrangement might be adopted—and the work conducted with much more economy. All these, however, are matters which the Americans will very speedily learn; every successive depression of trade will lead them more and more to see the necessity of managing every department of the business with the least possible expense; and as soon as they can equal the British in this, they will be able to compete with them, and that successfully too, in any market whatever.

Historical Sketch

a-a 141, 8–13: *Sole reliance on White's Slater,* a work replete with interesting information

b-b 150, 34

c 153, 23–26: who, in all probability, brought with him a knowledge of all the most recent improvements made by the English spinners.

d 153, 26: now flourishing

e-e 153, 32: they have been under the management of that gentleman in an uninterrupted state of improvement

f 154, 4: *On the interleaved page opposite this point Montgomery autographed the following instruction to the publisher,* The manuscript containing the continued history of the cotton manufacture, particularly of its introduction into Waltham and Lowell, commences a new paragraph on the opposite page. THIS PART OF THE MS IS MISSING.

g 154, 4–155, 21: Cotton spinning, according to the preceding statement, commenced in the then obscure village of Pawtucket in 1790, at which time only seventy-two spindles were put into operation. Since that time, the rapid extension of the business in this country has been equalled only by that of Great Britain. According to the report of a Committee, appointed by Congress in 1832, to inquire into the progress of spinning and of the manufacture of cotton goods,

The number of Mills in twelve States, were   795
Spindles in   do.   1,246,503

Looms in	do.		33,506
Males employed in the Manufacture		18,539	
Females do. do.		38,927	
Total employed,			57,466[4]

Previous to 1815, the whole weaving in the United States was done by hand looms, in many of which considerable improvements had been made, and great quantities of cloth were manufactured for home consumption. About 1814, a Mr. Gilmour landed in Boston from Glasgow, with patterns of the power loom and dressing machine, whom Mr. John Slater invited to Smithfield, and made known to him his wishes to construct these important machines; but not being able to prevail on the whole of the partners to engage in the business, Mr. Gilmour remained some time at Smithfield, employed as a mechanic, where he introduced the hydrostatic press, which proved to be of great advantage in pressing cloth, &c.

Judge Lyman of Providence had been endeavouring to construct a power loom, but failed in the attempt. On hearing of Mr. Gilmour, he, with some other gentlemen, entered into a contract with him, to build a power loom and dressing machine from the patterns he had brought from Great Britain, which he did to the great satisfaction of his patrons, from whom he received a compensation of 1500 dollars. These machines were soon after introduced into Pawtucket, where David Wilkinson commenced making them for sale. Gilmour was a man of great mechanical genius, but neglected to turn his talents and opportunities to the advantage of his family, and, consequently, on his death, they were left in poor circumstances.

The hand looms were soon superseded by the others, the introduction of which greatly aided in extending the business in this country, and has enabled the American manufacturers to compete with Great Britain in South America, India, and some other foreign markets.

Statistical notices

[a] 159, 9–160, 6: *After presenting the following table, JM remarked that the cotton industry workforce in New York state was just over three-fifths of a per cent of the state's whole population.*

STATEMENT *of the Cotton Manufacture in the State of* NEW YORK: *from William's Annual Register for* 1835.

COUNTIES	Number of Mills.	Capital invested.	Pounds of Cotton consumed annually.	Number of spindles in use.	Pounds of Yarn sold annually.	Yards of Cloth produced annually.	Num. of hands employed.
Oneida, ...	20	735,500	1,705,290	31,596	175,080	5,273,200	2,354
Rensselaer, ...	15	525,000	854,300	16,606	147,110	2,790,315	1,621
Dutchess, ...	12	445,000	833,000	17,690	185,500	1,952,000	1,974
Otsego, ...	11	304,000	618,543	15,344	56,000	2,322,000	1,077
Columbia, ...	7	218,000	559,000	13,266	199,000	1,150,400	1,265
West Chester, .	5	115,000	486,000	9,400	438,000	...	280
Washington, ...	5	100,000	168,800	3,606	33,500	717,650	275
Herkimer, ...	5	35,000	106,237	2,296	33,500	269,912	128
Saratoga, ...	4	144,000	270,000	5,752	...	1,210,660	460
Jefferson, ...	3	170,000	327,000	6,020	22,600	1,004,720	595
Ulster, ...	3	140,000	410,000	5,796	330,000	115,000	475
Orange, ...	3	135,000	251,000	4,200	4,000	740,000	460
Madison, ...	3	30,000	35,000	1,998	31,500	...	35

COUNTIES	Number of Mills.	Capital invested.	Pounds of Cotton consumed annually.	Number of spindles in use.	Pounds of Yarn sold annually.	Yards of Cloth produced annually.	Num. of hands. employed.
Tompkins, ...	3	28,000	55,500	812	1,000	199,063	97
Onondaga, ...	2	62,000	125,000	2,160	5,000	460,000	225
Monroe, ...	2	55,000	208,000	2,648	105,000	300,000	320
Clinton, ...	2	16,000	25,000	884	. . .	100,000	70
Rockland, ...	1	100,000	200,000	3,500	40,000	460,000	500
Schenectady, ..	1	77,000	118,000	2,000	20,000	416,000	200
Chenango, ...	1	75,000	200,000	4,474	. . .	800,000	225
Seneca, ...	1	70,000	190,000	4,000	. . .	550,000	150
Cayuga, ...	1	70,000	180,000	2,692	8,000	180,000	138
Franklin, ...	1	10,000	. . .	. . .	. . .	. . .	. . .
Suffolk, ...	1	10,000	36,000	576	33,000	. . .	30
Total, ...	112	3,669,500	7,961,670	157,316	1,867,790	21,010,920	12,954

b-b 162, 1

c-c 163, 1–166, 4

    cI 163, 3: *source of information was an account in the* Lowell Journal

    cII 163, 6: *at date of* Lowell Journal *account the town's population was 15,000*

    cIII-cIII 163, 7: eighteen

    cIV-cIV 163, 20

    cV-cV 163, 23–164, 14

    cVI-cVI 165, 15–19: 1824 into a town distinct from Chelmsford, and received its name from Francis C. Lowell, Esq., who was amongst the first to introduce manufactures in this place. There are now twenty-seven Factories

    cVII-cVII 165, 21–25: *room for twenty-seven more factories and for another 20,000 inhabitants to bring the population up to 40,000*

    cVIII 165, 30–34: There is also a steam boat which plies between Nashua (another manufacturing place) and Lowell, a distance of fourteen miles, which likewise co-operates with the railroad.

d-d 166, 3: at the beginning of 1839, about 9,000,000 dollars, equal to £1,875,000 sterling

e-e 166, 25: 37,984 throstle spindles

f-f 166, 26: 1,300 females and 437 males

g-g 166, 27: 220,000 yards

h-h 166, 28: 50,000 lb.

i-i 167, 1: 20,992 throstle spindles, 564 looms

j-j 167, 2: 830 females and 230 males

k-k 167, 4: 40,000 lb.

l-l 167, 10: 380 looms, and give employment to 470 females

m-m 167, 12: 40,000 lb.

n-n 167, 19: 5,000

o-o 167, 20: 154 cotton and 70 carpet looms

p-p 167, 23: 60,000 yards

q-q 167, 24: No. 12's

r-r 167, 28: 11,264 throstle spindles

s-s 168, 2: 404 looms

t-t 168, 3: 460 females

u-u 168, 4: 125,800 yards

v-v 168, 5: 34,000 lb.

w-w 168, 10: 31,000 throstle spindles, 910 looms

x-x 168, 12: 1,250 females and 200 males

ʸ⁻ʸ 168, 13: 200,000 yards of cloth, and use about 64,000 lb.

ᶻ⁻ᶻ 168, 19: 350 females, and 185 males

ᵃ⁻ᵃ 168, 20: 4,620 spindles, 38 broad cloth, and 92 cassimere looms

ᵇ⁻ᵇ 168, 22: 6,300 yards of cassimere and 1,500 yards

ᶜ⁻ᶜ 168, 23: 600,000 lb.

ᵈ⁻ᵈ 168, 28: 29, 248 throstle spindles and 830 looms

ᵉ⁻ᵉ 168, 30: 155,000 yards

ᶠ⁻ᶠ 168, 31: 53,300 lb.

ᵍ⁻ᵍ 168, 32: 50's

ʰ 168, 33–169, 2: This company spin the finest yarn on the dead spindle throstle of any that I am aware of in this country.

ⁱ⁻ⁱ 169, 3

ʲ⁻ʲ 169, 18–33: Lowell is not yet finished, but is still extending in wealth and population; nor can we conceive the extent to which it may yet be enlarged, as it is the opinion of many that there are still a sufficiency of power to propel double the machinery already in operation: at present it is the most important and extensive manufacturing town in the United States, and in all probability will continue to be so.

ᵏ 170 and 171: *These published statistics of Lowell manufactures dated 1 January 1843 replace those dated 1 January 1839.*

ˡ⁻ˡ 172, 29

ᵐ⁻ᵐ 174, 31

ⁿ⁻ⁿ 177, 7: 1st or 2d of December

ᵒ⁻ᵒ 177, 14

ᵖ⁻ᵖ 177, 23

�qq 178, 31: and comprises eighty-three dwelling houses, and twelve mercantile stores

ʳ⁻ʳ 180, 11–181, 8: *rephrasing apart from*

ʳˡ⁻ʳˡ 180, 6

ˢ 182, 12: The water power at this place is immense, and as yet scarcely begun to be occupied.

ᵗ⁻ᵗ 182, 18

ᵘ 185, 4: life and

ᵛ 185, 5: very

ʷʷ 185, 6–8: in every respect equal, and some of it superior to any thing of the kind made at the other

ˣ 186, 24: to tide

ʸ⁻ʸ 188, 10–24: When the Company first commenced operations, there were not more than ten houses at Paterson. In 1827 there were 6,236 inhabitants; 1,046 heads of families; 7 houses for public worship; 17 Schools; a Philosophical Society; 15 Cotton Factories, in which 24,000 spindles operate; 2 Factories of canvas; 1,644 spindles, employing 1,453 persons, whose wages are 224,123 dollars a year; extensive Machine Shops and Iron Works; there were 620,000 lb. of flax; 6,000 bales of cotton consumed annually, and 1,630,000 lb. of cotton yarn; 430,000 lb. of linen yarn spun in the same time; besides 630,000 yards of linen and duck; 3,354,500 yards of cotton cloth; and at that time, new Factories were still being built.

ᶻ⁻ᶻ 190, 14

ᵃ⁻ᵃ 194, 10

ᵇ⁻ᵇ 194, 32

ᶜ⁻ᶜ 195, 21: three

ᵈ⁻ᵈ 195, 22: ten

ᵉ⁻ᵉ 195, 23: is just now being

ᶠ⁻ᶠ 196, 31

ᵍ⁻ᵍ 197, 1

ʰ⁻ʰ 197, 5: four

ⁱ⁻ⁱ 197, 9: four

ʲ 197, 10: in a straight line

ᵏ⁻ᵏ 197, 21

^l 197, 22–25: is entirely appropriated as a common, and may be used by the inhabitants as a play ground, bleaching green, or for pasture.

^{m-m} 198, 7–11: Yet judging from the general appearance of the inside of the Factories, I do not see how they cannot be turned to very good account. The goods made are shirtings, sheetings, and printing cloth.

ⁿ⁻ⁿ 198, 23: ten or twelve

^{o-o} 200, 14: breadth 47

^p 201, 6–7: These are both 142 feet in length by 42 in breadth within the walls, and four stories in height besides the attics.

^{q-q} 201, 9–204, 22

> ^{qI-qI} 201, 9–12: There are in the three Mills 17,856 throstle spindles, and 568 looms in operation, consuming 39,000 lb. of cotton, and producing upwards of 105,000 yards of cloth per week.
>
> ^{qII-qII} 201, 12–202, 8: *at Saco a variety of striped and coloured goods, besides drillings and jeans are made. Dyeing the cotton in the fiber, rather than the yarn, state is the simplest and cheapest colouring method known to JM who thought it had not been tried in Britain. Giving shades unobtainable through yarn dyeing, it was employed by three companies in America: the Hamilton Manufacturing Co. at Lowell, a company at Philadelphia, and the York Manufacturing Co., at Saco. JM rates his firm's goods, especially its jeans, as the best made in the USA and its bedticking as the finest in the Boston market.* Samuel Batchelder, Esq., is the Company's Agent, under whose able superintedence the whole concern has been got up.
>
> ^{qIII} 202, 21–24 and 203, 4–33: *A correspondent of the* Portland Advertiser *calculates the relative transportation costs for a 5,000 spindle cotton mill, located 50 miles from its market and making 300 tons of goods per annum (or one ton a day) to be as follows: by land $4,200 p.a. ($7 per ton per 50 miles); by canal or railroad $1,800 ($3 per ton per 50 miles); and by sea $600 ($1 per ton per 50 miles). The differences of $3,600 and $1,200 are equivalent to the interest on capital of $60,000 and $20,000. So it would be better to pay $60,000 for the water power and mill site for a factory at tidewater falls in Maine rather than go 50 miles into the interior. And it would be better to pay $20,000 more for a Maine mill site than to choose a similar one requiring 50 miles of canal or rail transport.*
>
> ^{qIV} 203, 1–204, 11: building materials are inexhaustible—vast quantities of bricks are made for exportation, and there are valuable granite quarries, together with abundance of lumber, all of which offer a rare combination of facilities for building extensive works. The Eastern Mail Stage passes daily over the Island, and the Eastern Railway is intended to pass through the village, which will greatly facilitate the intercourse between this and Boston: the distance being 100 miles, it is expected the railroad cars will travel that length in five hours; so that leaving this place early in the morning, a person may go to Boston and do business for four or five hours, and return again in the evening.

^{qV-qV} 204, 14: six

^{qVI-qVI} 204, 15: four

^{qVII-qVII} 204, 19

^{qVIII-qVIII} 204, 20: two Newspapers published weekly

Miscellanies

^{a-a} 205, 20: on a small tube connected with the pulley B.

^{b-b} 207, 17

^{c-c} 209, 12–14: *Loom price (same) and dressing machine price ($425 or £88 10s 10d, only given.*

^d 212, 33: which is the prevailing colour for the wooden part of all the machinery here

^{e-e} 214, 3

^{f-f} 214, 30–32: £2 10s 10d Sterling, besides oil, tallow, stuffing, &c. This is about double the cost of the same power in Glasgow.

^{g-g} 214, 32

h-h 215, 2: at Lowell, Massachusetts, and Manayunk, Philadelphia
i-i 215, 6: THE MIDDLE SECTION OF THIS INTERPOLATION IS MISSING
j 216, 13
k-k 217, 14
l-l THE POSITION OF THIS LOOSE SHEET IS NOT CERTAIN
   Since it treats steam and waterpower costs, apparently in Britain, it has been placed at the end of the section on power costs.
m-m 217, 15–218, 29: *deleted; since 219 is unmarked, JM possibly intended some revision of the figures. In view of the ambiguity of JM's intentions, the whole section has been left in the text.*

EXPLANATORY NOTES

I *Plan and arrangement of the mills*

1 Double roof: one with a monitor or broken slope formed by clerestory lights which allowed the attic to be used for manufacturing.

2 Stirling & Beckton ranked as the fifth largest cotton manufacturing firm engaged in spinning and weaving in Manchester in 1833. V. A. C. Gatrell, "Labour, Power, and the Size of Firms in Lancashire Cotton in the Second Quarter of the Nineteenth Century," *Economic History Review* 2nd ser. 30 (1977) 100.

3 The four steam mills in Newport, R. I., included the Newport Steam Co. mill (25 h.p. engine), the Perry Manufacturing Co. mill (60 h.p. engine), and T. R. Hazard's Sea Isle Factory (15 h.p. engine). At Providence, R. I., there were two steam mills in 1838: Slater & Co.'s Providence Steam Mill (70 h.p.) and John Waterman's Eagle Mill (40 h.p.). The Newburyport mills with steam engines included William Bartlett's Mill No. 1 (15 h.p.) and the Newburyport Steam Cotton Co. (40 h.p.). All were high pressure engines. U.S. Congress, House, *Report on the Steam Engines in the United States*. House Document No. 21, 3rd session, 25th Congress. Washington, D.C., 1838 (hereafter cited as the *Woodbury Report*) 52, 85, 88.

4 Flat: floor or story.

5 British watch-keeping practices may have been more slack than American because of a greater reliance on fireproof construction which utilised iron framing. David Hartley (1732–1813) patented the use of iron plates, laid between joists and boards, as a method of fireproofing in 1773 (G. B. Pat. No. 1037), for which Parliament awarded him £2,500. By the 1820s the best constructed mills in Manchester incorporated iron frames. R. S. Fitton and A. P. Wadsworth, *The Strutts and the Arkwrights, 1758–1830. A Study of the Early Factory System* (Manchester: University Press, 1958) 196–217. Testimony of George Smith, Manchester cotton spinner, GB, *PP (Commons)* 1833 (690) VI, 549. For Hartley's work see Richard Warner, *Literary Recollections* (2 vols., London, no publisher, 1830) 2: 224–253.

   It is surprising that JM failed to allude to the new iron-framed fireproof mills in Britain, one of which, Henry Houldsworth's mill, had been built in Glasgow in 1804–1805. See Geoffrey D. Hay, "Houldsworth's Cotton Mill, Glasgow," *Post-Medieval Archaeology* 8 (1974) 92–100.

6 Main belt drives, pioneered by Paul Moody at the Appleton Mills, Lowell, spread to all industrial operations in the USA before the Civil War. Louis C. Hunter, *A History of Industrial Power in the United States, 1780–1930*, I. *Water-Power in the Century of the Steam Engine* (Eleutherian Mills-Hagley Foundation, 1979) 459–471.

7 Robert Montgomery, GB Pat. No. 6261 (26 April 1832).

8 The relative costs of belt and gear drives were, and remain, hard to ascertain. Ultimately, after 1870, American belt drives spread to Europe. Hunter, *Water-Power*, 466–472.

II *Notices of the various machines*

[1] James Lillie, Manchester engineer and partner of William Fairbairn (1789–1874), invented a willow which gained some contemporary note in the English cotton industry. See Andrew Ure, *The Cotton Manufacture of Great Britain* (2 vols., London: Charles Knight, 1836) 2: 8. Lillie is considered in A. E. Musson and Eric Robinson, *Science and Technology in the Industrial Revolution* (Manchester: University Press, 1969) 481–483.

[2] Samuel P. Mason, of Killingly, Connecticut, received an American patent for a cotton whipper on 8 July 1834. The reconstructed version is in NA, RG 241, Specifications 18, 507–508 and Drawings No. 8295, of which Montgomery's *CM* 1, Plate II is a good copy of figs. 3 and 4.

[3] Evidently the situation changed between the writing of the two editions of the *CM*; see text note eI.

[4] John Crane Whitin (1807–1882), founder of the famous textile machine making firm, patented his immensely successful picker and spreader on 20 July 1832. See Thomas R. Navin, *The Whitin Machine Works since 1831. A Textile Machinery Company in an Industrial Village* (Cambridge, Mass.: Harvard University Press, 1950) 26–34 and the patent in NA, RG 241, Specifications 14, 489–490 and Drawings No. 7178. A British scutching machine is illustrated in *CSMA* 1, Plate IV.

[5] William B. Leonard was agent of the Matteawan Co. at Fishkill, sixty miles up the Hudson from New York City between ca. 1819 and 1850. See *TIR*, 238, 243, 314, 335 and *Scientific American* 5 (1850) 253.

    Larned Pitcher (d. 1840), a British immigrant (ca. 1800), had a machine shop at Pawtucket. See George W. Browne, *Amoskeag Manufacturing Company of Manchester, N.H.: a History* (Manchester: Amoskeag Manufacturing Co., 1915) 197; James S. Brown to Zachariah Allen, 2 June 1876, Miscellaneous Papers Box, Allen Papers, RIHS.

[6] Hetherington's spreader appears to have been one of the minority of successful inventions in Britain not covered by a patent. Neither was it noted by Baines or Ure.

    John Hetherington & Co., machine makers of Store Street, Manchester, sent whole lines of cotton spinning and weaving machinery to the USA after the relaxation in the Board of Trade's machinery export licensing system of August 1842. See *Pigot & Slater's General and Classified Directory of Manchester and Salford* (Manchester: Pigot & Slater, 1841) 64, 76; *TIR*, 45.

[7] Joseph Brown and J. S. Wright were instrumental in completing the cotton mill built at the junction of the Quabaug, Ware, and Swift Rivers in Hampden County, Massachusetts, after the property was taken over by the Palmer Cotton Manufacturing Co. in 1831. The venture was started by the Three Rivers Manufacturing Co. in 1825. See Alfred M. Copeland, *Our County and Its People: a History of Hampden County, Massachusetts* (3 vols., Boston: County Memorial Publishing, 1902) III, 144–145; *McLane Report* I, 284–285.

[8] Oldham specialised in the manufacture of heavier cotton goods like velvets, velveteens, and moleskins, which were made with lower count yarns and given a surface finish. Double carding made a more uniform yarn than single carding and a combined machine was more economical than two separate ones. See Edward Baines, *History of the Cotton Manufacture in Great Britain* (reprint, London: Frank Cass, 1966) 418; *CSMA*, 72, and *CSMA* 3, 82, 84–85.

[9] The American patent for the railway drawing head for the cotton card was taken out by William B. Leonard of Matteawan (see note 5) on 16 September 1833. The reconstructed patent is in NA, RG 241, Specifications 16, 127–148 and Drawings No. 7761. While any connection has yet to be shown, Leonard's design was antedated by Johann Georg Bodmer's British patent No. 5016 (1824), 18–23 of the printed text. See *TIR*, 48, 305.

    Evidently the railway head was widely adopted in America because Leonard took the trouble to extend his patent for seven years, from 16 September 1847 (Specification, 127).

[10] Neil Snodgrass, engineer of Bridgeton, Glasgow, in the late 1830s who visited the USA in the mid-1830s. See *The Post Office Annual Directory for 1836–1837 of Glasgow* (Glasgow: John Graham, 1836) 200 and *CSMA* 3, 99.

[11] A practice unremarked in *CSMA* 1 or *CSMA* 3.

[12] The method of making pelisse wadding or batting was patented in the USA by the British immigrant James Beaumont on 3 December 1814. See D. Hamilton Hurd, *History of Norfolk County, Massachusetts* (Philadelphia: J. W. Lewis & Co., 1884) 948.

[13] Another example of the widespread American application of a stop motion device to save labor costs; it is similar to the one for the drawing frame, described below. For the use of stop motions in American textile machinery, see *TIR*, 198 and passim.

[14] Illustrated in *CSMA* 1, Plate VI.

[15] At Saco, the Saco Water Power Company, formed in 1839, opened a machine shop in 1842. Gibb, *Saco-Lowell Shops*, 125.

[16] Apparently Samuel Batchelder first applied a stop motion to the drawing frame (though this is not certain); certainly he achieved its transfer to Britain, through Daniel Treadwell, in 1835 (see Introduction).

The American patent for a drawing frame stop motion awarded to Lewis Cutting of Lowell on 15 May 1834 shows a mechanism very similar to that illustrated here. Cutting, however, claimed the spring bolt as his invention and this does not appear in the British patent for Batchelder's stop motion. A stop motion similar to Cutting's is found in the three twisting frames of the Merrimack Manufacturing Co., Lowell, surviving from the early nineteenth century (thus suggesting that they were made after 1834).

See NA, RG 241, Specifications 18, 197–198 and Drawings No. 8201; David J. Jeremy, "Innovation in American Textile Technology during the Early Nineteenth Century," *Technology and Culture* 14 (1973) 66–72.

[17] The Taunton speeder was invented by George Pitts Danforth (1791–1838) of Taunton, Massachusetts, and developed by the Taunton Manufacturing Co. Danforth secured the American patent on 2 September 1824. See John W. Lozier, "Taunton and Mason: Cotton Machinery and Locomotive Manufacture in Taunton, Massachusetts, 1811–1861," (PhD dissertation, Ohio State University, 1978) 161–162; *TIR*, 48, 109, 209, 216, 221, 243–246.

The Taunton or tube speeder was capable of producing a vast amount of coarse roving quickly, as Montgomery described in *CSMA* 1, 127–132. It was patented in Britain by Joseph Chesseborough Dyer: GB Pat. No. 5217 (16 July 1825). For Dyer, see *TIR*, 109, 242–250.

[18] The *Post Office Directory* for Glasgow for 1836–1837 lists C. Girdwood & Co. of Govan Street and H. Houldsworth of Cheapside Street among twenty-one machine makers and millwrights in the city that year, but not William Craig, who evidently valued the goodwill of the Girdwood name. Claud Girdwood & Co., founders, engine and machine makers, were a leading firm of textile machinery makers in the city in the 1820s. See *The Franklin Journal* 1 (1826) 107; "The Journal of John Smith," *Essex Institute Historical Collections* 106 (1970) 102.

[19] Montgomery visited the English manufacturing districts sometime during the first four months of 1836 for two reasons at least. Firstly, he wanted to incorporate English technical developments into the third edition of his *CSMA*, which was then retitled and enlarged. Secondly, with his American appointment in view, it would be prudent to familiarise himself with the state of his English competitors. See *CSMA* 3, Preface and JM to Daniel Treadwell, 6 January 1836, Treadwell Papers, Harvard University Archives.

[20] Gilbert Brewster, then of Poughkeepsie, New York, patented the Eclipse speeder on 18 April 1829. The machine is described in *CSMA* 3, 154–157.

[21] Sharp, Roberts & Co. was a leading machine making firm in Manchester; its preeminence derived in part from the mechanical genius of Richard Roberts (1789–1864), inventor of the best-designed self-acting mule, and of much else. See Musson and Robinson, *Science and Technology*, 478–479.

[22] Plate speeders seem to have been first built by Godwin, Rogers & Co., machine makers

of Paterson, New Jersey, in 1825. See *TIR*, 210. For Snodgrass's visit to America, see *CSMA* 3, 99–100.

[23] Charles Danforth (1797–1876), brother of George Pitts Danforth (see note 17), invented the cap spindle in 1828 when he was at Ramapo in the lower Hudson valley. The following year he moved to Paterson and remained here until his death, joining, expanding and taking over the firm of Godwin, Rogers & Co. See *DAB*.

[24] This form of the double speeder was illustrated in Rees, *Cylcopaedia*, s.v. "Cotton Manufacture," Plates VII and VIII.

[25] This throstle type of stretcher or extenser was possibly developed by Paul Moody for the Boston Manufacturing Co. at Waltham. See *TIR*, 187–188.

[26] William Higgins, a Manchester machine maker, patented two devices which improved the performance of the bobbin and fly frame: a spring presser for laying the roving "with a considerable degree of tightness," so inserting more roving into each cop or package; and a mechanism for forming conical ends to the cop. Both devices came from foreign sources, according to the specification. Higgins's fly frame was used in a large number of mills in New England and New York by the late 1840s. The major component in it was the differential gear, invented by Asa Arnold of Rhode Island but attributed to its English patentee, Henry Houldsworth. See GB Pat. No. 6639 (7 July 1834) in the name of William Higgins; J. S. Young, *Results of Investigations as to the Comparative Merits of the American Speeder and English Fly-Frame, as Machines for the Manufacture of Cotton Roving* (Portsmouth, N.H.: no publisher, 1849); Ure, *Cotton Manufacture* 2: 88.

[27] The dead spindle, communicated to him by a foreigner, was patented in Britain by Robert Montgomery, cotton spinner of Johnstone, Renfrewshire, Scotland: GB Pat. No. 6261 (26 April 1832). The dead spindle reached higher speeds than the earlier throstle spindle because flyer wobble was reduced by securing the flyer arms to a base pulley which was then powered.

[28] Charles Danforth patented the cap spindle on 2 September 1828. See NA, RG 241, Specifications 7, 231–232 and Drawings No. 5214.

[29] John Thorp's ring spindle patent of 20 November 1828 is in NA, RG 241, Specifications 7, 87–89 and Drawings, No. 5279.

[30] Henry Gore's light spindle, supported by tubes instead of collars, was popular in Manchester. See GB Pat. No. 6201 (22 December 1831) and Evan Leigh, *The Science of Modern Cotton Spinning* (2 vols., Manchester: Palmer & Howe, 1873) 2: 210–211.

[31] Estimates of the power requirements of the various spindle types could differ for several reasons: defects in measuring equipment; differences in the count of yarn being spun; confusion about what exactly was being measured, machines or mills. Two sets of experiments were recorded at Lowell in the early 1830s. The first, conducted by the Hamilton Manufacturing Co. in 1830, used measuring equipment devised by Warren Colburn (superintendent of the Merrimack Manufacturing Co.) which showed that one horse power might drive between 54 and 88 dead spindles (spinning machines only) or between 30 and 46 dead spindles (whole mill, including main gear). Montgomery, informed presumably by Batchelder who had been agent of the Hamilton Co. in 1830, reported that the dynamometer used at Lowell was inaccurate. The second set of experiments, performed in the Merrimack Manufacturing Co. mill No. 3, again with Colburn's equipment evidently, gave readings of one horse power per 80.7 spindles (spinning equipment only) or 52.7 spindles (whole mill, including drums), making a No. 28 yarn: a result corresponding to that cited here by Montgomery.

See Ithamar A. Beard, "Practical Observations on the Power Expended in Driving the Machinery of a Cotton Manufactory at Lowell," *Journal of the Franklin Institute* 2nd ser. 11 (1833) 6–15; *CM* 1, 205; Robert Israel to Lewis Waln, 30 September 1831, Waln Collection, Historical Society of Pennsylvania, a letter found by Steven Lubar who kindly brought it to my notice. For dynamometers, see the last chapter of *CM* 2.

[32] Ure, *Cotton Manufacture* 2: 131–133.

[33] With the backing of two New York capitalists, Morris Ketchum and Morris Grosvenor, Thomas Rogers (1792–1856) opened the Jefferson Works at Paterson, New Jersey, in 1832, after he broke from the older machine making firm of Godwin, Rogers & Co.

Levi R. Trumbull, *A History of Industrial Paterson* (Paterson, N.J.: C. M. Herrick, 1882) 111–113.

[34] See notes 5, 26, and 30 above.

[35] Larned Pitcher and James S. Brown; see note 5 above.

[36] For the Taunton interests of Samuel Crocker (1772–1853) and Charles Richmond (d. 1850), see Lozier, "Taunton and Mason," 78–140.

[37] The self-acting mule of James Smith, of the Deanston Works, Kilmarnock, Scotland, incorporated the "stripper," a winding-on mechanism patented by John Robertson of Crofthead, Renfrewshire, and Smith's own mangle wheel for changing the speed of the carriage during winding. Its simplicity of design made it popular, but only for low count yarns because of the strains it imposed. A Smith mule was illegally exported from Scotland by the son of William B. Leonard of Matteawan, who then secured its American patent for Smith, on 27 June 1838.

See GB, Pat. Nos. 6475 (21 September 1833) awarded to Robertson, and 6560 (20 February 1834), Smith's; Leigh, *Science of Modern Cotton Spinning* 2: 246; Lozier, "Taunton and Mason," 267–268.

[38] The Rocky Glenn Manufacturing Co. ran a cotton factory on Fishkill Creek, Dutchess County, New York. Started in 1836, the original factory was destroyed by fire and then rebuilt in 1841 and this may have provided the opportunity to install Smith's self-acting mules.

See Lorenzo Neeley, "Manufacturing Industry in the State of New York," *Hunt's Merchants' Magazine* 15 (October 1846) reprinted in *The Textile History Review* 4 (1963) 185.

[39] William Mason (1808–1883) of Taunton, Massachusetts, patented self-acting mule mechanisms on 8 October 1840 and 3 October 1846. The latter was perfected for and first used in the Essex Mill at Newburyport in 1843, so JM's remarks presumably relate to the earlier model of 1840. See Lozier, "Taunton and Mason," 271–274.

[40] Richard Roberts's self-acting mule patents are GB, Pat. nos. 5138 (29 March 1825) and 5949 (1 July 1830). The quadrant motion of the second patent, which elegantly solved the winding-on problem, eventually prevailed in all British self-actors.

Roberts's mule, like Smith's, was smuggled out of England and patented in the USA on behalf of its inventor (11 October 1841) by Rhode Island system manufacturers William C. Davol and his uncle Bradford Durfee, both of Fall River, Massachusetts.

See Leigh, *Science of Modern Cotton Spinning* 2: 244; Harold Catling, *The Spinning Mule* (Newton Abbot: David & Charles, 1970); Lozier, "Taunton and Mason," 268–269. For the Great Falls Co., see below chap. V, note 51.

[41] Mule patented in Britain by James Potter, a Manchester cotton spinner. See GB, Pat. nos. 7263 (21 December 1836) and 9366 (25 May 1842).

[42] For the Glasgow strikes, see GB, *PP (Commons)* 1837–1838 (papers 488, 646), VIII, "First and Second Reports from the Select Committee on Combinations of Workmen."

[43] Gilbert Brewster, a Rhode Island mechanic who moved to upstate New York in the early 1820s, invented several spinning devices and took out American patents between 1812 and 1831. Whether any covered more than a jenny is unknown. See Lozier, "Taunton and Mason," 255; *TIR*, 210, 221, 243–246.

[44] This was the self-acting mule invented by Ira Gay, machinist of the Nashua Manufacturing Co., between 1824 and 1827 and patented 10 April 1829. No specification or drawing has survived. See *TIR*, 212–213.

[45] For Moody's invention of the warping frame stop motion in 1815, patented 9 March 1816, see Jeremy, "Innovation in American Textile Technology," 64–66 and *TIR*, 183, 187, 195.

[46] The English warper described by Ure as automatic in fact relied on operative rather than mechanical responses to detect yarn breakages: Ure, *Cotton Manufacture* 2: 239–243.

[47] Orrell's factory was designed by William Fairbairn. Ure, who thoroughly described it, called it "a model of factory architecture" and frequently cited it for best-practice techniques. Ure, *Cotton Manufacture* 1: 296–304; 2: 90, 131, 323, 405.

[48] JM oddly (because Batchelder was his employer at Saco) repeats the incorrect date given in *CM* 1: Batchelder patented his steam dressing machine on 22 June 1832. See NA, RG 241, Specifications 14, 67–68 and Drawings No. 7120.

[49] Larned Pitcher and Ira Gay were partners in a Pawtucket, Rhode Island, machine shop until Gay went to Nashua, New Hampshire, in 1824. See note 5 above and *TIR*, 206, 212–213.

[50] JM seems to be referring to James Lillie's sizing machine which dressed yarn in the chain. See Ure, *Cotton Manufacture* 2: 249–253.

[51] The circular self-acting temple, patented by Ira Draper in 1816, was adopted by the larger northern New England cotton manufacturers in the 1820s. See *TIR*, 197–198.

[52] Amasa Stone, then of Providence, Rhode Island, took out an American patent for a warp let-off motion on 30 April 1829. If JM's memory was correct, Stone must have made two visits to England on patent business. In 1834, when resident at Liverpool (but describing himself as a machinist from Providence), he was awarded a British patent for a combined let-off and take-up motion which was applicable to silk as well as cotton, hemp, and woolen powerlooms. It looks a marked improvement on his American patent.

    See NA, RG 241, Specifications 8, 297–298 and Drawings No. 5474; GB Pat. no. 6704 (23 October 1834).

[53] James Kempton, a Connecticut manufacturer, told a Commons committee in 1833 that girls in American cotton mills tended four to six powerlooms each, but when Crockett visited Lowell in 1834 he found the girls tending no more than three looms each. Kempton's practice may have been exceptional or else there were wide differences in this respect between the Waltham and Rhode Island systems.

    See GB, *PP (Commons)* 1833 (690), VI, 167 and David Crockett, *An Account of Col. Crockett's Tour to the North and Down East* (Philadelphia: E. C. Cary and A. Hart, 1835) 94.

[54] Ure, *Cotton Manufacture* 1: 304–306.

[55] The only American patents by inventors living at Fishkill before 1837 were taken out by William B. Leonard (for a spindle in 1819, a powerloom take-up in 1827, and a railway carding head in 1833) and by William Fowler (for spinning cotton roving, 23 March 1836).

III *Comparative estimates*

[1] The framework adopted for calculating manufacturing costs is evidently derived from that used by JM to provide "Estimates of a Spinning and Weaving Establishment" in *CSMA* 3, 248–256.

[2] In the controversy which followed the publication of *CM* 1, JM stoutly affirmed that the two mills which he compared here were not hypothetical cases but actually existed and functioned. The point was important. If the two factories did not exist, then JM's calculations were pure supposition and his conclusions lost credibility.

    The dimensions and equipment of the British mill are identical to those of the Glasgow mill which JM cited in *CSMA* 3 in 1836: the latter was described as a "small manufacturing establishment" built of brick, measuring 90 by 38 feet within the walls and containing 2,100 throstle spindles, 2,400 mule spindles, and 128 looms. In the Justitia controversy, JM said that the mill was built in 1834 and that he possessed its plans. It is not unlikely that the mill in question belonged to his Scottish employers, MacLeroy, Hamilton & Co.

    The dimensions of the American mill tally with those of the two mills built for the York Manufacturing Co. at Saco, Maine, in 1835 and 1837: 142 feet long and 42 feet broad. Since JM became superintendent of the York factories in 1836; it is very likely that one of them provided the data for these estimates. Certainly it would be easiest for JM to present information on manufacturing costs to which he had the most direct access, and that would of course come from the mills which he himself managed. All the circumstantial evidence therefore buttresses JM's claims to have based his comparative estimates on functioning factories in Scotland and northern New England, in manufacturing districts specialising in the production of low number yarns and lower quality goods.

³ The blank spaces in this invoice presumably stood for Saco; Pliny Cutler; York. See Gibb, *Saco-Lowell Shops*, 749, n. 6.

⁴ Alexander Graham, *The Impolicy of the Tax on Cotton Wool* (2nd ed. Glasgow: Associated Cotton Spinners, 1836) 25.

⁵ A firm recorded in the Factory Commissioners' Reports: GB, *PP (Commons)* 1834 (167) XX, Section A1, No. 105.

⁶ J. Sommerville & Sons also completed the Factory Commissioners' returns in 1834: ibid., No. 110.

A John MacBride, manager of the Nursery Spinning and Weaving Mills, Hutchisontown, Glasgow, patented improvements in the jacquard loom in the early 1840s. See GB Pat. nos. 9032 (21 July 1841) and 10,259 (15 July 1844). By 1846 he had his own firm at the Albyn Works, Glasgow, engaged in cotton spinning and powerloom weaving. See GB Pat. no. 11,444 (12 November 1846).

⁷ Ure, *Cotton Manufacture* 1: xxxix.

⁸ Ibid. 1: xli.

⁹ James Kempton, a native of Philadelphia and a cotton manufacturer of Norwich, Connecticut, gave this information to a Parliamentary committee in 1833: GB, *PP (Commons)* 1833 (690) VI, 147; and, earlier that year to the Royal Commission on the Employment of Children in Factories, GB, *PP (Commons)* 1833 (450) XX, Section E 22.

¹⁰ Ure, *Cotton Manufacture* 1: xlii.

¹¹ Ibid.

¹² Ibid., 1:xlvii–xlviii.

¹³ Quoted in ibid., 1: xlix–1.

This schedule of wage rates, apparently from Paterson, New Jersey, was supplied by Benjamin Dawson of New York. Finlay, distinguished Scottish merchant and manufacturer with experience in the American trade, delivered the letter to the Parliamentary committee investigating manufactures, commerce, and shipping in 1833. See GB, *PP (Commons)* 1833 (690) VI, 70.

IV *Historical Sketch*

¹ This chapter derives from the badly organised life of Samuel Slater by George Savage White: *Memoir of Samuel Slater, the Father of American Manufactures. Connected with a History of the Rise and Progress of the Cotton Manufacture in England and America. With Remarks on the Moral Influence of Manufactories in the United States* (Philadelphia: the author, 1836) chiefly from chapters 1–3.

As noted in the Introduction above, White was one of the first to hail JM's *CSMA* 1 as a pioneering technical handbook.

Readers interested in the history of the early American cotton industry are referred to Caroline F. Ware, *The Early New England Cotton Manufacture. A Study in Industrial Beginnings* (1931; reprinted New York: Russell & Russell, 1966). For an entry to recent literature on the subject and for a study of British origins, see *TIR* and Anthony F. C. Wallace and David J. Jeremy, "William Pollard and the Arkwright Patents" *William and Mary Quarterly* 3rd ser 33 (1977) 404–425. For the Arkwright background see R. S. Fitton, *The Arkwrights: Spinners of Fortune* (Manchester: Manchester University Press, 1989). White exaggerates Slater's role as implanter of Arkwright technology into the

USA. For recent work revising Samuel Slater's role as employer, see Barbara M. Tucker, *Samuel Slater and the Origins of the American Textile Industry, 1790–1860* (Ithaca, N.Y.: Cornell University Press, 1984) and Jonathan Prude, *The Coming of Industrial Order. Town and Factory Life in Rural Massachusetts, 1810–1860* (Cambridge: Cambridge University Press, 1983).

[2] JM's hopes were widely shared and eventually realised. In the Middlesex Mechanics' Association exhibition held at Lowell in September 1851, G. C. Smith & Co. of Pawtucket, Rhode Island, displayed "One Slater spinning frame. One of the first frames built by Mr. Slater, in good state of preservation and running order, and so should be kept, a valuable relic." Five years later the Rhode Island Society for the Encouragement of Domestic Industry recorded its thanks to the heirs of Moses Brown for the donation of one of the first carding machines and one of the first spinning frames driven by water power used in the state, and voted to set up a sub-committee to get the machines repaired "and to take effectual measures for their preservation." Soon after, the two machines were reported "in first rate order for their preservation, and are now in the rooms of the society." In 1883 those same machines were donated by the Society to the United States National Museum, the Smithsonian Institution, chiefly because Brown University Museum (to whom they were given in 1880) could not or did not remove them from the Society's damp basement. Today Slater's first machines stand in the splendid Museum of American History, part of the Smithsonian Institution in Washington, D.C.

  See *Reports of the First Exhibition of the Middlesex Mechanics' Association, Held in the City of Lowell, September, 1851* (Lowell 1852) 8; *Transactions of the Rhode Island Society for the Encouragement of Domestic Industry* for 1856, 24; Smithsonian Institution, Accession files, no. 13,137, especially letter of Professor J. W. P. Jenks of Middleboro, Massachusetts, to Professor Spencer F. Baird, Secretary of the Smithsonian, 22 January 1883.

[3] Letter of Moses Brown to John Dexter (supervisor of revenue for Rhode Island), 22 July 1791, quoted in White, *Slater*, 89. The full text is found in Harold C. Syrett et al. (eds.), *The Papers of Alexander Hamilton* (26 vols., New York: Columbia University Press, 1961–1979) 9: 432–441.

[4] These figures are identical to those collected by the Friends of Domestic Industry, New York Convention, and printed in their *Report on the Production and Manufacture of Cotton* (Boston: J. T. and E. Buckingham, 1832) 16.

V *Manufacturing Districts*

[1] John P. Bigelow, *Statistical Tables: Exhibiting the Condition and Products of Certain Branches of Industry in Massachusetts, for the Year Ending April 1, 1837* (Boston: Dutton and Wentworth, 1838) 169.

[2] Preceding statistics from ibid. 202, 204, 209.

[3] Like the table published on p. 159 of *CM* 1, these figures come from Edwin Williams, *The New York Annual Register for the Year of Our Lord 1835* (New York: Edwin Williams, 1835) 150, 153. JM apparently did not copy his figures accurately: the figures in Williams, p. 153, are $350,011,629 (real estate valuation); $109,660,506 (personal estate valuation).

[4] Friends of Domestic Industry, *Report* 16, is the source of these figures though JM has inexcusably interpolated the New York statistics of 1835 into this table of 1831, and then failed to adjust the New York figure for pounds of cloth annually produced (not given in the 1835 data).

[5] This is also derived from ibid.

[6] This comes from the Sixth Federal Census: *Compendium of the Enumeration of the Inhabitants and Statistics of the United States, as Obtained at the Department of State, from the Returns of the Sixth Census* (Washington, D.C.: Thomas Allen, 1841) 361.

[7] See Gibb, *Saco-Lowell Shops* for a modern account of the early development of Lowell. Hunter, *Water-Power* chap. 6, provides an excellent description and analysis of the development of Lowell's water power system.

[8] See *DAB* for Kirk Boott (1790–1837).

[9] Statistics derived from *Statistics of Lowell Manufactures, January 1, 1843 Compiled from Authentic Sources* (broadsheet published at Lowell).
   For an early account of the Lowell corporations, see Charles Cowley, *Illustrated History of Lowell* (rev. ed. Boston: Lee & Shepard, 1868) 36–59.

[10] Oliver M. Whipple, by gaining control of water power sites on the Concord River, secured the interests of small, independent manufacturers at Lowell, which was otherwise dominated by the great textile corporations. Ibid. 61–63.

[11] *Statistics of Lowell Manufactures* was an annual broadsheet first published for 1 January 1835.

[12] Elisha Bartlett, *Vindication of the Character and Condition of the Females Employed in the Lowell Mills, Against the Charges Contained in the Boston Times, and the Boston Quarterly Review* (Lowell: Leonard Huntress, 1841) 21. Most of JM's other information on Lowell seems to have come from other sources. For the definitive analysis of female labour at Lowell, see Thomas Dublin, *Women at Work. The Transformation of Work and Community in Lowell, Massachusetts, 1826–1860* (New York: Columbia University Press, 1979).

[13] For the *Lowell Offering*, see Benita Eisler, *The Lowell Offering* (Philadelphia: J. B. Lippincott, 1977); Philip Foner, *The Factory Girls* (Urbana: University of Illinois Press, 1977); and Helena Wright, "Sarah G. Bagley. A Biographical Note," *Labor History* 20 (1979) 398–413.

[14] *The American Almanac and Repository of Useful Knowledge for the Year* 1837 (Boston: Charles Bowen, 1836) 10–33.

[15] JM refers to Britain's first effective Factory Act of 1833, 3 & 4 Will. IV, c. 103.

[16] Only Independence Day was a truly national holiday at this time: JM is describing the festivals and holidays prevalent in New England.

[17] This section is from White, *Slater*, 256–257.

[18] In ibid., the North East section is identified as Pawtucket.

[19] Ibid., 259.

[20] Ibid., 259–262.

[21] Section rI–rI does not come from White.

[22] White, *Slater*, 262.

[23] These comments on Slaterville, unlike those in the preceding paragraph, do not derive from ibid., 262–263.

[24] White, *Slater*, 267.

[25] Ibid.

[26] Paragraph t–t did not come from White.

[27] Section on Williamantic [Willimantic] comes from White, *Slater*, 267–268.

[28] Section on Greeneville from ibid., 268–270.

[29] Section on Paterson is based on ibid., 236–238, 383–384.

[30] Peter Colt (1744–1824), treasurer of the State of Connecticut and superintendent of the Society for Establishing Useful Manufactures, at Paterson, 1793–1796. See Joseph S. Davis, *Essays in the Earlier History of American Corporations* (2 vols., 1917; reprinted New York: Russell & Russell, 1965) 1: 460–497.

[31] Charles Kinsey and his associates, Crane and Fairchild. Ibid. 1: 505.

[32] But compare this figure with that given in the last chapter.

[33] Apparently not from White. The first of the companies reported was the Athens Manufacturing Co., organised in 1829 by William Dearing and four partners, including Captain John Johnson of Massachusetts. See Richard W. Griffin, "The Origins of the Industrial Revolution in Georgia: Cotton Textiles, 1810–1865," *Georgia Historical Quarterly* 42 (1958) 359–360.

[34] For the Saluda Manufacturing Co., incorporated in 1834, see Ernest M. Lander, Jr., *The Textile Industry in Antebellum South Carolina* (Baton Rouge: Louisiana State University Press, 1969) 38–39, passim.

[35] White, *Slater*, 271–272.

[36] Ibid., 272.

[37] Ibid., 278.

[38] Ibid.

[39] Ibid.

[40] Ibid., 279.

[41] Ibid.

[42] This factory, set up by Mr. B. Whitney and located at the corner of Tchoupitoulas and Roffignac Streets, was a vertically integrated plant producing twilled goods. The shortage of skilled labour (skilled workers had to be recruited from New England) led to the factory's early demise and no other cotton mill was set up in New Orleans before the Civil War. See Henry Rightor (ed.), *Standard History of New Orleans, Louisiana* (Chicago: Lewis Publishing Company, 1900). 532.

[43] In 1839 soon after the dam on the Kennebec River was built at Augusta, a heavy freshet destroyed part of the dam and, with it, ten mills just built to use the water power. See Hunter, *Water-Power*, 530n.

[44] The York Manufacturing Co. at Saco owned these mills (which JM superintended) while the Saco Water Power Co. managed the power and sold off mill sites. See Gibb, *Saco-Lowell Shops*, 104–149.

[45] For Amoskeag, see Browne, *Amoskeag Manufacturing Co.*; John B. Clarke, *Manchester: a Brief Record of Its Past and a Picture of Its Present* (Manchester, New Hampshire: J. B. Clarke, 1875).

[46] In order of construction, the first steam mills in Newburyport were the Newburyport Steam Cotton Co. mill (1835), Bartlett Mill No. 1 (1837), Bartlett Mill No. 2 (1841), and the James Steam Mill (1843–1845). See *Woodbury Report*, 52; Charles T. James, *Letters on the Culture and Manufacture of Cotton* (New York: George W. Wood, 1850) 15–16, 22–23.

[47] For the Newmarket Manufacturing Co., incorporated in 1822, see D. Hamilton Hurd, *History of Rockingham and Strafford Counties, New Hampshire* (Philadelphia: J. W. Lewis & Co., 1882) 397.

[48] At Dover the mills on the Cocheco River, at the Lower Falls, belonged to the Cocheco Manufacturing Co., incorporated 1827. See ibid., 818–819 and *TIR*, chaps. 6 and 9.

[49] JM derived this information from White, *Slater*, 401–404.

[50] The data in this table come from "Commercial Statistics . . . for the year ending on the 30th of Sept. 1839 from official documents," published in Freeman Hunt, *The Merchants' Magazine, and Commercial Review* III (1840) 453. I am grateful to Helena Wright for locating this reference.

[51] For the Great Falls Manufacturing Company, incorporated in 1823 and operating mills at the Great Falls on the Salmon Falls River, Somersworth, New Hampshire, see Hurd, *Rockingham and Strafford Counties*, 683–685.

[52] JM's observations on Saco were based on his own experience of living there and managing the York Manufacturing Company's mills.

[53] For the Saco Water Power Company, see Gibb, *Saco-Lowell Shops*, 112–115.

[54] This was the Portland, Saco and Portsmouth Railroad Co., incorporated 1837, which opened in 1842.

VI *Miscellanies*

[1] Pressure gauges and indicator diagrams, used by Boulton and Watt from the 1790s onwards, measured and recorded the power developed by steam engines, thereby giving an approximation of the horse power consumed by the whole of a production line. No notion of how much power individual machines absorbed was possible by these methods,

however. The first measuring instrument with this capability was described and published by a British engineer, James White, in 1822. It incorporated a differential gear and was almost exactly the same as the dynamometer built in the USA by Samuel Batchelder fifteen years later. However, British millwrights and engineers seem to have taken longer than their American counterparts to adopt the differential dynamometer, judging by the evidence of the encyclopaedias. Perhaps Americans were faced with harder decisions about alternative processing machines. Perhaps British engineers, employing steam rather than water power, could simply increase boiler pressures to meet new workloads and so not bother about changes in the power requirements of individual machines—within a range much wider than that available to American engineers, who faced the more rigid limitations of water power sites.

Warren Colburn's instrument, developed at Lowell in the late 1820s, seems to have been the first dynamometer employed in American cotton mills. Possibly it was of the Prony brake type. JM's doubts about it, presumably relayed to him by Batchelder who managed a Lowell company at the time, seem to be confirmed by the discrepancies evident in the published readings obtained from its use.

Batchelder's dynamometer, developed sometime before September 1838 (when CM 1 went to the printer), may have derived from White's design or else from Aza Arnold's differential, patented for the American cotton roving frame in 1823. JM's thorough description shows that it was a transmission type of dynamometer, cumbersome but rugged and suited to the low speeds of textile machinery. It earned Batchelder a silver medal at the second exhibition of the Massachusetts Charitable Mechanic Association in September 1839 and gained wide acceptance in both America and Europe in the middle decades of the nineteenth century.

See Richard L. Hills and A. J. Pacey, "The Measurement of Power in Early Steam-driven Textile Mills," *Technology and Culture* 13 (1972) 25–43; James White, *A New Century of Inventions, Being Designs & Descriptions of One Hundred Machines, Relating to Arts, Manufactures & Domestic Life* (Manchester: Leech & Cheetham, 1822) 15–25, plates 1 and 3; sources in chapter 2 above, note 31; *The Second Exhibition of the Massachusetts Charitable Mechanic Association, at Quincy Hall, in the City of Boston, September 23, 1839* (Boston: Isaac R. Butts, 1839) 2–3; William R. Bagnall, "Samuel Batchelder," *Contributions of the Old Residents' Historical Association, Lowell, Mass.* 3 (1885) 209; David J. Jeremy, "Technological Diffusion: The Case of the Differential Gear," *Industrial Archaeology Review* 5 (1981) 217–227. Edward W. Constant, "Scientific Theory and Technological Testability: Science, Dynamometers, and Water Turbines in the Nineteenth Century," *Technology and Culture* 24 (1983) 183–198, focusses on the Prony brake which, publicised in the same year as White's differential dynamometer, was simpler but much more inaccurate than its rival. Constant seems to have overlooked the differential dynamometer.

[2] Lowell machinery prices moved downwards, especially after 1822. See Gibb, *Saco-Lowell Shops*, 47: TIR, 188, 209.

[3] Rhode Island machinery prices, originally closer to cost than Lowell or Waltham prices, fell more slowly in the 1820s and 1830s. See *TIR*, 209.

[4] Otis Pettee's machine shop at Newton Upper Falls, about fifteen miles southwest of Boston, became a major supplier of machinery, as well as mechanics and key operatives, for Mexican cotton factories between 1837 and 1853, when Pettee died. See Gibb, *Saco-Lowell Shops*, 161.

[5] For Britain's prohibitory laws and their repeal see David J. Jeremy, "Damming the Flood: British Government Efforts to Check the Outflow of Technicians and Machinery, 1780–1843," *Business History Review* 51 (1977) 1–34.

[6] The Providence Machine Co. emerged from the Providence Steam Manufacturing Co. and the Providence Iron Foundry. Among its founding members in the late 1820s were Samuel Slater, David Wilkinson, and members of the Dyer family. See Providence Iron Foundry Minutes, 1819–1832, Slater MSS, Baker Library, Harvard University.

[7] A machine shop at Andover, started in 1828 by two mechanics from Worcester, was bought by Charles Barnes, George H. Gilbert, and Parker Richardson in 1832. They moved it to North Andover and it eventually became (through partnership changes

between 1841 and 1851) the well known firm of Davis & Furber. Sarah L. Bailey, *Historical Sketches of Andover (Comprising the Present Towns of North Andover and Andover), Massachusetts* (Boston: Houghton, Mifflin & Co., 1880) 599.

[8] Ure, *Cotton Manufacture* 1: 312–313. These figures were not the lowest quoted in Manchester: see ibid. 2: 442–443 for machinery prices that were about 15 percent lower.

[9] White, *Slater*, 320.

[10] Ibid.

[11] Described in Ure, *Cotton Manufacture* 1: 296–304.

[12] JM's data come from the Newburyport Steam Cotton Co. mill which was driven by a 40 h.p. high pressure steam engine, built in 1835 by the Providence Steam Engine Co. See *Woodbury Report*, 52.

[13] The first of these two mills may have belonged to the Perry Manufacturing Co. which installed a 60 h.p. high pressure steam engine built by Holmes & Hinkley in 1835. The second steam engine, and possibly mill, appears to have been built after the Congressional survey of steam engines of 1838. See ibid., 85.

[14] This odd size for a mill derived from the spindleage of the Boston Manufacturing Co.'s second mill built at Waltham in 1816–1818. See William R. Bagnall, "Sketches of Manufacturing Establishments in New York City, and of Textile Establishments in the Eastern States," ed. Victor S. Clark (4 vols., typescript, Washington, D.C.: Carnegie Institution, 1908; available in microfiche from the Museum of American Textile History) 2025.

[15] Robert Brunton, *A Compendium of Mechanics, or Textbook for Engineers, Millwrights, Machinemakers, Founders, Smiths &c. Containing Practical Rules and Tables* (1st ed., Glasgow: John Niven, 1824; 6th ed., same place and publisher, 1834; copy of 7th ed. not traced in British Library or National Library of Scotland).

[16] The Shaws Water Joint Stock Co. (incorporated 1825 with a capital of £31,000) built a reservoir on the moors at the source of Shaw's Water, southwest of Greenock. Engineered by Robert Thom and opened in 1827, the system provided power through two canals which descended just over 500 feet from the reservoir at Overton to the Clyde, three miles away. In 1845 the mills on the power canals paid £2–4 per horse power per annum, for water power, compared to not less than £30 per horse power per annum for the cost of steam power in Glasgow.

See *New Statistical Account of Scotland* 7: 432–438.

[17] A point JM made in the Justitia controversy, in his letter to the *Boston Courier* of 27 March 1841.

[18] Presumably related to Samuel Batchelder, though he is not noticed in Bagnall, "Samuel Batchelder."

PART II
THE JUSTITIA CONTROVERSY

I. INTRODUCTION

1. *The Justitia Controversy of 1841*

While Montgomery's *Cotton Manufacture* occasioned no comment in the leading British periodicals of the day,[1] it produced a stormy debate in the Boston newspapers. Montgomery's chief critic, writing under the pseudonym of Justitia, accused him of deliberately falsifying his data, partly in order to promote the sale of Saco waterpower.[2] Another opponent seized Justitia's Nativist insinuations about Montgomery's immigrant background and denounced the *Cotton Manufacture* as a foreign plot, designed to awaken British statesmen to the threat of American cotton manufactures in world markets.[3] Eventually the debate centered on the relative costs of water and steam power in America. Montgomery stood his ground against Justitia's impugnments: his motivation in writing the book, he reiterated, was simply to meet the request of British friends who wanted facts to counter contradictory reports circulating through transatlantic grapevines.[4] Against

---

[1] The following British journals were checked for notice of Montgomery's *Cotton Manufacture: Edinburgh Review, Gentleman's Magazine,* and the *Quarterly Review.*
[2] Justitia, 24 February 1841 (dates cited are those of the date of publication of the newspaper issue carrying the article concerned).
[3] Justitia, 16 February 1841; anonymous letter, 29 March 1841.
[4] Montgomery, 27 March 1841.

the charge of duplicity, he stood by his data, dispassionately explaining and amplifying it.

In the light of this debate Montgomery prepared his second edition of the *Cotton Manufacture*, and for this reason alone the so-called Justitia controversy must be examined here. In addition, the Boston newspaper articles significantly extended Montgomery's treatment of alternative power costs in America, a topic of recurrent interest. Since Justitia's views only have been subsequently heeded, because Justitia published them as a separate pamphlet,[5] this editor decided to collect the documents from both sides of the controversy and to present the complete text of the major ones here. The debate triggered by the first edition of the *Cotton Manufacture* can best be understood by looking first at its participants and then at its issues, as found in the following articles.

LIST OF DOCUMENTS IN ORDER OF PUBLICATION

Author	*Journal or newspaper*	*Publication date*
Anonymous review of CM	*Monthly Chronicle*	October 1840
Anonymous review of CM	*Boston Daily Advertiser & Patriot*	5 December 1840
James Montgomery, letter	"	6 January 1841
Justitia, letter No. 1	*Boston Courier*	16 February 1841
" No. 2	"	24 February 1841
" No. 3	"	2 March 1841
" No. 4	"	9 March 1841
" No. 5	"	12 March 1841
" No. 6	"	16 March 1841
'B', "Steam & Water, No. 1"	*Boston Daily Advertiser & Patriot*	26 March 1841
James Montgomery, letter	*Boston Courier*	27 March 1841
Anonymous letter, "The Boott Mills & Steam Cotton Mills"	*Boston Daily Advertiser & Patriot*	29 March 1841
James Montgomery, letter	*Boston Courier*	30 March 1841
Justitia, "Rejoinder"	*Boston Courier*	13 April 1841
" Appendix No. 1	*Boston Daily Advertiser & Patriot*	26 April 1841
'B', "Steam & Water, No. 2."	"	27 April 1841
" No. 3"	"	30 April 1841
Justitia Appendix No. 2	"	11 May 1841

[5] Justitia, *Strictures on Montgomery on the Cotton Manufactures of Great Britain and America. Also, a Practical Comparison of the Cost of Steam and Water Power in America* (Newburyport, Massachusetts: Morse and Brewster, 1841).

(a) The participants

Apart from Montgomery, the participants in the controversy concealed their true identities. However, three of the other combatants in the Boston newspaper columns left enough evidence in their letters for informed readers to make educated guesses about their authors.

Justitia, Montgomery's most heated and prolific antagonist, was almost certainly Charles Tillinghast James (1805–1862), engineer and pioneer of steam power in American textile manufacturing.[6] The attribution has been made before, though the grounds for it have never been demonstrated from the evidence of the articles themselves.[7] Certainly the experience and convictions of James closely matched the knowledge and polemics of Justitia. James, a native of Rhode Island, started as a carpenter, learned practical mechanics and then moved into mill engineering in the small, highly competitive Rhode Island mills and machine shops. Several steam-powered cotton mills started in New England in the late 1820s, most of them in Rhode Island where by 1832 four were under construction or in operation.[8] One was Samuel Slater's Steam Cotton Manufacturing Co. mill at Providence, set up in 1828, which used mules and powerlooms to make a relatively fine quality of cotton goods: $^7/_8$ and $^4/_4$ Sea Island shirtings made from No. 45 yarn.[9] Slater hired James to re-equip the mill in 1834 and this experience converted James to the cause of steam-driven factories. As James recalled in 1850, his next major assignment took him to the decaying seaport of Newburyport, Massachusetts, where a cotton mill was built to revive the town's economy. James installed machinery in this mill, Bartlett Steam Mill No. 1, which

---

[6] For James see *DAB*; Lozier, "Taunton and Mason," 375–388; Thomas R. Winpenny, "The Engineer as Promoter: Charles Tillinghast James and the Gospel of Steam Cotton Mills" *Pennsylvania Magazine of History and Biography* CV (1981) 166–181, concentrates on the impact of steam-powered cotton mills (notably the Conestoga Steam Mills), most promoted by James, in Lancaster, Pennsylvania, but misses James's involvement in the Justitia controversy. Further work is promised by Betsy Woodman and Elaine Tucker.

[7] For example the *National Union Catalog*; Peter Temin, "Steam and Water-power in the Early Nineteenth Century," *Journal of Economic History* XXVI (1966) 197; von Tunzelmann, *Steam Power*, 267n; Jeremy Atack, Fred Bateman, and Thomas Weiss, "The Regional Diffusion and Adoption of the Steam Engine in American Manufacturing," *Journal of Economic History* XL (1980) 289.

[8] *McLane Report* I, 951, 966–967, 970–973.
Carroll W. Pursell, Jr., *Early Stationary Steam Engines in America. A Study in the Migration of a Technology* (Washington, D.C.: Smithsonian Institution Press, 1969) 84–85.

[9] *McLane Report* I, 951. Samuel Slater considered that steam power was appropriate in New England only "under peculiar circumstances of location and business." Ibid., 927.

started in 1839; then he became company agent and the following year Mill No. 2 was built, "planned, constructed, and started by me, and run under my direction for a length of time," he recorded.[10] Justitia cited the Rhode Island steam mills and used the two Newburyport mills as his case studies for steam power; he also claimed fifteen years' work in textile engineering, "half in water, half in steam driven cotton mills."[11]

At the time of the controversy James had good reason to be proud of his No. 1 Newburyport steam mill. Its goods had won a silver medal in an industrial exhibition in Boston, in September 1839, and the citation read,

> This is an establishment lately erected upon the principle of the best Manchester Mills, the first of any importance that has been started in the United States, and must soon lead to correct estimates of the advantages of steam over water power. The goods here exhibited are of a very superior order, remarkably even and closely wove; and altogether of a better fabric than has ever been before produced in this country. They are in all respects equal to any British fabric of the kind that the committee has ever seen.[12]

Two years later goods from Bartlett Mill No. 2 took the gold medal.[13] One important argument that Justitia made for steam power was that it "is indispensable to the manufacture of a fine article of a superior quality, which is not the case with coarse goods."[14]

During the 1840s and 1850s James became a highly articulate consultant mill engineer, advocating in lectures and publications, and promoting through his professional capacities, the application of steam power to manufacturing. Because he urged Southerners to invest in steam cotton mills, he earned the enmity of northern New England's powerful textile capitalists who feared the emergence of rivals in the South.[15] Whether this later clash between James, the Rhode Islander and proponent of steam on the one hand, and the spokesmen for the northern New England cor-

---

[10] James, *Letters on the Culture and Manufacture of Cotton* (New York: George W. Wood, 1850) 15–16; idem, *Lecture on the Comparative Cost of Steam and Water Power, Delivered at Hartford, Connecticut, February 1844* (Newburyport, Massachusetts: Morse and Brewster, 1844), 25.

[11] Justitia, 11 May 1841; also 24 February, 12, 16 March, 13 April.

[12] James, *Letters*, 22.

[13] Ibid.

[14] Justitia, 2 March 1841.

[15] James, *Letters*; *DAB*.

porations, who stood for large-scale interests and reliance on water power, on the other, began before the *Cotton Manufacture* was published is unknown. Clearly Montgomery's book brought the arguments over water and steam power into the open for the first time, at a time when steam engines were being installed in the mills following the Waltham-Lowell system of manufacturing, though in far fewer numbers than in the Rhode Island system. While Justitia justified steam power largely on economic grounds, he also perceived a political dimension, complaining of the "power of corporations exercised through their agents."[16] Thus James's work in steam-powered cotton mills in Rhode Island and New-buryport and the public disputes he had with northern New England capitalists in the late 1840s dovetail with the experience and views of Justitia.

The identities of two other participants in the controversy may be guessed from even more slender evidence. "B" took Montgomery's side and defended him against Justitia's defamations.[17] "B" also regarded water power as more appropriate than steam for American manufacturers.[18] He was familiar with contemporary British and American technical authors, like Baines, Ure, Cleland, Brunton, Grier, and Allen, as well as the British Factory Commissioners' Returns.[19] He had experience as a practical manufacturer and knew about the difficulties in calculating manufacturing and power costs.[20] He was acquainted with the results of the dynamometer experiments at Saco, which unlike those at Lowell had not been published, for measuring the power requirements of dead throstle spindles.[21] And he rejected Justitia's charge of a conspiracy among the textile corporations.[22] All these clues point to Samuel Batchelder, Montgomery's superior in the York Manufacturing Co. at Saco, as being "B." Batchelder had been agent of the Hamilton Manufacturing Co. at Lowell before moving to Saco in 1834; he had invented a dynamometer which was used at Saco; and, as seen in the biography of Montgomery, he knew about the British cotton industry through the efforts of Treadwell and Montgomery.[23] But the evidence is only circumstantial.

[16] Justitia, 26 April 1841.
[17] B, 27 April 1841.
[18] B, 26 March, 27, 30 April 1841.
[19] B, 27, 30 April 1841.
[20] B, 26 March, 27, 30 April 1841.
[21] B, 27 April 1841.
[22] B, 30 April 1841.
[23] For Batchelder see *DAB*; Bagnall, "Samuel Batchelder"; and this editor's biography of Montgomery at the beginning of this edition.

The anonymous author of the letter dated 29 March 1841 obviously knew the Boott mills at Lowell better than Justitia and
quite likely he was the agent, Benjamin F. French,[24] or one of the
proprietors, Amos and Abbott Lawrence[25]: certainly it would have
been very appropriate for a company spokesman to correct any
errors of fact in Justitia's letters,[26] which treated the Boott mills
as an exemplar of water-powered manufacturing. If this fourth
critic were one of the Lawrences it would not be the last time that
James would clash with the family: in 1849–1850 James's calls to
Southern capitalists to take up steam-powered cotton manufacturing were countered by Amos A. Lawrence, son of Amos Lawrence.[27]

(b) The issues

The first American review of the *Cotton Manufacture*, which
consisted chiefly of a long summary of the book, levelled three
criticisms at Montgomery's comparisons of British and American
manufacturing costs. Montgomery's capital costs were thought to
be too low, by 5 percent in England and 6 percent in the USA
(prevailing interest rates, which had to be added to insurance and
wear and tear). His figures for British mill construction were regarded as excessively small. And his power costs apparently neglected charges for water power in the USA and for coal in
Britain.[28] After this review was reported in the Boston papers,
Montgomery wrote a reply in which he addressed only the question of comparative construction costs.[29] Justitia then entered the
debate. He questioned Montgomery's low figures for British mill
construction costs and launched the topic of power costs, which
drew other participants into the controversy.[30] Two major issues
therefore emerged. Because, the debate over power costs soon

[24] Cowley, *Illustrated History of Lowell* 56.
[25] *DAB*.
[26] Justitia, 26 April 1841, claimed that the Boott management, not surprisingly, had refused
him information on its mills' power costs. What data he used largely came from the
published annual *Statistics of Lowell*.
[27] See Thomas H. O'Connor, *Lords of the Loom: the Cotton Whigs and the Coming of the Civil
War* (New York: Charles Scribner's Sons, 1968) 68–69; James, *Practical Hints on the
Comparative Cost and Productiveness of the Culture of Cotton, and the Cost and Productiveness
of Its Manufacture. Addressed to the Cotton Planters and Capitalists of the South* (Providence,
Rhode Island: Joseph Knowles, 1849); idem, *Letters*.
[28] *Monthly Chronicle* (October 1840) 397.
[29] JM, 6 January 1841.
[30] Justitia, 16, 24 February 1841.

eclipsed that on construction costs, the latter can be quickly surveyed.

Montgomery marshalled several factors to explain his low figures for the building of a mill in Glasgow: foundations shallower than in America, permitted by less severe winters; cheaper prices for building materials; lower bricklayers' wages; and the minimal use of stone.[31] Justitia spotted discrepancies in Montgomery's calculations and cast doubt on the actuality of the mills cited: would an American factory with nearly 30,000 square feet of floor space really cost six times as much as a British mill with nearly 18,000 square feet?[32] Montgomery retorted that the Scottish factory was in fact larger than the *Cotton Manufacture* indicated (for it also had an attic) and that much space was lost in the American mill to water wheels, belting, and a less compact layout of machinery. But both mills existed and were running in 1841: Montgomery had the plans and specifications of the British one, built in 1834.[33] Justitia refused to accept that the Glasgow construction costs were as low as Montgomery claimed. On the authority of a "gentleman" who had superintended mill building in Manchester, Justitia believed that although stone, lime, hardware, and labor were cheaper in Manchester than in the USA, bricks and lumber were much dearer.[34] And there the matter stood insofar as public controversy was concerned. Both Montgomery and Justitia's informant may have been right about British brick prices: brick production, and therefore prices, displayed wide regional variations in the early nineteenth century, with building booms in Lancashire in the early 1830s peaking in 1836. It all depended on which dates and places were being compared.[35]

The bigger debate on power costs dealt with two subjects, Anglo-American power costs and, secondly, relative water and steam power costs in the USA. The first of these topics, started by the *Monthly Chronicle* of October 1840, raised three questions. How much did steam power cost in Britain and in the USA? Why did one horse power of energy drive less capital equipment in the USA than in Britain? And, what was the cost of water power in Britain? The first question received little attention because Justitia

---

[31] JM, 6 January 1841.
[32] Justitia, 16 February 1841.
[33] JM, 27 March 1841.
[34] Justitia, 13 April 1841.
[35] H. A. Shannon, "Bricks—A Trade Index, 1785–1849," *Economica* new ser. I (1934) 300–318; Matthews, *Study*, 113–118; A. K. Cairncross and B. Weber, "Fluctuations in Building in Great Britain, 1785–1849," *Economic History Review* 2nd ser. IX (1956) 283–297.

entered the lists and shifted the focus to steam versus water power in the USA. And the American disputants could only quote the outdated estimates of James Cleland, Zachariah Allen, Edward Baines, Jr., and Peter Barlow on British steam costs.[36]

That America's throstle spindles needed more power than Britain's throstle or mule spindles was commonly known among the technically informed. The debate over the *Cotton Manufacture* considered the latest estimates for current American spindle designs. Both Montgomery and "B" referred to experiments at Lowell and Saco (where an improved dynamometer was used) which showed that one horse power drove only 77 to 80 dead throstle spindles (the kind favoured in northern New England), compared to estimates of 100–180 live throstle spindles or 500 mule spindles per horse power.[37] However, Justitia, who ran mule spindles, persisted in attributing 20 percent more dead throstle spindles per horse power than his opponents' experiments warranted.[38]

Montgomery deflected the question of water power costs in Britain by arguing that, because ancient mill seats were originally purchased as land rather than as water power sites, rents would not reflect the true value of water power.[39] It simply was not true that all British water power sites were purchased in ignorance of their power potential, but no one chose to challenge Montgomery's judgement.[40] Far more important to American manufacturers, of course, was the question of relative water and steam power costs in the USA.

Montgomery freely admitted that in writing the *Cotton Manufacture* he had been unable to obtain "correct estimates of the average cost" of steam power in the USA.[41] His invitation to supply information was at once taken up by Justitia, in six letters to the *Boston Courier* in February and March 1841. Justitia's general arguments for steam mills sounded persuasive. Located in seaports, they incurred lower transportation costs; lower fixed capital costs, since they might be installed in existing buildings; and lower labour costs, given human inertia.[42] In addition they were well suited

---

[36] Justitia, 24 February 1841; for engineering authorities: Justitia, 13 April, 11 May 1841.
[37] JM, 30 March 1841; B, 27 April 1841.
[38] Justitia, 9 March, 13 April.
[39] JM, 27 March 1841.
[40] In Scotland alone for example, nearly two dozen out of 64 firms engaged in cotton manufacturing in the early 1830s employed water wheels, sometimes assisted by steam engines. See GB, *PP (Commons)* 1834 (167) XX, Section A.1.
[41] JM, 6 January 1841.
[42] Justitia, 2 March 1841.

to produce higher quality goods, because of the more regular mechanical motion of the power source, the higher humidity of the steam mill, the lower power requirements of fine yarn spinning, and the lower *ad valorem* transport costs of fine goods.[43] The profitable operation of such mills in Rhode Island and Newburyport, Massachusetts, whose Bartlett mills Justitia held up as models of successful steam power manufacturing, seemed irrefutable though his arithmetic was exceedingly careless (a defect which he shared with his opponents at times).[44] So too did his comparison of the Bartlett steam mills with the water-powered Boott Mills which had recently been started at Lowell.[45]

Montgomery, prudently perhaps, avoided any comment on this facet of the power question, until he came to revise the *Cotton Manufacture*. "B," however, subjected Justitia's arguments and figures to a close scrutiny, exposing serious flaws which were widened by the anonymous spokesman for Lowell. Firstly, Justitia had made no allowance for transportation costs for raw cotton, oil, starch, and manufactured goods between Boston and Newburyport.[46] "B" also reckoned that the Bartlett steam engines' fuel consumption costs were too low: the 3 lb. of coal per horse power and the 11 cents per horse power per day allowed by Justitia, should have been more like 10 lb. of coal per horse power and 50 cents per horse power per day at least, even allowing for improvements in steam engine design (such as high pressure boilers) and the American use of anthracite coal.[47] And if the Bartlett mills had been driving dead spindle throstles, such as the Boott mills ran, they would have required not the 25 per cent more power Justitia allowed, but at least 500 per cent more power (on the basis of the Lowell-Saco experiments). When an allowance of 8 looms per horse power was made, a 17,000 dead throstle spindle mill would need 250 horse power, not the 175 horse power of the two Bartlett engines combined.[48]

---

[43] Ibid. In fact a really smooth-running and economical high pressure steam engine was yet to be developed. Not until George H. Corliss had produced his automatic variable cutoff engine in the late 1840s did steam power really gain popularity in the textile districts. See Louis C. Hunter, *A History of Industrial Power in the United States, 1780–1930. Volume Two: Steam-Power* (Charlottesville: University Press of Virginia and Hagley Museum and Library, 1985) 251–253.

[44] Justitia, 24 February, 12 March.

[45] Justitia, 9, 12, 16 March 1841.

[46] Justitia, 12 March 1841.

[47] B, 27 April 1841.

[48] B, 27, 30 April 1841.

The margin of the advantage that Justitia claimed for his steam-powered mills narrowed and disappeared when his critics showed that the water-powered Boott mills' costs were lower than he had stated. His estimate for the cost of foundations was halved; that for the amount of coal for warming the mills was more than halved.[49] And the anonymous author speaking for the Boott interest showed that Justitia had made three more "egregious errors": over the Lowell method of selling water power; over the weight of cotton actually manufactured by the Boott mills; and over railroad freight rates.[50] Furthermore, "B" asked, if steam power really was so economical why had the city waterworks in Philadelphia, where Lehigh coal was much nearer and cheaper than in New England, switched from steam to water power?[51] None of these devastating criticisms deterred Justitia who stuck to the cause of steam power, quoting the evidence on his side of the argument: his engines and mills, his experience and his authorities, who were British.[52]

2. The Continuing Debate Over Water and Steam Power Costs in Nineteenth Century America

The data provided by both sides of the Justitia debate now allow the computation of another measure of steam and water power costs in the two specific locations, Newburyport and Lowell, Massachusetts in 1841. While a reading of the documents shows that prejudice surfaced more easily with Justitia than with his opponents, Justitia's data must be accepted for the steam power costs in the mills he managed. On the other side, we can now use the contributions of "B" and the Boott spokesman as well as those of Montgomery to calculate water power costs.

Justitia made a mistake by failing to calculate power costs separately from total manufacturing costs. By expressing the cost of power per yard of goods he fell into several accounting errors. He totally ignored labour costs, which immediately falsified his result for the cost of manufacturing cotton goods. His ratios of

[49] B, 30 April 1841.
[50] Anonymous letter, 29 March 1841.
[51] B, 30 April 1841.
[52] Justitia, 26 April, 11 May 1841.

manufacturing capacity to power for the various spindle types were inaccurate. And any savings he actually made and then attributed to the use of steam might in fact have derived from other savings in labour or capital costs. In his last letter to the Press, ironically, he acknowledged the importance of calculating the cost of power alone, regardless of how it was applied. If we take power costs alone, add transportation charges to Justitia's steam mill figures, regard Lowell water power charges as an operating rather than a capital cost, and make the other adjustments claimed by Justitia's opponents then a rather different picture emerges as the following two tables show.

WATER POWER COSTS AT LOWELL
(BOOTT MILLS 1–4), 1841

Capital costs

Waterwheels, gearing &c.		$ 85,000
Foundation		20,000
		105,000
	Interest @ 6%	6,300

Operating costs, per annum

Interest on capital	6,300
Water power for 29,248 spindles (@ $12 per h.p., p.a.)	5,341
Coal for heating (288 tons @ $6)	1,728
Transport of coal, cotton, oil, starch, goods (@ $1 to $1.50 a ton)	1,660
	$ 15,028

Horse power generated

At Lowell one "mill" of 3,584 spindles was reckoned as equivalent to 54.54 horse power. Therefore 29,248 spindles needed 445 horse power.

Cost per horse power

This can be calculated by dividing operating costs per annum by the number of horse power needed to keep the mill running:

$$\frac{15,028}{445} = \$33.77 \text{ per h.p., p.a.}$$

Sources: *CMI*, 215–216; [above, pp 213–215]; "B", Letter No. 1, 26 March 1841

STEAM POWER COSTS AT NEWBURYPORT
(BARTLETT MILLS 1–2), 1841

Capital costs

Total steam engines (169 h.p. combined)	$25,000
Foundations for engines	3,000
	28,000
Interest @ 6%	1,680

Operating costs

Interest on capital	1,680
Coal	6,045
Wages for engineer and firemen	1,137
	8,862

Horse power generated

At average pressure the two engines combined reached 169 h.p.

Cost per horse power

By formula previously used: $\dfrac{8,862}{169} = \$52$ per h.p., p.a.

Sources: Justitia, No. 5, 12 March 1841 and Appendix No. 2, 11 May 1841.

These figures confirm Temin's finding that capital costs formed a higher proportion of overall costs in water than in steam driven mills. But while the figure for steam costs (the same as Temin's) falls just below the wide range specified in Atack's simulation model for New England power costs, that for water power costs is just over half of the lower level of Atack's narrower simulated range. And at $34 per horse power per annum it is 30 percent lower than Temin's figure. The omission of depreciation and insurance, much higher for steam engines than for water-wheels, would have raised steam costs but these in turn would have been offset by the use of larger engines. Clearly water power costs at Lowell have been seriously overestimated, largely because a rate of $4 per spindle, which included rent on four acres of land as well as water power, was quoted by Justitia and has been repeated

[53] Temin, "Steam and Water Power," 197.
[54] Jeremy Atack, "Fact in Fiction? The Relative Costs of Steam and Water Power: a Simulation Approach," *Explorations in Economic History* XVI (1979) 409–437; Atack et al., "Regional Diffusion," 281–308.

uncritically ever since. When he came to prepare the second edition of his *Cotton Manufacture*, Montgomery assembled costs that were nearly double these present ones for steam power at Newburyport and just under half the present ones for water power at Lowell. But without knowing the method by which he reached these figures, it is impossible to tell how much prejudice was built into his cost accounting.

II. DOCUMENTS IN THE JUSTITIA CONTROVERSY

1. LIST OF DOCUMENTS IN THE SEQUENCE OF WRITING AND ARGUMENT

American reviews of the Practical Detail and Montgomery's response

Anonymous review	*Monthly Chronicle*	October 1840
Summary of this review	*BDAP*	5 December 1840
Montgomery's response	*BDAP*	6 January 1841

Justitia's letters attacking the Practical Detail and Montgomery's responses

Number 1	*BC*	16 February 1841
2	*BC*	24 February 1841
3	*BC*	2 March 1841
4	*BC*	9 March 1841
5	*BC*	12 March 1841
6	*BC*	16 March 1841
Montgomery's response, No. 1	*BC*	27 March 1841
" No. 2	*BC*	30 March 1841
Justitia, Number 7	*BC*	13 April 1841

Attacks on Justitia's arguments in favour of steam power and Justitia's responses

"B," Letter No. 1	*BDAP*	26 March 1841
Anonymous letter	*BDAP*	29 March 1841
Justitia's reply (*Strictures*, Appendix 1)	*BDAP*	26 April 1841
"B," Letter No. 2	*BDAP*	27 April 1841
" No. 3	*BDAP*	30 April 1841
Justitia's last letter (*Strictures*, Appendix 2)	*BDAP*	11 May 1841

Justitia published his nine letters as a pamphlet, *Strictures on Montgomery on the Cotton Manufactures of Great Britain and America, also, A Practical Comparison of the Cost of Steam and Water Power in America* (Newburyport, Mass.: Morse and Brewster, 1841) 75.

2. DOCUMENTS

Review of James Montgomery's *Cotton Manufacture* in the *Boston Daily Advertiser and Patriot*, 5 December 1840

THE COTTON MANUFACTURE OF THE UNITED STATES, COMPARED WITH THAT OF GREAT BRITAIN.

The last number of the Monthly Chronicle contains an analysis of an important work lately published at Glasgow, by "James Montgomery, Superintendent of the York Factory at Saco, Maine." The author is an Englishman, and is well acquainted with the state of the cotton manufacture in Great Britain, by many years practical acquaintance with the business and he has also become acquainted with the state of the manufacture in this country by four years experience and observation, under the most favorable circumstances. His work is therefore full of facts of great interest, and which it is important for the manufacturers of this country to know. The article in the Monthly Chronicle, to which we refer, gives an analysis of the more important facts, with tables abridged from the work, giving the comparative prices of each description of machinery, and each branch of labor. It is too long for publication in a newspaper, but we here give the general recapitulation, of the estimates there presented, showing the cost and produce of an American and an English Cotton Mill of the description there given, with the general remarks of the Chronicle, which we recommend to the attention of those who are interested in this subject.

[Then came a summary of the review printed in the *Monthly Chronicle* (October 1840) 385–399. It included a verbatim reprint of the critical sections of the *Monthly Chronicle* review, as follows.]

It will be observed that the foregoing estimate embraces no computation of the cost of working power, or of the comparative expense of steam and water power. The estimate of $7\frac{1}{2}$ per cent. for the wear and tear of machinery and buildings also, if intended to embrace besides wear, the interest on capital, seems to be inadequate, and perhaps hardly more than sufficient to cover the charge of wear and repairs, in which case a further allowance of 6 per cent. in this country, and 5 in England, should be made for interest.

We can hardly imagine however that there is not some error in the estimate of the comparative cost of buildings in the two countries. The sum of $4,608, appears to be a very small sum for the cost of a building of the dimensions described, and we can hardly suppose it to be adequate to the erection of such a building in a style of strength and durability, bearing any comparison with buildings used for the same purpose in this country; since the difference of cost stated is evidently much greater than can be accounted for from the greater cheapness of labor and materials in Great Britain.

The cost of water wheels and gearing, and also of a steam engine, is given in the estimate above quoted, but no estimate is made of the cost of water power, or of coals for producing steam. These are important items, and the question of their comparative cost is one of considerable interest. We regret that it is overlooked in this work.

It will be observed that these computations apply only to those branches of the cotton manufacture, to which the advantages of mechanical power can be applied with the greatest effect, and which the experience of our countrymen has enabled them to prosecute most successfully, and not at all to the finer and more complicated manufactures, which require the application of a greater amount of manual labor, and of skill which is the result of continued experience. Yet this result presents a most encouraging view of the prospects of this important manufacture in this country. It shows that in the manufacture of those descriptions of cloth for which there is much the greatest demand, the mills of this country are at this moment able to sustain a competition with the most favorably situated establishments of Great Britain, and to supply the articles produced at equal prices in markets foreign to both countries. But this is not all. The whole comparison shows that those items in the computation which produce a result in our favor, are of a permanent character, secured to us by our national position, and of which we cannot be deprived; while a portion of those, in which we labor under a disadvantage, will change in process of time, and render the general result of the comparison still more favorable to us. This remark will apply particularly to deficiencies in economical management, deficiencies in certain portions of the machinery, and the want of equal experience and skill in a portion of the hands employed.

These considerations must relieve those who are interested in the American cotton manufacture, from any serious apprehensions of permanent decline of this branch of industry in this coun-

try. An ample pledge, for the continuance of a demand for the
products of this manufacture, is to be found in the universal want
of the civilized world, of a material which can be in no other way
so cheaply supplied. If this material could be furnished by foreign
laborers on better terms than by our own, the American manufacturer might well feel, that he held his command over the market, even of his own country, by an uncertain tenure. But if it be
proved that the advantages of his position are such, that notwithstanding the materially higher cost of labor in this country than
in Europe, he can still furnish his products at a price at which he
cannot be underbid, even in a foreign market, by the manufacturers of any other country, his position is as safe and independent, as in the nature of things, the emoluments of any occupation
can be.

Letter of James Montgomery dated 16 December 1840, printed
in the *Boston Daily Advertiser and Patriot,* 6 January 1841.

To the Editor of the Daily Advertiser

The Monthly Chronicle for October last, contains a review of
a work lately published by me on the cotton manufacture of the
United States and Great Britain; from which you published extracts in the Semi—Weekly and Daily Advertisers of the 5th ult.
To some of your remarks in the article referred to, I deem it
incumbent on me to give the following explanation.

After copying the estimates (given in the work,) of the cost of
buildings, machinery, &c. for a cotton factory in this country and
in Britain, together with the amount of wages paid per fortnight,
and the estimated quantity of cloth produced in the same time,
from each mill respectively; from which are deduced comparative
estimates of the cost of manufacturing in each country,—you then
proceed to remark:

"We can hardly imagine however, that there is not some error
in the estimates of the comparative cost of buildings in the two
countries. The sum of $4,608 appears to be a very small sum for
the cost of a building of the dimensions described, and we can
hardly suppose it to be adequate to the erection of such a building
in a style of strength and durability, bearing any comparison with
buildings used for the same purpose in this country, since the
difference of the cost stated is evidently much greater than can
be accounted for from the greater cheapness of labor and materials in Great Britain."

In reply to the above it may be stated, that the estimated cost of building a cotton factory in Great Britain of the dimensions described in the work referred to, was originally made out from specifications—now in my possession—of a mill erected in Glasgow in 1834. The same estimates were also published by me in another work, entitled "The Theory and Practice of Cotton Spinning,"[1] which though reviewed in various periodicals in Glasgow, Edinburgh and Manchester, no objections were ever made to the estimated cost of buildings. Previous to their first publication they were laid before a master builder in Glasgow, who considered them as correct as could be made out in theory, and that the sum of £960–$4,608, was amply sufficient for building a factory complete, according to the plan and dimensions specified.

In order to ensure the utmost accuracy before publishing them a second time, that part of the manuscript was submitted to the inspection and correction of gentlemen in this country, whose practical experience rendered them perfectly qualified to detect any error in the estimated cost of building a factory in America. They were also submitted to the examination of several manufacturers in Glasgow, whose practical knowledge eminently fitted them for correcting any inaccuracy in the estimates of building a factory in that country. If therefore they are not correct, I know not what other means could have been selected for having them so.

One principal item in the cost of building a factory in this country, arises from the very deep and expensive foundations necessarily required on account of the severity of frost in winter; this however, is an expense that is almost entirely saved in Glasgow, where I have never known the frost to penetrate twelve inches below the surface; and taking the country as a whole, the average *greatest* range of the temperature is from 84° to 8°, and it is very rare indeed, that frost continues more than four weeks at any one time without a change, and equally rare for the cold in winter to descend as low as 20 fahrenheit. So that two feet in depth for a stone foundation, is perfectly sufficient for a mill of five stories in Glasgow. Nor is any centre wall required to sustain the centre pillars, as a flat stone laid on the surface of the earth below each pillar, is quite sufficient to support a column of five stories. And for a mill of that height, the walls of the first story might be 22 inches thick, the second, third and fourth, 18 inches, and the fifth only 14 inches. The cost of such brick work in Glasgow is about £5.45–$27.60 per rood of 272 square feet 14 inches thick, including materials of good quality, and best workmanship, the walls faced with hard brick on the outside, and finished in the best style; while the contractor engages to remove all the rubbish from the premises, together with the earth thrown out from the foundations. The above is equal to 101.4 cents per square foot, 14 inches thick. Brick work in Lowell of the same style, costs $10 per 1,000 bricks, or about 40 feet square one foot thick, equal to about 25 cents per foot; making a difference of 14¾ cents per foot in favor of Glasgow. In regard to the difference in the cost of foundations, I am informed, that the foundations of a mill in this country, driven by water, of that part of the building below the level of the ground, would cost on an average, $5,000; while the foundations for a mill of the same dimensions in Glasgow, I should think could not exceed $500, thus making a difference in favor of the latter of $4,500.

Lime is obtained within a few miles of the city, and sold by the load or chaldron, at a much cheaper rate than in this country, while bricks are made in the suburbs.

The cost of plastering	in Glasgow	in Lowell
For one coat,	4½ cents	10 cents
" two coats,	5½ "	12½ "
" three coats	7 "	

Cost of slater work per
   rood of 272 sq. ft $16.08

Cost of slater work per 1,000 brick, or 40 sq. ft. $10 or about 6 cents per square foot in Glasgow, and 25 cents in Lowell. Thus in plastering work there is a difference of over 50 per cent., 75 per cent. in slater work in favor of Glasgow. Besides this difference of slater work, the contractor in Glasgow binds himself to uphold the roof in good order for two years, and at the end of that time leave it in a perfect state of repair.

The above prices paid in Glasgow for brick, plaster and slate work, include materials of the first quality, and best workmanship. They are copied from documents now in my possession as the ordinary average rates paid for such work when it was in a prosperous state.

I have no means at present of comparing carpenters' and joiners' works, nor of the cost of stone work; but in regard to the latter it may be mentioned, that except for foundations to steam engines, very little stone work is required for a Mill in Glasgow, and what is used is entirely free stone; in squaring and finishing which, one man can do as much work as twelve men upon granite.

Your remarks in regard to the comparative cost of steam and

water power in Britain, shall not be forgotten; but I regret that I have never been able to obtain correct estimates of the average cost of steam power in America. It is presumed that the calculations of the cost of water power at Manayunk and Lowell—as given in the work referred to—are correct, while water privileges at the same cost as sold at Lowell, are now offered for sale by the Saco Water Power Company. But it would certainly be interesting to manufacturers generally, if a correct comparison could be given, as to the actual cost of steam and water power in this country. And if any of your readers or correspondents will furnish the requisite estimates of the average cost of steam power, I will take proper means to ascertain the actual average cost of steam and water power in Britain.

J. MONTGOMERY.

Saco, Dec. 16th.

Justitia, No. 1   *Boston Courier,* 16 February 1841

THERE has recently appeared in the United States a work which claims as its author "JAMES MONTGOMERY, *Superintendent of York Factories, Saco, in the State of Maine.*" This book purports, on its title-page, to contain "A Practical Detail of the Cotton Manufacture of the United States, and the state of the Cotton Manufacture of that country, contrasted and compared with that of Great-Britain; with Comparative Estimates of the Cost of Manufacturing in both Countries." Mr. Montgomery is a foreigner; but, as that fact should not in the least degree prejudice his work in this country, in the minds of any American citizen, so neither should it, from any feeling of mistaken courtesy, exempt it from impartial criticism, or its errors, whatever they may be, from detection and exposure.

The field which the author of the above-named work has undertaken to explore, is one of vast magnitude and extent, and presents a very numerous assemblage of details and facts, all of them of importance to the friends of American Manufactures; and particularly so to those engaged in the business; as their success depends very much on the facilities with which it can be most productively managed, at the smallest expense. To collect these facts and details, and to arrange and compare them in a proper manner, so as to arrive at a correct result, requires a mind of no ordinary intellect, accompanied with an uncompromising spirit of candor and impartiality, and entirely unembarrassed by prejudices growing out of national pride, or personal interest. How far Mr. Montgomery has shown himself competent to the task he has undertaken, I propose to show by a brief review of some few of the statements and estimates contained in his work.

The writer is fully aware of the prevailing disposition to excuse all errors in publications, because it should not be expected that one man should know everything. As relates to trifling errors on unimportant subjects, such an apology may be well enough. But on a subject of such vital importance to the Cotton Manufacture of the United States, and the comparison and contrast between that and the Cotton Manufacture of Great Britain, the public mind should not be misled; and no man should undertake to spread the facts in the case before the world, unless fully qualified to state every point of the least importance, without resort to hypothetical estimates, or bare supposition. If not thus qualified, the public have the right to hold him to a rigid responsibility for all his errors. Any man who may undertake such a work, is supposed to know all about it. If he misleads the mind of the reader, either from ignorance or design, the effect is the same; and no sufficient apology can be pleaded in justification; and more especially when facts are within his reach, of which he might avail himself if he would.

In Mr. Montgomery's work, which he has deemed of sufficient importance to cause it to be transported across the Atlantic for publication, there is doubtless embodied a considerable amount of useful information. But it is so intimately blended with matter that is entirely incorrect and deceptive in itself, as to be directly calculated to mislead the mind of any one not intimately acquainted with the subject; and no one who can detect the errors and misrepresentations can place implicit confidence in other statements, that rest entirely on the responsibility of the author. Such a work is but a poor guide for the uninformed to rely on, and can be of small service to others.

The writer has no feeling of hostility to Mr. Montgomery, nor would he, in any case, knowingly withhold the need of praise from any one to whom it may be justly due. But from careful perusal of the work above alluded to, he is perfectly satisfied, and further than that, he knows, and can make it appear, that there are departures from truth in its pages, in important particulars, which

ought to have been known to the author, and an exposition of
which goes to demolish some of his labored estimates, in which
more than one New-England manufacturer is deeply interested,
if Mr. Montgomery himself be not. It is difficult, even with a full
measure of charity, to understand how these errors should have
originated in ignorance of the facts; or to avoid the suspicion that
something besides public good inspired the pen that indited them.
If, however, to enlighten the public mind were the original intention, it has signally failed of its object, at least, in some important particulars, and those of a character, information on which
was at hand, and which might have been had without difficulty.
I proceed to point out a few of them.

At page 114 [above, p. 136] commences a table of what purports
to be "comparative estimates" of the cost of building factories,
machinery, &c. in Great-Britain and America.—In order to sustain
the statements I have made, let us take the first estimates in that
table, comprised under the head of "Building and Gearing." It
will strike every one, that among the multiplicity of factories in
the two countries, one, two, or more, in actual operation, might
have been selected in each country, nearly similar, and the true
cost of which might have been readily ascertained. But demonstration, by means of known facts, does not appear, in this instance, at least, to have suited the purpose of the author. On the
contrary, he *supposes* two cases—one, a factory in America, 142
feet long, by 42 feet wide, four stories high, with an attic; the
other in Great-Britain, 90 feet long, by 38 feet wide, five stories
high, flat roof and no attic. What supposition is this to begin with,
for a "comparative estimate" of cost! The American *supposed* factory is to contain, on its five floors, 29,820 square feet, while the
British factory, on the same number of floors, (for it has no attic)
is to contain only 17,720 square feet, a difference of 12,720 square
feet; precisely as though, in Great-Britain, this difference in dimensions and superficial contents would occasion no difference
in cost.

Still more unfairly, Mr. Montgomery has supposed, in Great
Britain, a steam-mill, and in America, a water-mill, precisely as
though there were no water-mills in G. Britain, and as though the
cost of erecting a factory, to be operated by steam, was necessarily
equal to one to be operated by water. In fact, the general tenor
of his work goes to carry the impression, that steam-power cannot
be advantageously used in America, though much cheaper than
water-power in Great-Britain; and hence that water-mills must be

built here, at least till *"coals become cheaper."* But more of this here-after.

In the table referred to, the cost of building alone, for the American factory, is set down at $25,000; while that of the British factory is stated at no more than $4,408—less than one-fifth the cost in America. Is there a rational being that could believe, for a moment, that the expense of erecting a building in the United States was nearly six times greater than for a similar building in Great-Britain? Yet such is the necessary conclusion from the "Comparative Estimate," unless acquainted with the subject. Let it be remembered then, first, on most sites for the use of water-power, the preparation for building, in excavating, laying foundations, &c. costs quite as much, and in many cases even more, than the building itself. Hence, the building itself would probably cost, on his estimate, not more than $13,000. In the ratio of comparative size, the English building would cost over $7,000; and the circumstances of location, and the *different qualities* of the two buildings, would probably account for most of the remaining difference in comparative cost. In the second place, a factory to be operated by steam, may be erected on any selected spot, with but little more cost for excavation, foundation, &c. than is required for other buildings of equal weight. On the whole, we do not believe there would be found any very material difference between the cost of a building in the United States, and one precisely similar in Great-Britain.

Another statement requires explanation. The comparative *supposed* factories are *supposed* to contain 128 looms each. Yet that in Great-Britain is *supposed* to require a steam-power equal to only twenty-five horses, while in the American factory, a water-power is required equal to eighty horses! If there be this *supposed* difference between steam-power and water-power, Mr. Montgomery would do well to recommend the former, without waiting for *"coals to become cheaper."* If there be such an amazing difference between the processes of the two countries in manufacturing, that the power equal to that of twenty-five horses will do as much work in Great-Britain as eighty will in the United States, he would confer a great obligation on the American manufacturer by developing the why and the wherefore. Possibly he may *suppose* how it is done. Probably no one else can.

These remarks may appear of but little consequence to many, but they serve to exhibit, in some degree, the unfairness of certain estimates, and to give more weight to statements which follow, having for their object to correct some misstatements and evasions of facts in the work alluded to, and which have a direct bearing on the progress of cotton manufactures in this country.

Justitia, No. 2   *Boston Courier,* 24 February 1841

I COME now to view some of the comparative estimates made by the author of the book, of the cost of manufacturing in Great-Britain and the United States. The estimate is for a fortnight, in the table before referred to. In the American mill, his estimates foot up, $1,954 45. In the British mill, with the same number of looms, the footing is $1,040 59; making a balance of $913 86 in favor of the British mill, for two weeks, or of $23,760 36 per annum. But this estimate must be taken with some good degree of allowance. It will be recollected that his *supposed* British factory is a steam-mill, the capital stock of which is but about $54,000; while his *supposed* American water-mill has a capital of $110,000. The difference, therefore, in interest and insurance, is not far from $600 per fortnight, on his calculation, and would, if the capitals were equal, cut down the difference in cost, and make them nearly equal. This would be the case, provided both mills were operated by steam. In another place, he says, water-power in Britain costs four times as much as in America; and then again, that steam-power in the United States costs twice as much as in Glasgow. If, then he wished to make fair comparative estimate, why did he not furnish examples of both in each country, on a fair average, and compare them with each other?

That factory labor in Great-Britain is somewhat cheaper than in the United States, is well known. But recent improvements in machinery in this country, and especially in steam-mills, where economy in the expense of power is a great desideratum, have very much reduced the amount of labor required to produce a given result. And the fact is, as can be shown to the satisfaction of any one, by experience, and on good authority, the aggregate cost of labor performed in some such, exceeds but very little, that required for similar establishments in England. Besides in Mr. Montgomery's supposed American water-mill, by his own show-ing, the product exceeds that of his supposed British steam-mill, by 16,100 yards per fortnight, amounting to 418,600 yards per annum. The saving in the cost of labor in the English mill, then, compared with the product of the American mill, cannot be very great. The author, however, for some reason, seems anxious to show, after all, that the American manufacturer has the advantage of the British manufacturer. Consequently, though he gives, in the estimate above referred to, a balance of 19 per cent, he says afterwards, that the difference in the import charges, &c. on the raw material, more than counterbalances that advantage, and gives a balance of 3 per cent in favor of the American manufacturer.

As another evidence of fairness, the author confines himself to low numbers; 18 being the highest number he has assumed; precisely as though Americans either did not, or could not, manufacture any thing finer.

To show another instance of fairness in estimates, Mr. Montgomery takes as the basis of a general estimate the cost of steam-power in this country, and to prove it double the cost of steam-power in Glasgow, "a mill in Massachusetts" containing, as he says, 3,700 spindles, operated by an engine of the power of forty horses, at an expense of $12 20-100 per day. At the very moment he penned this statement, it can hardly be doubted that he must have known it deceptive. There were other mills in Rhode-Island and Massachusetts, then running by steam-power, at a cost much less. I will name one—the mill of Providence, R.I. Steam Cotton Manufacturing Company, which, at an expense hardly exceeding $14 per day, operates 10,500 spindles, with all the necessary preparation for making cloth, besides the lathes, turning engines, circular saws, &c. in a machine shop, in which, from 50 to 70 hands are employed in the manufacture of machinery. This is a well-known fact; and the above mill is about a fair average of the steam-mills in Massachusetts and Rhode-Island.

The writer has already said more than was at first anticipated; but seeing I have proceeded thus far, I must crave the privilege of taking a glance at the subject of steam-power, by way of comparing its cost, in this country, with Mr. Montgomery's estimates, and also with his estimates of the cost of water-power. In doing this, I shall not refer to suppositions or hypothesis; but shall base my estimates on facts as they exist, and which are capable of ocular demonstration.

For some reason, perhaps the author of the work alluded to in these numbers may be able to define it, he appearing very anxious to dispose of water-power at Saco, he seems to manifest no friendship for steam-power in the United States, connected with cotton mills; though a fair investigation was due to the subject in such a work as his purports to be, in order for the public to know whether it could or could not be profitably employed, for the benefit of our northern towns and cities. But he merely states that a few mills to be operated by steam, have been erected and put in operation, that one or two more are about to be, and that, in one mill in Massachusetts, steam, of the power of forty horses, costs $12.20 per day. This it must be confessed, is rather a limited and condensed view of a subject so important, by one who professes to go into detail, and to make comparative statements and estimates, and at whose command might and would have been, if called for, all the information on the subject, that could have been given. I will endeavour, in measure, to make up the deficiency.

The statement already alluded to, of the "one mill," is, as every one must see, very unfair. And how the author can reconcile such a mode of procedure with his assumed candor, it is difficult to conceive, without a greater stretch of credulity than the writer is capable of exercising. In the first place, no one acquainted with the high degree of excellence to which steam engines and machinery have been brought in the United States, can for a moment believe, that, as he seems to imagine, a British steam-engine, calculated at the power of twenty-five horses, can perform twenty-five percent. more work than a good American engine of the estimated power of forty horses. Yet, just this is the inference, and the only inference, to be drawn from his "comparative estimate," and his account of the steam-mill spoken of, in Massachusetts. And then, again, it is equally mysterious how his Massachusetts steam-engine, of the power of forty horses, should operate a mill with one hundred looms, while, in his, "comparative estimates," a water power equal to the power of eighty horses, should be required to operate one hundred and twenty-eight. These conflicting statements he is left to reconcile. I cannot pretend to do it.

The American people are proverbial for their enterprizing spirit, and inventive powers; and it is extremely doubtful, if any form of the steam-engine, or any invention of improvement on that noble machine, at all useful, has been brought forward for a quarter of a century, or more, that has not been applied to use in this country. Not only so, but the *Yankees* have made several useful improvements and appendages themselves; and which have been found of much value, in the course of operations by means of steam-power.

It has ever been the aim of manufacturers of steam-engines in this country, as well as of scientific men, and others interested in the subject, to obviate difficulties in construction and operation, and to adopt and apply to use, every available improvement, for the three-fold purpose of promoting durability, facility of operation, and economy in fuel. That they have not been altogether unsuccessful in their attempts, is fully evinced by the speed of our steamboats and locomotives; many of which are quite equal to the best English specimens, and not a few superior, on a comparison, side by side. There are also in operation in this country, at the present moment, in some of our manufactories of cotton, engines which are allowed to surpass the best British engines, in form, construction, and economy of fuel, as well as in style of workmanship. As to our cotton machinery, in nearly, if not in all, its various departments and modifications, Mr. Montgomery to the contrary, notwithstanding, it is believed to be quite equal, if not superior to the English machinery adapted to the same fineness of manufactures. Under all these considerations, and the writer is fully confident that they are correct, he would respectfully ask of Mr. Montgomery, how or why an American steam-engine should not be quite as effective as a British one? We would ask of him why the manufacture of cotton cannot be carried on by steam-power, contiguous to navigable waters at the North, or elsewhere, as cheap as by water-power in the interior? And as he has appeared, by design or otherwise, to avoid this part of the subject almost altogether, I will give my comparative estimates in my next and last number.

Justitia, No. 3   *Boston Courier,* 2 March 1841

IN my last number it was proposed to give, in No. 3, a comparative estimate of the cost of steam power and water power, and with that to close the discussion. The writer, however, has changed his plan, so far as to waive that subject in the present number, for the purpose of adding a few further remarks.

From the first attempts to manufacture goods in the United States, by means of steam power, most people have been, or appeared to be, quite skeptical as to the practicability of the profitable application of that power for such a purpose. It has been the general, and almost universal impression, that steam could not, by any possibility, be employed at a cost so low as water. With people in general, on this, as on all other practical subjects, argument is of no avail, even if sustained by known facts, and proved true by the incontrovertible laws of nature. Practical illustration is demanded in each individual case. Persons when comparing steam and water power, seem to forget that the latter costs any thing, either originally or in its application; and would appear to think it as free for manufacturing as for domestic purposes, and as cheaply applied.

People either forget, do not know, or do not take the trouble to consider, that steam-mills may be erected and put in operation, at one half the expense, generally, which is required for water-mills, exclusive of machinery, and that thus a large amount of interest on the original investment of capital is annually saved. Nor yet do they appear to take into the account the well-known fact, that steam-mills erected and operated contiguous to navigable waters, save all the heavy expense of inland transportation. Without now entering into details on this particular item, which is a very important one, I will here refer the reader to a note in Montgomery's work, quoted by him from the Portland Advertiser. It may be found on page 202 (above p. 230); and is brought up by Mr. Montgomery to show the superior advantages of the water power of Saco, Me., contiguous to navigation, over water power in the country remote from navigation. The Portland writer allows transportation from Boston to Maine, for "*mills situated on or near the tide-waters,*" at one dollar per ton. For other mills, fifty miles from Boston, where transportation can be had by railroad or canal, he allows three dollars per ton. In his estimate, he includes the transportation of cotton only, to the mill, and that of the manufactured article back again. From his estimate he omits the heavy cost of transportation of oil, iron, starch, flour, &c. Yet for a mill of 5,000 spindles, he makes the difference in the cost of transportation between the locations in Maine, and others accommodated with railroads and canals, $1,200 per annum on cotton and cotton goods only. He allows $600 per annum for the transportation between Boston and the ports in Maine, for mills situated in the latter places, for the same articles. This is unnecessary to the steam-mill in a seaport, and hence, the writer admits by his statement, that the difference in favor of the steam-mill would be $1,800 per annum. This amount would pay the interest on a capital of $30,000; which is a very pretty item in the scale of manufacturing operations, with a mill of 5,000 spindles. But, in the sequel, I think I shall be able to swell this difference to a much greater amount.

I wish it still to be borne in mind, that, as to the comparative cost and profits of steam power in maritime locations, and water power in the interior, very much depends on the quality of goods manufactured. For obvious reasons, the cost of manufacturing coarse goods, by steam power, would approximate nearer to the cost attendant on manufacturing a similar article by water power, than would be the case with those of a finer texture.

1. The cost of transportation depends very much on the fineness of the goods manufactured;—2. Coarse goods require a greater amount of power to manufacture them, than is required for the manufacture of fine goods;—3. Steam is indispensable to the manufacture of a fine article of a superior quality, which is not the case with coarse goods.[2] But cloth of a superior quality, I have not the least hesitation in saying, of number thirty, and upwards, can be made with the steam mill on navigable waters, at much less actual cost, all things taken into the account, than by the interior water mill. But of this more hereafter.

And now should the advocate of water power say, he can manufacture coarse goods by means of that power, as cheap as by steam, it will not materially affect the question. He has his choice, to manufacture coarse or fine, at his option. So has the proprietor of the steam mill. On coarse goods, water power can only claim equality with steam power. On fine goods, steam power has greatly the advantage, and hence has manifestly the advantage on the general scale.

After all, the reader is requested to recollect, that the writer does not predicate his cost of steam power on the example furnished by Mr. Montgomery, as the basis of a general estimate. In that case, he gives a single example, where he says, a steam mill, with 3,700 spindles, was driven by an engine of the power of forty horses, at the expense of $12 20 per day for coals, engineer, and fireman. I shall furnish examples of a far different character from this, and show that double the number of spindles can be run, and more than double the work done, and that such is now the fact, with a less cost than he has set down in the case he has given.

One more item of the advantage of location to the steam mill. Proprietors of water mills, generally, in the interior, are under the necessity of purchasing land, and erecting dwellings, for the accommodation of their operatives. This requires a very considerable amount of capital, which is thus diverted from other objects. In seaports, this necessity does not exist. Dwellings are generally found in abundance, already erected; and any deficiency is readily supplied by the owners of real estate themselves. Added to this, there is another consideration. There is, in general, to be found in maritime places, an abundance of help, of nearly all descriptions wanted in the mill. A vast proportion of these persons either cannot or will not leave their homes, to labor in distant establishments, without the inducement of high wages; and many not even for that inducement. But they readily and gladly go into mills in their immediate vicinity, will work for less wages than would command their services abroad, and in fact can well afford to do so, as they can live as well with their own families and friends, at much less expense. This is a very important consideration; and the difference it makes in favor of the steam mill, is, in many cases, sufficient to pay the daily expenditure of the power to drive it. And finally, to all this is to be added the vast difference in the original cost of the water mill and the steam mill, making a corresponding saving to the latter, in the annual interest on the original outlay.

I have now given the reader some of the data which are to serve as the basis of my estimates and comparisons. In them it is believed a case will be made out, sufficiently plain and clear for every one to understand, and sustained by proofs sufficiently strong to satisfy even the most incredulous; and those proofs drawn from actual results, capable of ocular practical demonstration.

The writer has not asserted, nor is he about to assert, or attempt to show, that steam power has ever, in all cases, been applied to the manufacture of cotton goods, with a profitable result, in this country. In some cases, he is aware it has not; and the same may truly be said of water power. But in neither case has the power been in fault. Other circumstances have operated unfavorably. In the application of steam-power, a great deal depends on the judicious construction of engines and other machinery, and a proper adaptation of the means to ends; and very much, indeed, to strict attention and judicious management. When all these particulars are attended to, and the principles they suggest fully carried out, steam power can be as profitably employed, to say the least, in our maritime towns, as water power can be in any part of New-England. In order to make good this proposition, he will go into a full and explicit comparative estimate. And that estimate shall not be made on *supposed* results, from the *supposed* operations of *supposed* mills. It shall be based on actual data, which cannot be controverted.

In making this statement, after the remarks I have heretofore offered, it is perhaps incumbent on me, and if not, it is at least my wish to say, I have no particular partiality for steam power over water power; and that I would not wish to distort a single fact, nor to make a single statement contrary to fact, to create a favorable impression in regard to one, or an unfavorable impression in regard to the other. I have long been practically acquainted with both, as applied to manufacturing purposes; and my remarks, intended only, as far as my feeble efforts may go, to set the public right on the subject, are results of my own personal experience, founded on my best judgment. On their merits, others must and will decide for themselves.

I have promised to make my estimates from known data. Consequently it will be necessary to do what otherwise would not be done—name particular establishments. For this course, my apology is, that the nature of the case demands it, and that it will be pursued without disrespect to any one, or the least possible intention or desire to inflict even the most trifling injury. It is presumed no one will doubt this, when told that the water mills I shall select are those establishments at Lowell, called the Boott Mills, unquestionably equal to any others, and superior to most in the United States, operated by water power.

Justitia, No. 4   *Boston Courier,* 9 March 1841

IN my last number a comparative estimate was promised of the cost of water power and of steam power; the former located in the interior, and the latter on navigable waters. I now proceed to redeem that promise; and, in doing so, shall take, as the basis of an estimate of the cost of water power, the Boott Mills at Lowell; using as guides the official statistics of the Lowell manufactories,[3] and the statements of Mr. Montgomery. Those establishments, it will be conceded, are as favorable specimens of manufacturing by water as our country can furnish, as has been before hinted.

The Boott Mills are four in number. They contain, together, 29,248 throstle spindles, and 830 looms; the number of spindles in each mill varying from 7,000 to 8,000. Our business now is to ascertain their actual cost and expenses, for the comparison. In doing this, however, it will be unnecessary to bring into the account items which would be equal in water mills and in steam mills, as they would not affect the relative results, and as the object is merely to ascertain the difference, if any, and to strike the balance. The cost of the building, for instance, above the foundation, not being necessarily different, whether for a steam or water mill, need not be included in the estimate; and so of some other items.

First, in the list of expenditures, I take the article of fuel; the amount of which, consumed in the four Boott Mills, per annum, according to the statistical account alluded to above, is Anthracite coal, 750 tons—wood, 70 cords. The cost of coal at Boston, I set down at $6 per ton, and the cost of wood at Lowell $6 per cord; which would make the original cost of these articles $4,920. This, it must be recollected, is fuel for warming the mills—a purpose for which steam mills require no purchase of fuel, they being warmed in the best and the most effectual manner, by the steam itself, after it has performed its office as a moving power.

Second, transportation. Here, as to the cost per ton between Boston and Lowell, I take Mr. Montgomery as a guide. He states it at $2 per ton. 1. The transportation from Boston to Lowell, of 750 tons of Anthracite coal, as $2 per ton, is $1,500. 2. The transportation of cotton from Boston to Lowell, and of the manufactured articles from Lowell to Boston. As the amount of transportation of these articles for the manufacture of coarse goods is greater than for fine goods, I shall call it for the Boott Mills on the scale of No. 40; the same fineness as those usually manufactured in steam mills, to which I shall refer, to make out a case as favorable as can be for the water mills. Say, then, the above named 29,248 spindles turn off, each 4 skeins of yarn per day, No. 40. The aggregate number of skeins for the four mills will be 116,992 skeins per day, and 701,952 skeins per week. At 40 skeins per pound, the weight per week would be about 9 tons; and per ann. 468 tons.[4] As however the estimate of waste is 14 per cent., that amount must be added to the weight of yarn, which will give the amount of cotton transported from Boston. This will give the amount at 533 1/2 tons nearly per annum; to which add 468 tons transportation of the manufactured article to Boston, and you have 1,001 1/2 tons nearly; which at $2 per ton, will cost $2,003 per annum. 3. The Boott Mills consume 7,100 gallons of oil per annum, and 100,000 pounds of starch. Allowing the oil to weigh 8 pounds per gallon, that and the starch together make up 78 tons[5]; the transportation of which will cost $156 per annum. Thus it will be seen the annual cost of transportation of coal, cotton, cotton goods, oil and starch will amount to $3,659 1/2[6] per annum. And it must be recollected that all this expense is saved to mills located in places contiguous to navigation.

Third. Cost of water power at Lowell is $4 per spindle. At that rate, the power to drive 29,248 spindles would cost $116,992; but, as throstle spindles are supposed to require one fourth more power than mule spindles, and producing an equal quantity and quality of yarn, and as the latter are usually adopted in steam mills, I deduct one fourth from the above cost at Lowell, and estimate it at $3 per spindle; which, for 29,248, will amount to $87,744.*

Fourth. Cost of application of power. I take Mr. Montgomery's estimate for this. According to his estimate for one mill of about 5,000 spindles, and deducting one fourth from the cost, as in the case of water power as above, the water wheels for the four mills, together with shafting, belting, gearing, &c. will cost $85,000.[7]

Fifth. Cost of foundations for four mills, on water privileges, together with wheel pits, flumes, race ways, gates, &c. $40,000.

The buildings themselves, as before observed, above the foundations, are not taken into the account, there being no necessary difference in cost, whether for steam mills or water mills. According to the above estimates, which are believed to be within the limits of truth, the sum total of the cost of water, wheels, shafting, belting, gearing, foundations, wheel pits, &c. &c., is $212,744.[8]

The annual interest on that sum is $12,764 64; which, added to the cost of fuel and transportation ($8,589 50)[9] makes $21,354-14.[10] This amount constitutes the annual expense of the water power, for articles of consumption not required for the purpose of steam mills, for which they are indispensable in water mills; for expenses not incurred in maritime places; and for the cost of power, and the means of application, in which foundations are included. The above amount must therefore be taken as the basis of succeeding estimates, in order to determine which runs at the least expense: the water mill at Lowell, or the steam mill in a maritime location.

In making the foregoing calculations, it has been my object to circumscribe them within the range of well-known facts, as learned by practical experience, and made public by means of official data, and the admissions of persons most interested in water power, of whom Mr. Montgomery is one. In doing this, it is conscientiously believed, that, for the purpose of satisfying all concerned, and obviating every objection on their part, the above amount has been reduced, by several thousand dollars, below what actual results would have borne me out in stating; but yet the estimate is near enough to truth for all practical purposes, as will be fully tested and made to appear, on the comparison.

* This additional power is required only as far as the spindles are concerned, other machinery being equal; so that the difference is not so great as might at first be imagined.

The case is now a plain one for the reader. It is evident that, to give the water mill the advantage over the steam mill, or to place both on a footing of equality, it must be shown, that, all other things being equal, the steam mill must expend, in the course of the year, the above $21,354 14, for the same purposes to which it is appropriated in the water mill, or for some others for which the water mill does not require it, to realize the same, or a similar result, as to work or profits, or both. If steam power on the sea-board cannot do this, it must yield the palm to water power at Lowell. But if, on the other hand, steam power can operate without exceeding that limit, it will prove itself equal to water power. And if, at the expiration of the year, any portion of the aforesaid amount shall remain unexpended, or, what is the same thing, shall have been saved, the steam power will claim that advantage, as to actual expense. The advantages, however, accruing either to steam or water, must at the same time be qualified, increased or diminished, in the ratio of the comparative difference of profits realized from the quantity and quality of work performed. This ground of calculation is a fair one, because every one has his option to employ either water or steam as he may elect to do; and because the proper criterion, by which to judge is, to determine by which mode the greatest profit can be made in a given time, with an equal amount of cost, all other things being equal.

I am fully aware it may be said that, in some parts, water power may be obtained much cheaper than at Lowell. That probably is the case: while, in other parts, it costs more than in Lowell. The cost at the latter place has been assumed as a fair average. There is much unoccupied water power in our country, and those who own it, anxious to dispose of it to advantage, sometimes are willing to sell mill-sites, at *first* at very moderate prices, to have operations commenced, as an inducement to others to purchase. In some instances, for ought I know, privileges may be disposed of, to first applicants gratuitously, or nearly so. But operations once commenced, water power increases in value; and very soon, it will probably be found impracticable to purchase it at less than the Lowell rates. The difference between the cost at Lowell and Man-ayunk, near Philadelphia, is, as estimated by Mr. Montgomery, as follows: At Lowell, the cost is $200 per horse power; at Manayunk it is $1,016—more than 5 times as much as at Lowell. Thus prices vary in different locations, from one extreme to the other; and where the price is very low, as a general rule, it may be presumed there are drawbacks in other matters, to counterbalance. Lowell I take as a fair average, and have used it accordingly.

Having now said what I had contemplated, by way of estimate of the expenditures of water mills, I dismiss that part of my subject, and shall, in another number, bring forward calculations and estimates of steam mills, as to cost, expenses, &c., and carry out the promised comparison, in order to ascertain what difference there may be, and on which side is the advantage.

Justitia, No. 5   *Boston Courier*, 12 March 1841

FROM the estimates made in my last communication, I proceed, as promised, to compare with them my estimates of the cost of steam power on navigable waters, for manufacturing purposes. In doing this, as in the former case, items of cost and expenditure common to and equal in both, will not be taken into the account, they being immaterial in the comparison. I shall select for my purpose the steam mills called the Bartlett Mills, at Newburyport, Mass.

The Bartlett Mills, are two in number—Mill No. 1 is 150 feet long by 48 feet wide; four stories high, with an attic; with a picker room, engine, and boiler room detached from the main building. This mill contains 6,336 spindles, and 144 broad looms; and the machinery is operated by an engine of 75 horse power. The cost of foundation was $1,000. The cost of engine, shafting, gearing, belting, &c., was $10,000—making $11,000 in all. Mill No. 2 is a building about 200 feet long by 49 feet wide. This mill contains 10,664 spindles, and 224 broad looms; a large proportion of the looms for weaving goods 9-8 and 5-4 in width. The foundation of this mill cost $2,000. Cost of engine, shafting, gearing, belting, &c., $15,000. It is five stories high, and has a boiler room detached from the main building.

Thus the cost of foundation, engines, shafting, gearing and belting, in these two mills, is $28,000. The number of spindles in both is 17,000—less by 12,248, than the number in the four Boott Mills, which have been selected for the comparison; and more than one half that number, the average for two of those mills, by 2,376.

It must now be recollected, that, in the steam mills, the foundations, engines, shafting, gearing, belting, &c., are to be offset against the original outlay for water power, foundations, water wheels, wheel pits, flumes, raceways, shafting, gearing, belting, &c., in the water mills; as, for steam mills, the former constitute a practical equivalent.

The Bartlett Mills being two in number, and the Boott Mills four, the expenses of the latter must now be divided, and one half assumed as applicable to two of them. It will be remembered that a deduction of 25 per cent. was made from the cost of water power, &c., on account of the supposed greater power required for throstle spindles than for mule spindles. The amount left was $212,744. The annual interest on that amount was $12,764 64. One half that amount, to be applied to two mills, is $101,372.[11] The annual interest on it is $6,382 32. The outlay on engines, foundations, &c., for the two Bartlett Mills, we have seen to be $28,000; the annual interest of which is $1,680. This amount, abstracted from $6,382 32, as above, leaves $4,702 32—a balance to the credit of steam power, towards the payment of expenditure for purposes other than those required for water power.

The actual cost of fuel for the four Boott Mills, together with the cost of transportation of coal and other articles, as already seen, is $8,598 50.[12] One half that amount, for two mills, is $4,299 25. Steam power prevents the necessity, and even the utility, of purchasing fuel to warm mills operated by it, and for which purpose it is used at Lowell. A location on navigable waters prevents the necessity of inland transportation. Hence, the above expense of $4,299 25 per annum is saved to the Bartlett Mills. This, added to the amount $4,702 32, before brought forward makes $9,001 57. One more item remains to be added. The Bartlett Mills contain 17,000 spindles. That is more than half the number contained in the four Boott Mills by 7,376.[13] This difference is about one-seventh; for which add $1,284[14] to the last named sum, and we have $10,285 57 per annum to be applied to the purposes of generating and managing steam power, with which to operate the Bartlett Mills. In other words, the case stands thus. The amount of capital invested for the two water mills, and that for the two steam mills, are supposed equal or in proportion to the difference in the numbers of spindles. The capital of the water mills has been mostly expended, and the interest on the residue appropriated; and the mills are in complete operation. The steam mills lack nothing but power to operate them, which requires nothing but fuel and attendance, to generate and control it. For this purpose, they have remaining so much of their capital as will yield an annual income of $10,285 57. The only question now to be solved is, can the requisite steam power be furnished for that amount? That question I will now attempt to answer.

I have said that the Bartlett Mills contain 17,000 spindles. This requires some qualification. Mill No. 1 has been in full and complete operation for two years, with its full complement of 6,336 spindles, and 144 looms. Mill No. 2 is as yet only in partial operation; having now in use about one half its machinery. But from the average actual results of two years in Mill No. 1, the cost of power for that establishment has been accurately ascertained. The engine in Mill No. 2 has now been running about three months; and repeated experiments to test its power enable me to give a result which will be found equally accurate.

In Mill No. 1, the average consumption is 1 1/3 tons per day. In Mill No. 2 it will not exceed 1 2/3 tons—making an aggregate of 3 tons per day, or 930 tons per annum for the two mills, containing together 17,000 spindles, and 368 broad looms. I allow the coal at Newburyport to cost 6 1/2 dollars per ton; though for the Lowell Mills I estimated it to cost but 6 dollars in Boston. At 6 1/2 dollars per ton, 930 tons for the Bartlett Mills would cost $6,045, the expense of fuel per annum. To that amount must be added the wages of one head engineer, at $2 per day, and of two firemen, each 5 shillings per day. These come to $1,136 67[15] per annum, and added to the cost of fuel makes up $7,181 67. Deduct this from the balance of $10,285 57, and you still have a balance left of $3,103 90, and the steam mills in full operation; and which it will be seen is sufficient to increase the steam power at least 50 per cent. This excess is equal to the interest of about $52,000, and that sum would purchase at Lowell (if Mr. Montgomery's estimate of 4 dollars per spindle is correct, as I suppose it to be) mill sites and water power for two mills, with each 6,500 spindles. So that, on the whole, it would be of no benefit to the Bartlett Mills to be located at Lowell, even if water power were furnished free of expense; provided, and which it is firmly believed, the foregoing estimates are within the limits of truth.

As has before been said, so it is respectfully solicited that the reader will bear in mind, these estimates and comparisons are not the results of any particular partiality of preference for steam power over water power; nor are they intended, by any means to promote the interests of one, at the expense of the other. Their only design is to exhibit to the public a fair and impartial statement of facts, for the information of those who may require it. I make no pretension to infallibility, and therefore may have committed errors in the foregoing calculations and estimates. If so, I shall be happy to be set right though I am fully persuaded they will be found essentially correct.

In addition to the advantages of steam power at Newburyport, as already given, when compared with water power at Lowell, and other inland locations, an experiment has been made, and proved successful, to reduce the quantity of fuel necessary, according to the statement already given. Steam pipes have been laid under ground and surrounded by substances which are bad conductors of heat, by which the steam may be conducted from the boilers in either mill, to the engines of both. Mill No. 2, has enough of its machinery in operation to make up full 13,000 spindles, and a proportionate number of looms, besides its shafting, gearing, &c. for the remainder. The machinery in operation is therefore nearly equal to that of two of the Boott Mills. On the next day after the laying of the connecting pipes had been completed, before, of course, the substances which surrounded them had been heated at all, or the joints perfectly cemented, the experiment was tried to run both mills by the power of the steam generated in the boilers of Mill No. 2. It was commenced at the usual hour of starting in the morning. Both mills were operated with the usual speed during the day. And both mills, finishing rooms, offices, &c. warmed with the "*exhaust steam.*" And the whole was done with two tons of anthracite coal. There is no doubt, therefore, that both establishments, with their 17,000 spindles and 358 broad looms, may be operated with 2 1/2 tons of coal per day, were they connected; and which would reduce the quantity required, per annum from 930 tons to 775. Thus the steam power for the two mills can be generated with a quantity of coal 25 tons less than is required to be used in the four Boott Mills at Lowell.

The great error in popular opinion relative to the employment of water or steam power for manufacturing purposes, is the supposition that the power itself is the great desideratum in the list of expenditures, and which is to be looked at as a paramount consideration. This is a fallacy; as the foregoing calculations fully show.

The mere cost of water power or steam power, that is of the elements of power, as seen, constitutes but a small proportion of outlay. The interest on the cost of water power for two of the Boott Mills is only $3,509 76 per annum; and the actual cost of steam power for the Bartlett Mills is but $7,181 67; a difference of but $3,671 91 per annum. And we have seen how readily that difference is sunk, and a balance found against water power in favor of steam power, by a difference in location. The mere original cost of a moving power is, therefore, of secondary consideration: and its relative profits, or merits or demerits, the quality and quantity of goods, however, taken into account, will depend altogether on location.

I have now concluded what I had to say on this part of my subject, and submit it to the consideration of those who may feel an interest in it. To conclude finally my remarks, I shall close with one more number, in which I shall present the question in a different form, with comparative estimates, *pro rata*, of expenditures and profits.

Justitia, No. 6   *Boston Courier,* 16 March 1841

IN the preceding numbers, I have examined Mr. Montgomery's work, as far as steam power is concerned, when compared with water power, and drawn out some practical and comparative estimates on the subject, as suggested and confirmed by my own practical experience. In the close, I shall now proceed to make a brief practical comparative estimate, founded on what is believed all will concede to be correct principles. In doing this, I shall compare the cost of power of the Boott Mill No. 2, which spins and manufactures yarn No. 40, with the Bartlett steam Mill No. 2, which spins and manufactures yarn of the same number. As the basis of this calculation, I shall take the cost of power and its requisites, and endeavour to show and compare the cost per yard, in each establishment, and the proportionate cost of power, &c., to the relative value of goods in market. In doing this, there being a difference in the width and quality of the goods in the two establishments, I shall give, as the most equitable mode of calculation, the cost per cent. of the market value of the manufactured article from each establishment respectively. And this, it is believed, must strike every one as a fair practical rule.

If the proprietors of the Boott Mill No. 2, run the throstle spindle, it is their own choice to do so; and it is to be presumed that, it being driven at great speed, they find an equivalent for the extra power required, in the quantity of work performed, or that they would not adopt it. And if it can be shown that, to manufacture by steam power, in the Bartlett Mill No. 2, is no greater cost per yd. taking into view the market value of the article produced, it is presumed every one must be satisfied, however skeptical.

According to the best statistical information I have been able to obtain, and which is believed to be sufficiently correct for the comparison, Boott Mill No. 2, including water wheels, shafting, gearing, belting, &c. &c. cost 100,000 dollars, and contains 8,640 spindles. The interest per annum on that amount is $6,000. At $4 per spindle (according to Mr. Montgomery) the water power for 8,640 spindles would cost $34,560, the interest on which is $2,073 60. This added to the above amount, makes $8,073 60. Cost of fuel for heating mill, in proportion to the amount required for the four mills $1,600; making $9,673 60. About the average amount spun by this mill is four skeins per day per spindle, making 266,600 pounds[16] per annum, of cotton to be transported from Boston to Lowell, and the corresponding weight of cloth to be returned to Boston. This, together with oil, starch, waste, &c. will make an aggregate amount of about 300 tons, which, at $2 per ton, is $600. This added to the foregoing, makes $10,273 60. Cast away the cents, I shall assume the amount 10,273 dollars as the aggregate cost per annum, for the Boott Mill No. 2.

The cloth turned off per annum I set down at 1,500,000 yards, and which is very near the truth; the quantity not exceeding. That number of yards, compared with the above aggregate cost, ($10,273) will give the cost per yard at 69-100 of one cent, or to avoid the use of fractions, about seven mills per yard.[17] This, therefore, is the cost of water power, wheels, shafting, gearing, building, &c. of the Boott Mill No. 2, per yard of cloth.

Let us now take an estimate of the Bartlett steam Mill No. 2, and then compare the results. This mill cost, for building, engine, shafting, belting, gearing, &c. $60,000; the interest on which is $3,600 per annum. Cost of transportation, nothing, the mill being located on the bank of navigable waters. Cost of fuel to warm mill, nothing, the steam being applied to that purpose, after having performed its work as a moving power. Cost of one and two thirds tons of anthracite coal per day, at $6 50 per ton, for the engine, amounts to $3,354 per annum.[18] As one head engineer has charge of the engines in both mills, and is paid two dollars per day for his services, I charge one dollar per day to Mill No. 2, for engineer, and 5 shillings per day for a fireman, which makes $568 50 per annum.[19] These items of expenses for the Bartlett steam Mill No. 2, practically equivalent to those enumerated for the Boott Mill No. 2, amount in all to $7,522 50. Casting away the cents in this case, also, the difference in the aggregate expenditure, for the purposes named, is $2,751, to the credit of the Bartlett mill, in the gross amount.

The Bartlett Mill No. 2, has 10,664 spindles, and 224 broad

looms, 5-4, 9-8, and one yard wide. Taking the results of Mill No.
1, and those of No. 2 thus far, the quantity of cloth per annum
will be about 1,000,000 yds. less by 500,000 yards, than the quantity produced by the Boott Mill No. 2. The difference in length
is however counterbalanced, and more too, by the difference in
width and quality. The amount of cost per annum, of the Bartlett
mill, $7,522, being compared with the number of yards, viz.
1,000,000, it will be found that the cost per yard is 7 1/2 mills—
one half mill higher than the cost of the same one yard in *length,*
at the Boott mill, without regard to width or quality.

The writer believes the above comparative estimates to be
founded on the true and only correct basis; that is, showing the
cost of the power and its requisites, per yard, as compared with
the market value, respectively of the articles produced. This rule
must and will hold good, in all cases in which mills are employed
on similar numbers, provided the mills turn out about an equal
number of pounds, and that the goods are of about the same
market value. In one particular, the Boott mill has the advantage
in this estimate; for as the Bartlett Mill No. 2 turns off the greatest
number of pounds, the aggregate value in market will be greater,
even if there be no difference in quality. This had not been taken
into the account; and the estimates have been made on the ratio
of the cost of power, &c. as though the number of pounds turned
off in both mills, was equal; which is not the case. To show, then,
the actual relative results of the two mills, per annum, it becomes
finally necessary to compare the cost of power, &c. as shown before, with the market value of the products of both mills respectively.

Taking the quantity of cloth produced by the Boott Mill No. 2,
at 1,500,000 yards per annum, as stated, at the price it commands
in market, the cost of power, as given before, is something more
than six and a half per cent. on the full amount. On the other
hand, take, as stated, 1,000,000 as the aggregate annual product
of the Bartlett steam Mill No. 2, at the price it commands in
market, and the cost of its power, &c. as before given, is four and
a half per cent. of its market value.

In this latter estimate, the correct principle is involved; and the
result is, a difference in favor of the steam mill, of $3,300 per
annum. This is more by $540, than was found by estimating the
cost per yard.[20]

Thus, I have given my estimates as promised; and in doing
which, I have aimed to be as correct as the nature of the case

would admit. If I have erred in any point, and possibly I may have done so, it has been unintentional, and I stand ready to be put right. Had health and business avocations permitted, I might have gone more into detail, but as it is, it is presumed no material error will be found, to affect the general result. No particular establishments would have been named, had not the circumstances absolutely demanded such a course. But in naming them, the writer believes he has done injustice to none, inasmuch as he merely commented on what the public knew before.

One other remark I have here to make, and then close the subject.

Mr. Montgomery, under the head of "Cost of Steam Power in the United States," as before noticed, takes a single establishment as the criterion. And why he makes that selection, probably, is best known to himself. There are, and long have been, many steam cotton mills, or several at least, in "the United States," running at a much less expense. The mill he selects, he says, has 3,700 spindles, 100 looms, &c. and a steam engine of 40 horses' power. The cost he sets down as follows: 1 1/4 chaldrons of bituminous coal, at $8 per chaldron, per day: wages of engineer $1 33, and a fireman 87 cts. making $12 20 per day—$3,806 40 per annum. This, he says, is about double the cost of the same power in Glasgow. Thus he leaves it to be inferred, that steam power in the U. States costs double what it does in Glasgow, of necessity. The Bartlett steam Mill No. 2, it has already been seen, has an engine of 100 horses' power, and expends fuel to the amount of only $3,354 per annum, to drive 10,664 spindles, and 224 looms; while the mill he has selected consumes, for an engine of 40 horses' power, 3,700 spindles, and 100 looms, fuel to the amount of $3,120 per annum. The fuel, then, for the Bartlett Mill No. 2, does not cost half as much as that for the other steam mill named above, in proportion to the power of its engine, and the number of its spindles and looms. And though engine, &c. in the Bartlett Mill No. 2, will probably require less fuel to drive them, than almost any others in the country of like magnitude, yet the cost is quite as near the general average, to say the least, as the specimen selected by Mr. Montgomery. Whatever that gentleman's views on the subject may be, the truth is, there is not the great difference he appears to imagine, between the cost of steam power in Glasgow and the United States, at the present time; and I have, I think, fully shown, that it may be quite as advantageously used in seaports, and on navigable waters, as water power in the country provided that engines and machinery are constructed on the best models and best principles, and arranged and managed in the most judicious manner.

In conclusion, permit me to add, these numbers are not the production of any feeling inimical to Mr. Montgomery. His work I consider erroneous on many important particulars, as to steam power, and calculated to mislead those who may refer to it for information. It is the book, and not the man, with which I have to do; to the book only, have I intended to direct attention.

Letter of James Montgomery dated 17 March 1841, printed in the *Boston Courier*, 27 March 1841.

THE COTTON MANUFACTURE OF GREAT BRITAIN AND AMERICA CONTRASTED. No. 1.

Mr. Editor—Several articles appeared in some of your late numbers, purporting to be a review of a work lately published by me, bearing the above title. The writer of those articles, under the signature of Justitia, states, "that he knows and can make it appear, that there are departures from truth, in important particulars," in the work referred to; and further, "that it is difficult to understand how these errors could have originated in ignorance of the facts." Now whatever consideration may be due to the anonymous writer, who thus charges me with wilful falsehood, some reply is certainly due to the public. And I doubt not you will allow me a fair opportunity to do so, through the medium of the same columns; and also to inquire, how far he has succeeded in supporting the above charges. It may be proper, however, that I set myself right with the public, as to the origin and design of the above-named publication.

In consequence of numerous contradictory reports circulated in Britain, regarding the practical state of the cotton manufacture in America, the author was requested by his friends to write some account of these matters, upon which they could rely; and, particularly, to make out some comparative estimates of the cost of manufacturing in both countries. For this purpose two mills were selected, one in each country; and, as fortunately happened, they contained 128 looms each. The one mill was making heavy shirtings, the other fine drilling; but as the size of the yarn and weight of the cloth were the same in both, they were considered fair subjects of comparison. The British mill was built in 1834, the plans and specifications of which are now in my possession, and have been examined by gentlemen in this country well acquainted with such matters, who are perfectly satisfied regarding their authenticity and correctness. The American mill was built in 1835, and, as already stated, both mills contain 128 looms each, with all the subordinate machinery suited to the general practice, style of building, and manner of arranging the machinery, *in each country* respectively; *and both mills are now* (March, 1841,) *in full operation.* These facts *are perfectly susceptible of proof*, whenever such proof is wanted by the public generally, or any respectable gentleman individually.

The comparative estimates of the cost of manufacturing in the two countries, after being inspected by gentlemen in this country, were sent to the author's friends in Glasgow, that those statements which referred to the cost of manufacturing in Britain might be corrected, and accompanied by the practical details and other matters embraced in the work, they were deemed not unworthy of being laid before the public. And be it remembered, that the author arrived in this country in June, 1836, and the whole MS was sent off in September, 1838; so that it cannot be supposed he could collect, in that short period, all the materials which a longer experience, and better acquaintance with the country, would have enabled him to do, especially as he had to collect the materials himself, no writer having gone over the same ground before. The work was likewise published in Glasgow, where it was impossible for the author to correct the press; and though some little inaccuracies may have escaped his notice in the manuscript, it is presumed there are nothing more than may be found in the first editions of any other work.

The historical sketch of the rise and progress of the cotton manufacture in the United States, was chiefly compiled from "White's Memoir of Slater."[21] The author has since learned that that part of the work is not so complete as could have been wished; inasmuch as the progress and present state of the business in Massachusetts are not treated so fully as might have been expected; some of the manufacturing villages are not brought forward with that prominence to which they were entitled; and certain *respectable* individuals, who took an active part in establishing the cotton manufacture in Lowell, are not so much as named in the work. These omissions arose entirely from ignorance of the facts; and the author would now most gladly avail himself of any information on the above subjects, which any person could furnish.

These things premised, the communications of your correspondent may now be examined in detail.

He begins his first article, by informing your readers that I am a foreigner, but cautions them not to harbor any prejudice against me on that account. He will be pleased, therefore, to accept my best thanks, for thus taking me under his very gracious protection. I am not aware, until now, that such a stigma rested upon the intelligence of your readers; nor am I prepared to admit that they are capable of such illiberality, as to entertain any prejudice against any work, merely because the author is not a native-born citizen. But, of course, your correspondent knows better, or why caution them against it?

After various general denunciations of the work, as "intimately blended with matter that is entirely incorrect, deceptive, and calculated to mislead," and of "interested motives" on the part of the author—language which is certainly very honorable to an anonymous writer—he then proceeds to particulars, and says—

"At page 114 [above, p. 136] commences a table of what purports to be comparative estimates of the cost of building factories, machinery, and &c, in Great Britain and America." He then takes the first table, which comprises building and gearing, and proceeds thus:—"It will strike everyone, that among the multiplicity of factories in the two countries, one, two, or more in actual operation, might have been selected in each country, nearly similar, and the true cost of which might have been readily ascertained. But demonstration by means of *known facts* does not appear, in this instance, at least, to have suited the purpose of the author!! On the contrary, *he supposes two cases!!!* One a factory in America, 142 feet long by 42 wide, four stories high, with an attic; the other in Great Britain, 90 feet long, by 38 feet wide, five stories, *flat roof and no attic*. What a supposition is this to begin with for a comparative estimate of cost!"

Here the reviewer asserts, without the least shadow of evidence, that I have *supposed* two cases; two hypothetical, or fictitious cases. He made this public assertion, not knowing whether it was true or not. But from what has already been stated, it will be seen that my comparison of expense, both for building and operating factories, in this country and Great Britain, was founded on factories *actually built,* and now in operation; but as I did not feel myself at liberty to publish the concerns of others, I referred to them as two factories, supposed to be of certain dimensions, and containing a certain number of looms, &c. It is therefore certainly no fault of mine, that these factories should differ somewhat in their construction and size. They are fair specimens of the style of building and arranging machinery in each country; and I thought myself particularly fortunate, in being able to obtain all the particulars of two factories so *exactly* alike, in the size of the yarn and weight of cloth manufactured, which thereby furnished the best means of making a fair and correct comparison of the cost of manufacturing (as it is usually conducted) in the two countries generally.

I would just remark, that your correspondent distinctly states that the British factory referred to has a *flat roof* and *no attic.* This statement he afterwards repeats. Now, if he will only turn to page 113 [above, p. 135] of the work he undertakes to review, he will find it there plainly stated, that the British factory *has an attic,* which is used for the "picking and scutching." And as to the *flat roof,* there is not such a word in the whole book.

With a design to show the incorrectness of my estimates, your correspondent next adverts to the difference in the superficial contents of the two factories. But all he makes of this is merely to show that there is a difference; what that difference is, or whether it is right or wrong, he cannot tell. He says, indeed, that the superficial contents of the American factory are 29,820 square feet, and that the British factory contains 17,100 square feet, making a difference of 12,720 feet. Now here he is lamentably wrong, in a very simple calculation. The American factory is 142 feet long by 42 wide, making the superficial contents of one story 5,964 square feet. Four times that number gives 23,856—to this is to be added the attic, which is only two thirds of one of the other stories—3,976; this added to the former gives 27,832 square feet, as the superficial contents of the whole mill; whilst the British factory including the attic, contains 19,380 square feet—making a difference between the two mills of only 8,452 square feet: and which is accounted for as follows:—

The basement story of the American mill contains the water wheels; whilst the 128 looms, with all the subordinate machinery, are contained in the three upper stories, with the attic—the whole superficial contents of which are only 21,868 square feet.

In the British factory, the 128 looms, with all the subordinate machinery, are distributed through the whole five stories, besides the attic—the whole superficial contents of which, as already stated, are 19,380 square feet; thus leaving a difference of the space occupied with machinery, of only 2,538 square feet. And be it remembered, that the American factory is driven by large belts passing up through the floor, which of itself occupies a considerable space; while the British factory is driven by an upright shaft, ascending in a recess, outside the main building; and from which crossed shafts enter into the different rooms, about two feet under the ceilings, so that there is not a single inch of any of the floors occupied with the gearing. Besides, the process of manufacturing is much more condensed, and the machinery more compactly arranged in the British than in the American factory. As in the latter, two sets of machines called speeders and extensers, are employed to do the same work; which in the former is done on one set of machines called fly frames.

Your correspondent proceeds: "Still more unfairly, Mr. Montgomery has supposed in Great Britain a steam mill, and in America a water mill." It may be a satisfactory answer to this to say, that in making a fair estimate of the general cost of manufacturing in the two countries, it seemed to be proper to found the calculations upon the mode of operation which is usual and customary in each. Had I done otherwise, I should have justly exposed myself to the charge of unfairness. The proportion of steam mills to that of water mills, in this country, does not amount to one percent; and perhaps the same remarks might be applied to Britain.[22] It is known, however, that the water mills in the latter country are of long standing; I know of none that has been built within the present century, and the ground upon which they stand was originally purchased, not as water privileges, but as private property. Hence as regards the cost of water power in Britain, no correct data can be furnished as the basis of a general estimate. In my late publication, I therefore made no allusion or reference whatever to the cost of water power in that country.

Your correspondent next refers to the difference in the cost of buildings, and says, "In the table referred to, the cost of building alone, for the American factory, is set down at $25,000; while that of the British factory is stated at no more than $4,408,[23] less than one fifth of the cost in America." "Is there a rational being," he exclaims "that could believe for a moment, that the expense of erecting a building in the United States, was nearly six times greater than for a similar building in Great Britain." Now, my statement, instead of referring to a similar building, refers to one which your correspondent has just said in the preceding paragraph, measures only 17,100 square feet on the floors, while the American building (as he says) measures 29,820. And accuses me of unfairness on the other side of the question, in assuming that "this difference in dimensions, and superficial contents, would occasion no difference in the cost."

After deducting this difference, and making a proper allowance, say $5,000, for the extra cost of the foundations in this country, on account of the depth of the frost and security from undermining the walls in mills driven by water; there will still be a difference in the cost of such buildings, in favor of Great Britain, which, surprising as it may appear, is fully warranted by the documents in my possession.

Notwithstanding the exclamation of your very sapient correspondent, he proceeds very gravely to account for the difference in the cost of the two buildings. He makes out that the American factory would, exclusive of the foundations, cost about $13,000; and in the ratio of comparative size, the British factory would cost over $7,000, foundations included. In this he proceeds upon the supposition, that the cost of building materials and workmanship, are the same in this country as in Britain. Now, from the documents referred to above, and other means, I *can prove*, that labor and materials are over 58 per cent cheaper in Glasgow than in Lowell. Statements to this effect were published in the Daily Advertiser, (3d January)[24] and well known to your correspondent. Hence, if his $7,000 (his supposed cost of the British mill) be reduced by 58 per cent, it will then stand, upon his own showing, only $2,940. Possibly *he* can explain how a mill could be built for that sum. Probably *no one* else can.

This communication being already too long, I will take up his other articles in my next. And having examined the specific charges of your correspondent so far, I may be permitted to enquire, what "departures from truth" have been shown, the detection of which should impair the confidence in other statements, that rest upon the responsibility of the author? What errors and misrepresentations have been pointed out by your correspondent? NONE! The comparative estimates of building, labor, machinery, &c., comprises a detailed statement extending to twelve octavo pages. It would not have been strange if in some of these I should have been mistaken. But your correspondent has pointed out *no error*, in regard to the rate of wages, the amount of labor required in conducting the business, the quantity of goods produced, or any of the expenses or estimates, that would affect the *results* of the comparative cost of manufacturing in Great Britain and America.

JAMES MONTGOMERY.

Saco, March 17th.

Letter of James Montgomery dated 20 March 1841, printed in the *Boston Courier,* 30 March 1841.

THE COTTON MANUFACTURE OF GREAT BRITAIN AND AMERICA CONTRASTED. No. 2.

Mr. Editor,—In my former communication, I exposed the fallacy of your correspondent, Justitia, in asserting that I had supposed two hypothetical cases, for my comparative estimates of the cost of manufacturing in this country and Great Britain. I also flatter myself that I fully explained, to the satisfaction of your readers, some of the comparative statements contained in the work, the title of which stands at the head of this article, in reference to which your correspondent attempted to fasten upon me the charge of duplicity and wilful falsehood.

The next subject in order upon which your correspondent animadverts, is in relation to the difference of *power* required in the two factories which formed the subject of comparison. He says— "The comparative *supposed* factories are *supposed* to contain 128 looms each. Yet that in Great Britain is *supposed* to require a steam power equal to only twenty five horses, while in the American factory a water power is required equal to eighty horses;" and proceeds—"If there be such an amazing difference between the processes of the countries in manufacturing, that the power equal to twenty five horses will do as much work in Great Britain as eighty will in the United States, he would confer a great obligation on the American manufacturer, by developing the why and the wherefore. Possibly he may *suppose* how it is done. Probably no one else can."

Now, I would just enquire what error your correspondent has pointed out in the above statement? He found, in the first place, that there was a difference in the superficial contents of the two mills; and second, that there was a difference in the cost of the buildings; and here, now, he finds a difference in the *power*. But does he say that any of these are errors, or does he give any indication to show that he *even knows* what *is error* from what *is truth*? NONE WHATEVER!!

Your correspondent knows very well that the two mills are *not* compared as doing the same *quantity* of work; that it is expressly stated that the produce of the British mill is 35,200 yards per fortnight, and the American 51,300 yards in the same time. He is also aware, if he has examined the estimates he undertakes to condemn; that in the British mill, the filling, or weft, is stated to be spun on mules, and the warp upon the live spindle throstle frames; whereas, in the American factory, the whole is spun on the dead spindle spinning frame, which requires nearly three times the power of the former.

I would further observe that I have no where stated, as your correspondent would have his readers suppose, that the power for driving the British mill was equal to twenty-five horses; nor do I make any estimate of the necessary power for driving either of the mills. At page 72 [above p. 101] it is simply stated that "the weight of the dead spindle spinning frame, compared with the common throstle, (live spindle) is considered to be as eighty of the former is to one hundred of the latter." In the tables to which your correspondent refers, no comparison is made of the power, *but of the cost* of buildings, machinery, labor, and &c.; and had he been disposed to honorable criticism, he would have confined his remarks entirely to the latter. This, however, would have led him greatly beyond his depth.

Amongst the various items connected with the British factory, a condensing engine of twenty-five horse power is enumerated, and the cost of which is set down. Now, if I am right, as regards the *cost* of the engine *actually* employed, together with the *cost* of all the other items included in the comparative estimates, it will make no difference as to the *result*, whether the engine was stated to be twenty-five, or thirty-five horse power. The result, or conclusion fairly drawn from my comparative estimates was, that the British manufacturer had an advantage over the American, equal to 19 per cent. in the cost of manufacturing. But this was more than counter balanced, by the advantages which the latter had over the former, in regard to the impost, charges, &c. on the raw material which left a clear balance of 3 per cent. in favor of the

American manufacturer. This conclusion, for very obvious reasons, your correspondent does not even *attempt* to controvert, though he ought to have known, that if a conclusion fairly deduced be *right*, the premises cannot be materially *wrong*. When your correspondent insinuated that I was in an error regarding the power required for the two factories, why did he not show the extent of the error, and apply the correction? No man has a right to charge another with error, unless he can demonstrate what is truth. Instead of his empty bombast about setting an American and English steam engine side by side, had he set his *truth* side by side with *my error* the public could then have judged impartially between us. Seeing he has not done so, your readers are left to draw their own inference.

The above may be regarded as a satisfactory reply to the statements of your correspondent. But it is evident he regards this difference in the power allowed for the two mills, as his strong point, upon which he seems to stake the issue of his case, seeing as he returns to it with peculiar emphasis in his second article. I will therefore take the liberty of entering a little more minutely into the subject.

The common estimates of power required to operate cotton factories in Glasgow are as follows:

One horse power calculated at a medium, will operate
100 throstle (live) spindles with preparation
500 mule spindles with preparation
12 looms with warping, dressing, &c.
See Greer's Mechanics' Calculator, page 325, 2d edition, Glasgow.[25]
See Brunton's Compendium of Mechanics, 4th and 5th edition, Glasgow.[26]
See also Allen's Science of Mechanics, page 147, Providence.[27]

The two first named gentlemen—Greer and Brunton—were long connected with the large machine and engineering establishment of the late Claude Girdwood & Co., Glasgow,[28] where they had ample opportunities for testing the power required for cotton factories; not by means of dynamometers, but by the ordinary amount of machinery operated, by various steam engines of determined powers. It is also to be remembered, that their estimates include the mill gearing. Dr. Ure's experiments agree with the above, as to the mule spindles with preparation, but makes one horse power equal to 180 throstle spindles, preparation included, and 10 looms, dressing, &c.[29]

Now, the British mill mentioned in my comparative estimate, having 2,160 throstle (live) spindles, 2,400 mule ditto, with 128 looms, if tried by the preceding estimates of power, would require a fraction over 21 horse power for the throstles, and a fraction less than 5 horse power for the mules, making 26 horse power for the spinning, with preparation; and over ten horse power for the looms, equal 36 in all; or, according to Dr. Ure's estimate, only 30 horse power. It is probable, however, that the latter does not include the mill gearing, for which five or six additional horse power will be amply sufficient, in consequence of the neat, light style, in which such gearing is fitted up in Glasgow.

Experiments were made at Lowell by means of a dynamometer, to ascertain the weight of American machinery; and according to the results of these experiments, as well as the results of experiments tried at Saco, by Batchelder's dynamometer, it was found that 80 spindles of the dead spindle spinning frame, with preparation, requires full one horse power; and that an equal power is required for 8 or $8^{1/2}$ looms with dressing &c.[30] Therefore, as the American factory mentioned in my comparative estimates, contained 4,992 dead spindles, with 128 looms, if tried by these results, will be found to require a fraction over 62 horse power for the spinning, with preparation, and over 15 for the weaving, dressing, &c., (taking the later at 8.5 looms to the horse power,) making the whole machinery equal to the power of 77 horses, exclusive of the mill gearing; and in consequence of the large heavy belting, wooden drums, and the speed at which the gearing are driven, we cannot suppose the latter to require less than 12 or 13 horse power, which would make the whole mill to require a power equal to 90 horses. Making a difference of power required to operate the two mills of something over 50 per cent.; and however incredible this may appear to some of your readers, to me it is nothing at all surprising, having had experience of the machinery in both countries. Nor does it appear that the Americans have paid so much attention to the saving of power, as has been done by the British; which may no doubt arise from the fact, that the steam power of the latter costs about (if not over) four times the cost of water power of the former.

Your correspondent states as "another instance of fairness in estimates," "Mr. M. takes as the basis of a general estimate of the cost of steam power in this country, and to prove it double the cost of steam power in Glasgow, a mill in Massachusetts, containing, as he says, 3,700 spindles, operated by an engine of the power of forty horses, at an expense of $12.20 per day. At the very moment he penned this statement, it can hardly be doubted that he must have known it deceptive."

A person who is thus so liberal in his charges of deception and falsehood on others, might have been expected to show some *little respect* for truth himself; but of course, he being a native, it is not supposed his statements shall be very scrupulously examined. The steam mill referred to by your correspondent, is a small mill in Newburyport,—not over one hundred miles from his own residence. The original statement which I published is yet in my possession, dated December, 1837. It was very obligingly furnished to me by the then superintendent of the factory; a gentleman, who for honor and respectability of character stands far above your correspondent. The mill then contained 1,584 mule spindles, 2,116 throstle (dead) spindles, with 100 looms; making light cloth from No. 30s yarn. And if we allow 100 throstle spindles to the horse power, 500 mules ditto, and 10 looms, with all the necessary preparation (which is certainly a fair estimate, considering the fineness of the yarn)—we shall then have 34 horse power as the weight of the machinery, leaving six for the gearing, making just 40 horse power, as stated to be the estimated power of the engine.[31]

And yet your correspondent is perfectly puzzled to understand how these different mills should require such a difference in the propelling power. He concludes a long statement with these words:—"And then, again, it is equally mysterious how his Massachusetts steam engine, of the power of 40 horses, should operate a mill with 100 looms, while in his comparative estimates a water power equal to the power of 80 horses should be required to operate 128. These conflicting statements he is left to reconcile; I cannot pretend to do it."

And why, I would ask, did your correspondent undertake to condemn, as deceptive and erroneous that which he knows nothing about? Why did he undertake to criticise a subject, of which he is so palpably ignorant? Had he possessed the practical knowledge of a common overseer in our factories, he would have known there was nothing irreconcilable in my statements—which I presume have now been satisfactorily explained to your readers.

The next, and only remaining subject referred to by your correspondent, as requiring any explanation from me, is in the beginning of his second article, where he refers to the difference in the amount of expense incurred, per fortnight, for conducting the two mills. In the American mill, the expenses foot up $1,954.45; and in the British mill, $1,040.59, making a balance in favor of the latter, of $913.86. This balance he endeavours to explain away by a quibble, perfectly in keeping with all his other statements. He finds there is a difference in the capital stock invested in the two establishments. The capital for the British being $54,000, and for the American, $110,000, which he says "would make a difference in insurance and interest of not far from $600 per fortnight." And this, "If the capital were equal, would cut down the difference in the charges per fortnight, and make *them* nearly equal! This would be the case provided both mills were operated by steam."!!

To show the ridiculous absurdity of the above, I will just enumerate the different items, in which the British manufacturer has an advantage of the American.

The cost of building materials and labor, are at least 58 per cent. cheaper in Glasgow than in this country. The price of machinery is also 48 per cent. cheaper. Insurance on Manufacturing stock is 7–8 per cent. in Glasgow; in this country it is from 1 to $1^3/_4$ percent. Interest on capital, 5 per cent. in Britain, 6 per cent. in America. Besides all these—the credit given on manufactured goods, sold in Glasgow,—and I believe in Manchester, too,—is from one to two months; whereas, in this country, it is from six to eight months. Yet, notwithstanding these various items in favor of the British manufacturer, your correspondent will prate about the capitals and expenses being equal, and says "that such would be the case, provided both mills were operated by steam." Bah!

Having now not only examined the charges of wilful falsehood, on the part of your correspondent, but the grounds on which his charges were founded, I hope I have shown to the satisfaction of your readers, that these grounds do not warrant the application of ignorance, and deceptive motive, to the author of the work referred to; but, on the contrary, that the comparative estimates reviewed by your correspondent are perfectly consistent with facts; and that the mistaken views which he took of them arose from practical ignorance, or misapprehension, on his part. I now leave him to the very unenviable position in which he stands before the public.

JAMES MONTGOMERY.

Saco, March 20.

Rejoinder

Justitia, No. 7   *Boston Courier,* 13 April 1841

AS Mr. Montgomery has replied to my foregoing numbers, I solicit the privilege of making a rejoinder. In doing this, I shall endeavour to be as brief as the circumstances of the case will permit. Delicacy might perhaps prevent me from entering the field again, as an *"anonymous writer"* after the significant sneer of my antagonist, were it not that he soon appears to have forgotten himself, and determined to place himself on a level with me, by entering the list in his own defence. Under these circumstances, I feel no repugnance to breaking a lance with him.

In the outset, Mr. Montgomery has charged me with having attributed wilful falsehood to him. Whether this charge is made with the intention of enlisting public sympathy for himself, as a persecuted man, or for some other purpose, I am not aware; but certainly I have not laid any accusation of *"wilful falsehood"* against Mr. Montgomery. In speaking of the errors in his work, it is true, I said it was difficult to understand how they could have originated in ignorance of facts. I repeat that remark; predicated on the supposition that Mr. Montgomery supposed himself possessed of all necessary information on this subject. But my communications went to show that he was not. If he was, I have no qualification to make to the remark. If he was not, it needs none. Mr. Montgomery seems to entertain some slight misgivings himself, as to the complete accuracy of his book; for he, in his reply to me, casts himself on the indulgence of the public, and, admitting that "little inaccuracies" may have crept into it, attempts to palliate them, by pleading the short period occupied in the preparation of the work, and the fact that the work was published in Scotland, so that the author, being in the United States, could not superintend the printing.

All this comes with an ill grace from an author who publishes to enlighten the public mind on an important subject. Why so much haste? The world would not have been convulsed in consequence, if Mr. Montgomery had taken four years to prepare and perfect his work, instead of driving it through, half made up, in two years.

And had it been published in the *United States,* instead of being sent to *Scotland* for that purpose, the author would have had it under his own eye, to correct and revise it during its progress through the press. His apologies are futile. Shortness of time, when more time could be had, and the publication of his work at the distance of three thousand miles, when it could have been as well done within the length of his own shadow, afford no apology for errors. Errors there are in Montgomery's work, and however they may have found their way into it, whether by his own pen, or the *pen* or press in *Scotland,* he is accountable for them. As a specimen, I would now particularly cite the following case, to which he and I have both referred before, and which is not only an error, but evidence of unfairness.

At page 214 [above p. 211], of his book, Mr. Montgomery has an estimate, headed "COST OF STEAM POWER IN THE UNITED STATES." Any person of common sense, on reading this, would expect to find, in what is to follow, a fair actual average of the cost professed to be stated. If unacquainted with the subject, he would, on perusal of the article, suppose that Mr. Montgomery had made due enquiry, and found that the cost of steam power was $12 20 per day, as a general average, for an engine equal to 40 horses. No other inference can be drawn, because, though he gives (undoubtedly the correct result) of a solitary engine, he tells you it is the "*Cost of Steam Power in the United States.*"

In the first place, this is a gross error; for it is well known, that, in general, steam power in the United States costs less than he has stated, by at least 25 per cent., and in many cases, by 50 per cent. It is grossly unfair, in the second place, because, though he has given the cost of water in various parts of the United States, he has passed by ten thousand steam engines, neglected every means of information he might have derived from Pittsburg, Philadelphia, New-York, and elsewhere, and taken as his stand, and for the "*Cost of Steam Power in the United States,*" an engine which runs at a greater expense of fuel than almost any other one he could name, in proportion to its power. I have never doubted the correctness of the statement of expenses relative to the engine to which Mr. Montgomery refers, and he knows it. Why then his mean and dastardly insinuation that I question the veracity of the gentlemen who gave it? But I said, and I repeat it, that Mr. Montgomery knew that statement deceptive, as published by him as a standard of the "Cost of Steam Power in the United States." The statement in that respect is untrue—untrue as a general average, which Mr. Montgomery makes it, though true as the cost of steam power for the establishment it refers to. As to epithets, innuendoes, and comparisons of private personal character, Mr. Montgomery is welcome to them. Wherever he may find gentlemen who "*stand far above me*," instead of envying, I shall honor them; and hope Mr. Montgomery will profit more by an association with such persons, than, by his writings, I should suppose him heretofore to have done.

I notice here, another of his "*little inaccuracies.*" In his estimates of the cost of machinery, he states the price of mules, by the Providence Machine Company, at 2 dollars and 75 cents per spindle. The actual price is 2 dollars and 42 cents; making a difference of about 14 per cent. from his statement, while some other companies sell at prices 25 per cent. less than he states. In the payment of bills for machinery, such an "inaccuracy" would not be considered very small.

If Mr. Montgomery supposes me to think my countrymen entertain "illiberal prejudices" against foreigners, he is mistaken. On the contrary, I am fully aware that a very numerous class of them are readily overcome with the empty boasts of foreigners without merit, and duped by foreign works without intrinsic value. It is the interest of foreign adventurers, among whom I do not by any means include all foreigners, to keep up this feeling among the Yankees, by imposing on our credulity, and inducing us, if possible, to bow the knee in humble adoration of John Bull and his numerous offshoots. It was a work of supererogation in me, to remind the American reader that Montgomery's book was a work of foreign origin, having the Glasgow imprint, having been sent there probably to be *revised* as well as printed, and bearing on its title page "*United States of America,*" in small sized capitals, and "*Great Britain*" emblazoned with *large ones*. Besides this, the labored effort, in almost every department, to show Great-Britain as our superior, is sufficiently characteristic of its birth. Even John *Bull's sheep* skins are made to act a conspicuous part; for, in a note at page 210 [above p. 209], the author says, "The sheep skins used in Great-Britain are equal in quality to calf skins used in America, and are fully double the size." He does not condescend to inform us what calf skins he means, whether of *John Bull's* calves who congregate here, or some others. He forcibly reminds me of a newly imported Englishman who, on viewing an apple tree in a gentleman's garden, pronounced the apples to be of tolerable size, but not near so large as those in England. The tree had been loaded for the occasion, with half grown pumpkins! Thus, in Mr. Montgomery's view, we have little or nothing in our manufactur-

ing establishments to be compared with those in Great-Britain—
more properly Glasgow—even our operatives have not the merit
of being so nearly starved!

To return—Mr. Montgomery takes me to do for *supposing* that,
in his "comparative estimate," he had introduced two "hypothetical and fictitious cases;" instead of two factories, one in Britain
and the other in America, as he now says he did; "*actually* built,
and *now* in operation;" and says I have misrepresented him. I cry
the gentleman's mercy. I presumed that Mr. Montgomery wished
his readers to understand him as he wrote, and to *suppose* when
he *supposed*. Thus I read and thus I *supposed*. The fact is, the
gentlemen has, in no place in his book, claimed that the factories
on which he founds his "comparative estimate," were "*actually*
built, *now* in operation," or ever had been, or ever would be.
Whatever he might have had in his mind, he wrote thus—"In the
following estimates of the cost of buildings, machinery, &c. of
manufactories in the United States and in Britain, each factory is
SUPPOSED to contain, &c." "In estimating the cost of a factory
in the United States, the extent of the mill is SUPPOSED to be,
&c." "The factory in Britain is SUPPOSED to be," &c. This is his
own language—this the manner in which he introduces his mills.
It is all *supposition* in his book—all "*hypothetical* and *fictitious*" to his
readers, whatever it may have been to him; and having thus *supposed* his cases, he might as well have required his readers to tell
where his mills were located, and who owned them, as to know
that they existed at all. His book gives not the least intimation of
their existence or operation; and he knows it. How then have I
misrepresented him?

Mr. Montgomery next quarrels with me because I gave to his
British factory a "*flat roof.*" Well, I will attempt to account for
having done so. In his introduction, Mr. Montgomery *supposed*
the American factory to be four stories high, *besides an attic;* and
his *supposed* British factory he supposed to be five stories high,
besides an attic. But when he came to make up his estimate in
figures, he set down his American factory with the attic, and his
British factory *minus* the attic. I read him precisely as he wrote. I
found no supposed British factory with an attic, in his estimate,
though I found an American one; and hardly knowing how to
dispose of the attic-*less* building without a roof, I *supposed* a "*flat
one,*" rather than leave the fifth story open, exposed to the
weather. But as Mr. Montgomery has *supposed* his *supposed* attic to
be restored to his *supposed* British factory, I *suppose* I may be con-

tent to take away my *supposed* "flat roof," and to let it sink into the proper insignificance of a "*hypothetical and fictitious case.*" Alas, poor Yorick!

As to my error in calculation of the superficial contents of the floors of the two factories, it does not exist, except in the case above referred to; and for that, I am not accountable. In the *supposed* American factory, there are five floors, including the attic, each floor being 142 by 42 feet. Their aggregate area, as before stated, would be 29,820 feet. Including the *supposed* attic in the *supposed* British mill, there are six floors, each 90 by 38 feet; and the aggregate area is 20,520 feet. The difference, 9,300 feet. Mr. Montgomery's wit, bluster, and "*depth,*" to the contrary notwithstanding, the mode of calculation is correct; and all the error in the former one, originated in his own blundering supposition. And as to the manner in which the different stories may be occupied, or whether they be occupied at all, makes no difference to the point, the question involving only the cost of the building, and not the arrangement or sizes of belts and machinery. But our author cannot permit an opportunity to pass, without sounding the praises of John Bull; and hence, in this place, he condescends to exhibit the superior sagacity of the old gentlemen, in his factory arrangements, compared with Jonathan's clumsy habits.

Mr. Montgomery attributes to me the "*supposition*" that "the cost of building materials and workmanship are the same in this country as in Britain. This *supposition* is one of his dreams—it is not mine. He says that I made out the cost of the American factory, exclusive of the foundations, (*mark that*) about $13,000. Very good. Now he says, if I deduct from that [the amount of $13,000] the sum of $5,000 for difference in the cost of foundation in Britain and this country, and the cost of difference in size in the two factories, I shall bring the cost of the British factory down to $7,000. No doubt of it. But why deduct any thing for difference in cost of foundation? If Mr. Montgomery wishes to deduct his $5,000 on that account, it must come from his estimate of $25,000, in which the cost of foundation is included. My estimate of $13,000 does not include that item. All he can deduct from this sum for foundations is the cost of foundation for his steam mill in Britain. Here it would cost but $1,000. There, according to this account, it would cost less than $500. Thus, instead of reducing the cost of the building in Britain, as he vainly supposes, according to my calculations, to $2,940, he would find it about $5,000—even admitting, which I do not believe, that materials and workmanship are cheaper by 58 per cent. in Britain than they are in the United States, as he says.

After having become conscious of the very gross error exposed in the foregoing paragraph, perhaps Mr. Montgomery may not be quite as much disposed to exult over the supposed mistakes of others; and he may listen with somewhat more patience to what little I have to say. I have said, I do not believe the comprehensive statement he makes with regard to the comparative prices of materials and workmanship, in the two countries. I have it from a respectable source, from a gentleman who has been employed at Manchester as a master, to superintend the erection of factories, and setting up engines, that labor is about 33⅓ per cent. cheaper there than here—that stone, lime and hardware are cheaper, but that bricks and lumber are very much dearer. On the whole, he considers that Mr. Montgomery's supposed *brick* factory, in Britain, would cost nearly if not quite as much there as here.

On the subject of the engine of 25 horses' power which he connects with the estimate of the British factory, he scolds a while, and then says: "I have no where stated, as your correspondent would have his readers suppose, that the power to drive the British mill was equal to 25 horses; nor do I make any estimate of the necessary power to drive either of the mills."

Well, what does Mr. Montgomery wish *his* readers to *suppose*? He *supposes* his establishments, and feels himself very much outraged if others do not consider them real substance—factories "*actually* built, *now* in operation." In connection with those factories, we find set down an engine of 25 horses' power, its cost carried out and made up into the footing, or aggregate amount, and the daily expense of coals to operate it, carried forward in the items of current expenses. But lo! when Mr. Montgomery finds his power quite unequal to his work, he wheels about and says, Oh! I did not wish to have it understood that my supposed engine would drive the mill! Oh no—the gentlemen is quite willing to have a "*hypothetical* or *fictitious*" engine in his *bona fide* mill, but my difficulty is, to understand how, with such an engine, that mill can be "*now* in operation." But his flurry about the business is all fudge. His statement was a blunder, either of himself, or some one else; and no one who reads the estimate, can doubt that he intended the engine to drive the mill, and supposed it sufficient for the purpose. Every one who reads it must so understand it; for it is set down as one of the fixtures of the mill; and its cost, which is but about 50 per cent. of the cost of an engine of the requisite power, is carried into the estimate of aggregate cost. It does not exhibit a spirit of superior fairness and candor in Mr. Montgomery to treat the subject as he has done, in a jesuitical manner. How much reliance can be placed on the statements of an author, who to avoid the honest confession of an error, will reduce himself to such pitiful shifts, and involve himself in such preposterous absurdities?

Having placed himself in an awkward position, by weaving into the "comparative estimate" what he appears to know nothing about, first, Mr. Montgomery disclaims all intention to make an estimate, and then proceeds immediately in an attempt to make one, in the hope of extricating himself from his difficulty. That estimate, when completed, exhibits another of his "*supposed* hypothetical, *or fictitious cases.*" It is not necessary to notice this estimate in detail, as its errors can be exposed without it. Suffice it to say, he adds 11 horses' power to his former 25, making up 36; and seems to conclude that will answer very well for his *supposed* British factory. Let us now turn to Grier, and see how far he sustains him.

Grier allows one horse power for 100 throstle spindles and preparation. The British mill is *supposed* to contain 2,160 throstle spindles; requiring, according to Grier, a power equal to 21 60/100 horses. For mule spindles, and preparation, spinning filling No. 18, the power of 8 horses is required to drive 2,400, the number *supposed*, in the British mill; and the looms, 128 in number, with dressing, &c. require a power equal to that of 10 66/100 horses. The aggregate power therefore required for the machinery in Mr. Montgomery's calculation, would be, according to Grier, equal to that of 40 60/100 horses. This is for the machinery alone; but then, about four tenths of the power generated, is lost in overcoming the friction of the engine, shafting and gearing; and which require an additional power equal to 16 10/100 horses, making up an aggregate of 56 70/100. Besides all this, as an engine cannot long work with its maximum power, its nominal power should exceed its actual required power by 25 per cent.[31] And therefore, to drive Mr. Montgomery's British mill, he should have an engine of at least 66 horses' power, instead of 25 or 36. Perhaps, if the subject be not "*beyond his depth,*" he may, by this statement, detect the cause of his own error.

To substantiate the foregoing statement, I submit the practical result of a mill of about the same amount and description of machinery, as that set down by Mr. Montgomery; observing however, that this mill manufactures cloth No. 30, and requires of course less power than to manufacture No. 18, as it does in the British mill. There are 5 cylindrical boilers, each 28 feet in length, and 30 inches in diameter. The aggregate fire surface is 490 feet. Allowing 9 feet per horse power, this gives a power equal to 54 4/9 horses. Again—area of main cylinder, 154 inches—piston 5 feet stroke, 30 double strokes per minute. Pressure of steam acting against the piston 37 1/2 pounds per square inch. Thus, area of cylinder 154 x 37 1/2 x five feet stroke x 60 strokes per minute— $\frac{1,732,500}{33,000}$ —52 1/2 horses power.[32]

   This is no supposed case, but a practical illustration, and shows conclusively that the actual power required to overcome friction, and drive the machinery, is not less than that of 52 1/2 horses. Did the mill manufacture No. 18 instead of No. 30, the power would require to be increased about 20 per cent.

   I have thus noticed all that Mr. Montgomery has advanced, that I deem of the least importance. As to criticism on his book, I undertook nothing more than to point out a few of his numerous errors; and as to following him "beyond my depth," I have little apprehension of drowning where he can manage to keep his head above water. I shall not quarrel with him about ignorance. If he is the possessor of superior knowledge, I shall be happy to see some of it displayed;—as it is, I believe the world is but little indebted to him on that score. I confess I am too ignorant to reconcile his errors with truth, or his contradictory statements with each other; and have not yet arrived at that degree of gentle-manly refinement, and the powers of mimicry, to enable me to assimilate myself to a sheep, and to cry "BAH"! Profound in knowledge, however, as he may fancy himself, he is challenged to reconcile his statements with each other, as to the difference of power required by his showing for the Massachusetts mill with 100 looms, and the British mill with 128, by any thing he has said, or can say. And the darkness and obscurity in which he has left the subject in his book and in the Courier, show that he was too ignorant or too careless to be fit to write a book to enlighten or inform the public mind. As to his hodge-podge remarks relative to financial subjects, and comparisons between affairs in Britain and America, I consider them as mere quackery, and unworthy of notice. They may be partially correct—they may not. He prob-ably does not know. They are a part of his "*hypothetical* or *fictitious*" castle in the air, deserving little credit, except such parts as are gleaned from other authors. And now, perfectly satisfied with my *"position before the public,"* however *"unenviable"* it may be deemed by any profound ignoramus, foreign or domestic, I dismiss Mr. Montgomery, his book, and his communications; and shall consider neither as worthy of further notice.

Letter of "B" printed in the *Boston Daily Advertiser and Patriot,* 26 March 1841.

STEAM AND WATER.

Mr. Hale[33]—As the comparative cost of steam and water power for driving factories has from time to time been a subject of discussion, I was glad to see that a correspondent of the Courier, under the signature of Justitia, promised, a week or two since, to come forward with such statements as would set the matter at rest. I have accordingly examined with some interest the preliminary articles, and at length find in the Courier of the 11th and 15th, the promised calculations. I propose to make a brief examination of those articles, and to give you my views on the subject, which differ materially from those of the writer referred to.

He takes for the subject of his comparison the Boott Mills at Lowell, and the Bartlett Mill at Newburyport, and proposes to leave out of the calculation, all such expenses as would be the same in mills driven by water, as in those driven by steam, and to state an account of such expenses as differ in the two modes of operation, so as to come to a fair conclusion on the subject. A summary of this account, according to his manner of stating it, would stand as follows:

Four Boott Mills, containing 29,248 throstle spindles, and 830 looms.
The expense of warming these mills is estimated as follows:

750 tons coal, at $6 per ton,	$4,500
70 cords of wood, at $6 per cord,	420
Transportation of 750 tons coal, at $2	1,500
Making the cost of fuel	6,420

The next item of expense is for transportation from Boston, of 533½ tons of cotton, at $2 1,067

468 tons of goods to Boston, at $2	936	
83 tons oil and starch, at $2	166	
Transportation of the above articles	2,169	
These two sums added together make the expenditure for warming the mills, and for transportation of cotton, manufactured goods, oil and starch, the sum of		8,589
He then puts down for purchase of water power for 29,248 spindles, at $3		87,744
Water wheels and gearing for four mills		85,000
Foundations for four mills, being an extra expenditure on account of driving by water, for wheel pits, &c.		40,000[34]
Making an expenditure of capital of		$212,744

The annual interest on which is $12,764.64.[35]

The writer than proceeds to the corresponding expenditures in the two Bartlett mills, containing 17,600 mule spindles, and 368 looms, and estimates the cost of foundation

for No. 1 Mill,	$	1,000
Engine and gearing for same,		10,000
Foundation for No. 2 Mill,		2,000
Engine and gearing for same,		15,000
Making an expenditure of capital	$	28,000

The annual interest on which is $1,680.

He then says, "the Bartlett mills being only two in number, and the Boott mills four, the expenses of the latter must now be divided, and one half assumed as applicable to two of them." Accordingly he sets down half the amount of interest on capital expended, as above stated,

on Boot mills	$ 6,382	32
And deducts interest on Bartlett mills	$ 1,680	00
Leaving for balance of interest in favor of Bartlett mills,	4,702	32
To this he adds one half the amount of expenses stated above, for fuel and transportation,	4,294	50
Making the sum of	8,966	82

He then says, the Bartlett mills containing 2,376 spindles more than half the number in the Boott mills, it is proper to add in that proportion   <u>1,284 00</u> [36]

And we then have in favor of the Bartlett mills, to be used for generating
steam power, the sum of $10,280 82

He then proceeds to state the cost of
steam power, as follows: 930 tons of
coal at $6\frac{1}{2}$,   6,645 00[37]

1 Engineer, 310 days at $2   620 00

2 firemen, at $6 per week each,   516 67[38]

Total Cost of steam power,   <u>7,181 67</u>[39]

Which deducted, leaves a balance in
favor of steam mills of $3,099.15

In examining and comparing these statements, the first particular that will be noticed, by any one acquainted with the business,
will probably be an item put down in the expenditures of the
Boott mills, for transportation of manufactured goods, cotton, oil
and starch, amounting to $2,169.50, against which there is no
corresponding charge whatever, in the expenses of the Bartlett
mills. Is it to be inferred from this, that no expense is incurred
by the mills at Newburyport for transportation of those articles?
Are the goods sold without bringing them to Boston? If a part of
the cotton is received directly from Southern ports, does not a
part, at least, pay freight from Boston? Whatever difference there
may be in the cost of transportation, for mills situated in Lowell
or Newburyport, should fairly be taken into the account, in this
view of the matter, and no more. So far as the transportation from
Newburyport should be by railroad, the expense would be as much
as to Lowell, the distance being something more. The freight of
cotton by water, part of it received, perhaps, direct from the south,
might be something less than one dollar per ton. We certainly
shall not estimate the whole transportation at too high a rate, by
calling it half as much as between Boston and Lowell, or one dollar
per ton, making the item $1,084.75.

The writer also sets down for warming the four Boott mills, 750
tons of coal. His authority for this is what he calls the "official
statistics of the Lowell factories." He probably refers to a printed
sheet containing a statement in round numbers, of the average
quantity of goods manufactured, number of bales of cotton used,
and quantity of starch, oil, coal and other materials. This, though

no doubt correct as to the average quantity of coal *purchased*, would not be considered, without knowing how it was used, or whether it was used at all, the most satisfactory data from which to calculate the cost of warming the mills; a matter which, in this argument, involves more than the whole amount of the difference stated by this writer in favor of the use of steam. But suppose we take his statement to be correct, let us inquire whether his argument will not prove too much. He sets down the cost of fuel for warming the four Boott mills, at $6,420 00

One half that sum, for two of the mills, is   3,210 00

Now as fuel for warming the mills is only wanted for half the year, take for the corresponding six months, half the cost of fuel set down for driving the steam mills for the year, which would be   3,022 50

There would remain a balance of $187 50

According to this, all the steam for driving the mills, during one half the year, would be insufficient to warm them!! It is not long since I was requested to examine a project for manufacturing by steam, in which the estimates and calculations involved the same absurdity, and there seemed to be something of a kindred character in the whole scheme.

I believe the most accurate estimate of the expense of warming a mill may be derived from an experiment, the result of which, I believe, was published in your paper, two to three years ago. It stated the actual weight of coal used for one week, during which the state of the thermometer was noted from hour to hour, night and day.

The average temperature of the mill was   67 07

Temperature of the external air,   14 05

Therefore the temperature was raised   53 02

The quantity of coal consumed was 7,484 pounds. Assuming ten weeks during the Winter at this temperature, which was in fact the coldest week of a cold Winter, and twenty weeks more, for Spring and Fall, requiring half the quantity of fuel, we have 149,680 pounds. The mill in which this experiment was tried, was of similar construction to the Boott mills, with the same number of windows and doors, and of the same dimensions, except two feet less in length, being 145 feet long. Therefore, add for this difference 10,322 pounds—160,002 pounds, or nearly 72 tons for one of the Boott mills. It would therefore require for the four mills, 288 tons,

Which would cost $1,728 00
Transportation at $2 per ton,   576 00

And this it should be recollected, is for warming the mills the whole twenty-four hours, while those mills that are warmed by the exhaust steam from the engine, have the benefit of it only during working hours, and consequently the temperature for some time in the morning is much too cold for manufacturing operations, or for comfort, unless other means are provided, no estimate for which is included in the statement.

I would make but one other correction in the estimate for the expense of Boott mills. He states the cost of foundations at $10,000 per mill. This is much too high for ordinary situations. About $5,000 would be a fair average.

In the present view of the matter I do not think it necessary to go into any examination of the expense, as stated for steam engines. I would only remark that this is *estimation* and not *practice*. It is not even *experiment* for though in the beginning one would infer there were 17,000 spindles in operation in the Bartlett mills, in the end it seems "this requires some qualifications," and it is stated that "mill No. 2 is as yet only in partial operation; having now in use about one half its machinery." I propose to take the estimate for the steam mill as stated, with the following very obvious correction. Justitia says "The Bartlett mills contain 17,000 spindles. That is *more* than one-half the number contained in the Boot mills, by 2,376. This difference is about one-seventh, for which *add* $1,284," &c. Now the number of looms in the Bartlett mills is 368, which is *less* than half the number in the Boott mills by 47. I therefore correct the statement by a deduction accordingly. It will not be disputed that the number of looms affords a more correct comparison than the number of spindles, particularly as the estimate in the one case is for *mule* spindles, and in the other for throstle spindles—and the writer cannot object to this, unless he can show that his looms produce more cloth than those in the Boott mills, which I think he will not attempt.

I have thus, without going into detail more than was absolutely necessary, prepared the way for making up a corrected account of the comparative expense in the two cases stated, as follows:
For warming four Boott mills, 288 tons of Coal, @ $6. $1,728.00
Transportation of same, at $2*   576.00

* The Rail Road freight for coal to Lowell is but $1 a ton, exclusive of loading, or for cotton, $1 50.

Difference of expense of transportation of manufactured goods, cotton, oil and starch, as already stated,	1,084.75
Making the amount of these expenses,	$3,388.75
Purchase of water power for 29,248 spindles, at $3,	87,744.00
Water wheels and gearing for 4 mills,	85,000.00
Cost of wheel pits and foundations,	20,000.00
Making an expenditure of capital of	$192,744.00
The annual interest on which would be	11,564.64
Half this amount for the two mills, is	5,782.32
to this add half the other expenses, as above	1,694.38
Makes the expenses of two Boott mills	$7,476.70
Deduct for difference of 47 looms less in Bartlett mills, than in 2 Boott mills,	846.00
Leaves for expense of Boott mills of equal capacity of 2 Bartlett mills,	$6,630.70
Against this sum we take the expenses of the Bartlett mills, as set down by Justitia, without alteration, being for the cost of coal, engineer and 2 firemen.	$7,181.67
Interest on cost of engines, foundations and gearing	1,680.00
Which makes the actual expense of Bartlett mills,	$8,861.67

In the foregoing statements, I have followed the steps of Justitia, not because his mode of stating the matter gave the most perspicuous view of it, but because my object at present was to show, from his own premises, and in a manner to which he could not object, that he had placed the balance on the wrong side of the account. And I trust it is now shown, as the result of his own statements, corrected in a few obvious particulars, taking into view all the items that entered into his calculations, and excluding all others, that instead of a balance of $3,099.15, in favor of the Bartlett mills, there is a balance of $2,239.97 against them. But even this is very far from what the author professed and promised to do. That is, to show the comparative cost of *steam and water power*. So far from it, he does not calculate, nor even attempt to estimate the *power* to drive the different mills. He also takes it for granted that there is no difference in the expense of mule spinning and throstle spinning, and assumes that mills driven by water must of necessity be subject to the heavy expense he puts down for transportation, and that those driven by steam require no such expenditure. In relation to these matters I propose to make some further observations. B.

Anonymous letter printed in the *Boston Daily Advertiser and Patriot*, 29 March 1841

THE BOOTT MILLS AND STEAM COTTON MILLS. The following communication was received before the article in Friday's paper, on the same subject, was published, though after that article was in the hands of the printer. It may perhaps be thought that that article is a sufficient reply to the essays referred to, but as the corrections here made are not embraced in the article, and as they apply particularly to statements in reference to the Boott Mills, we here insert this communication.

MR. HALE—A friend has called my attention to a series of essays that have recently appeared in the Courier—purporting to be a review of a work published in England by Mr. Montgomery, of Saco, on the state and prospects of the Cotton Manufacture in the United States.

With the reviewer's estimate of that work I in the main concur. I think that with a good deal of valuable information are mingled up some serious errors; and that the book was written with a design not American but English—to awaken the attention of statesmen in Great Britain to the impolicy of the duty on Cotton.

I must confess, however, that I like the review still less—because, without the valuable information, it contains to say the least as serious errors—and is moreover written with a design alien from its apparent object—a design to create in the public a confidence in Steam Cotton Mills in general, and in the Bartlett Mills at Newburyport in particular.

It is not my intention to enter at large into the argument. Experience and all the best authorities on England, have settled the question that water power is *in itself* cheaper than steam. That it should afford any rent whatever, is conclusive of this question. Water power, Mr. Montgomery states to cost four times as much in Britain as here, while steam power in the United States costs twice as much as in Glasgow. Why, in the name of common sense, should the English Manufacturers pay such prices for water power, unless it were in itself far cheaper than steam.

The author claims peculiar credit for the authenticity of his statements. "They shall be founded," he says, "on incontrovertible data," and he selects the Boott Mill No. 2, at Lowell, as a standard of comparison. Now would you believe it, Mr. Editor, he is not furnished with one single correct fact with respect to that mill except the number of spindles.

I shall illustrate this by showing three egregious errors that he has committed, and then leave him to what credit he deserves.

First—Cost of water power. He says "at $4 per spindle, (according to Mr. Montgomery) the water power for 8,640 spindles, would cost $34,560, the interest on which is $2,073.60."[40] He tells us elsewhere that coarse goods require a greater amount of power than fine goods—and everybody knows that they require fewer spindles. Does he then believe, or not believing does he wish to persuade others that the price of water power at Lowell or anywhere else on the face of the earth is or ever was sold by the running *spindle*, so that a fine mill requiring half the power, shall yet pay more for that half, than a coarse mill does for the whole? The *fact* is that power at Lowell is and always has been sold by the number of cubic feet of water to be drawn in a second of time in a given fall. The price paid moreover is for land and water. The whole land required for mills, cotton houses, machine shops, counting rooms and boarding houses goes with the power, and has in practise been worth all or nearly all the price paid. For Boott Mill No. 2 the sum paid for land and water was about $15,000. This first blunder or misstatement then is one of his incontrovertible data, is about one hundred and twenty-six per cent.

Second. Product of the mill. Our reviewer states that "about the average amount spun by this mill is four skeins per day per spindle, making 264,000 pounds[41] of cotton to be transported." The *fact is*, that the quantity of cotton used in Mill No. 2, for the year ending in July last, (and it is now using somewhat more) was 388,000 pounds, showing an error òr misstatement in another essential point, of about forty six per cent.

Third. Expense of transportation. This the reviewer states on the authority of Mr. Montgomery, at $2 per ton. Any person connected with either the railroad or any of the factories at Lowell, could have told him that the freight on coal is $1, and on merchandise $1.50 per ton; add expense of loading and unloading, 25 cents per ton; and it leaves an error in the one case of sixty, and in the other of fourteen per cent in another of the incontrovertible data of his calculations.

I could easily add other examples—they occur in every line; but I have selected the above from a single paragraph, because they are easily understood and readily verified. "This, therefore," he triumphantly sums up, "is the cost of water power, wheels, shafting, gearing, building, &c. of the Boott Mill No. 2, per yard of cloth."

"Let us now," he continues, "take an estimate of the Bartlett Steam Mill No. 2, and then compare the results."

In this, Mr. Editor, I do not mean to follow him—being reminded by it of a remark of a friend of mine, when it was proposed in an argument to assume as true one half of certain calculations put forth by one notorious for his non-adherence to fact. "The two halves," said my friend, "being precisely alike, I see no reason for believing the one more than the other."

I will only add, for the writer's edification and that of your readers, a few sentences quoted from his own introductory article.

"On a subject of such vital importance as the cotton manufacture, the public mind should not be misled; and no man should undertake to spread the facts in the case before the world, unless fully qualified to state every point of the least importance, without resort to hypothetical estimates or bare supposition. If not thus qualified, the public have a right to hold him to a rigid responsibility for all his errors. Any man who undertakes such a work is supposed to know all about it. If he misleads the mind of the reader, either from ignorance or design, the effect is the same; and no sufficient apology can be pleased in justification."

Further Letters from Justitia
No. 1   *Boston Daily Advertiser and Patriot*, 26 April 1841

There appeared in the columns of the Boston Daily Advertiser, a few weeks since, two communications, one signed 'B', and the other without a signature, containing strictures on my communications in the Boston Courier on the subject of Steam and Water Power. On those strictures, I would wish to make a few brief remarks, which want of leisure has heretofore prevented. In doing this, I would state, in the commencement, my object is neither to write up the Bartlett Mills, nor to write down any others, as seems to have been supposed. The Bartlett mills require no writing up. They are well known, and speak for themselves. Their stock is all taken up, and the demand for their goods in market is much greater than can be supplied. I have no interest in writing down any other mills; and my only object has been to show to capitalists in sea-ports, that they may employ their capital at home as well as abroad, and occupy their own tenements, as well as to go into the country with their business.—And notwithstanding the counter-statements above referred to, I think my attempt has been successful.

The communication of 'B' requires but few comments, which shall be brief; as the entire force of the argument rests on some two or three errors of his own, the correction of which will sweep away his entire fabric.

'B' says that, by my calculation, I make the amount or rather the cost, of fuel, to warm two the Boott mills for six months, greater than that required to generate steam power for the two steam mills for the same period. What then? If it be true, why not? He admits the *quantity* of fuel stated for the Boott mills, to be correct; that is, as *"purchased,"* though he intimates that it is used for various purposes. Suppose it is. The coal to generate steam in the steam mills, furnishes heat for all necessary purposes. Hence the argument of 'B' amounts to nothing. I have stated *actual quantities* of fuel—of fuel *used* in each establishment. The *cost* has nothing to do in this case, because, to the price of coal at New-buryport, is added the cost of transportation from Boston to Lowell, which I reckoned at $2 per ton, or $750 for six months. The amount of *fuel* therefore, and not of *cost*, is the criterion; and I am fully prepared to prove, that, the amount stated for the Bartlett mills, is within the truth. The amount stated for the Boott mills, is the amount stated in the published statistics as being *used*. I have allowed one hundred and fifty tons for the Bartlett mills, per annum, above the quantity used by the Boott mills. I would ask, how does this speak for the knowledge of 'B'! I am not at all surprized, after pursuing such statements from his pen, at any he makes afterwards. If he examined with the same tact "the project for manufacturing steam," no wonder he thought it absurd. But the author of that *"project"* is ready to come under bonds, notwithstanding 'B's" judgement on the subject, to carry it out in practice.

With his project for warming mills, I have nothing to do. My basis was the Lowell statistics, which purport to give the actual amount of coal and wood *"used"* per annum. And as to the warmth of steam mills in the *morning*, 'B' evidently knows nothing about it; for a steam mill warmed by steam, has the atmosphere at a temperature sufficiently high, at any time.

'B' prudently declines to estimate the expenses of steam mills, so we are deprived of the benefit of his knowledge and *experience* on that subject. But he supposes my estimates *mere* "estimation, and not practice;" because one of the Bartlett mills was not in full operation. What a sage conclusion! Mill No. 1 has been in full operation for more than two years. Suppose Mill No. 2 was not yet built; would not Mill No. 1 furnish practical data for a correct estimate to include thousands of others? Mill No. 2, however, now has 6,000 spindles in operation; and the result far exceeds my former calculation. But 'B' thinks I should allow for transportation of many heavy articles between Boston and Newburyport. The case did not require or admit it. Though the proprietors of the Bartlett mills might choose to transport coal, &c. from Boston, and back again, that would be nothing to the purpose. My estimate was for steam mills contiguous to the ocean—on navigable waters—where owners might, if they pleased, import for themselves, and avoid the heavy cost of inland transportation. The Bartlett mills were introduced to make out a case; and their proprietors may or not avoid that expense. It matters not in the estimate of what may be done. In fact many steam mills do avoid that expense, in toto; and some of them are so located that even the cost of importation is less than at Boston. If some steam mill owners, on navigable waters, do not avail themselves of all the facilities afforded by their locations, it does not vary the result as to what others do, and what all may do. The Bartlett mills and Boott mills were cited as examples of the relative cost of steam power and water power, under circumstances most favourable to each, and which might be applied to all. It was not intended to confine the estimate either to them, or to Lowell and Newburyport. As a general result, my estimate was correct.

To show how accurately 'B' has conned over my former communications in the Courier, read the following. "He" [*Justitia*] "takes it for granted, that there is no difference in the expense of mule spinning and throstle spinning; and assumes that mills driven by water must be subject to the heavy expense he puts down for transportation, and that those driven by steam require no such expenditure." This is not true. I deducted 25 per cent. from my estimate of the cost of water power, &c. for the "*difference between mule spinning and throstle spinning*." I have further said, "the difference [in the original cost] is sunk, and a balance found against water power, in favor of steam power, *by a difference in location*." He therefore attributes to me the reverse of what I have advanced or assumed. Besides, my statements were founded on water mills in the interior, where inland transportation is necessary, and steam mills on navigable waters, where it is not necessary. After all, 'B' has paid an unintentional compliment to the Bartlett mills; for after having, as he supposes, saddled them with additional heavy expenses, beyond those of the Boott mills, they still go on and prosper, and successfully compete with the water mills.

Not satisfied with my estimates and comparisons, 'B' offers others more congenial with his views—Finding the half of the looms in the four Boott mills more than equal to the number in the two Bartlett mills, by 47, he concludes that the number of looms, and not the number of spindles, in the establishments is the proper ground of calculation. Here again, the falsity of his premises vitiated his inferences, and sweeps away his argument. Did he recollect, when he made his careless statement, that the Boott mills manufacture cloth less than one yard in width, while the Bartlett mills manufacture, in one establishment, cloth full one yard wide; and in the other cloth from $1\frac{1}{8}$ to $1\frac{1}{4}$ yard wide? This fact at once disposes of his 47 extra looms, and leaves him at perfect liberty to enter on a new calculation. So much for 'B'; and now, Mr. Editor, I have a few words to say to your other correspondent.

Your other correspondent promises to "illustrate some of my egregious errors," and then to leave me to what credit I deserve. First, then, he accuses me of the commission of an error, in my statement of the cost of water power. On that subject, I followed Mr. Montgomery; and, from subsequent information, I am satisfied of his correctness. And therefore, sir, I am not yet convinced that there is an error. Your correspondent says the water and land cost *about*—mark that—*about* $15,000; and that the land is worth nearly all the first cost. That *about* is a very indefinite and ambiguous phrase. Pray why not tell precisely the cost? Your correspondent no doubt could do it. And as to the land—pray what might it be worth without the water? Why precisely as much as so many square feet of any other frog swamp. The value of the land is made by the water power; and without it, is worth next to nothing. But suppose the water always has been sold by the cubic foot, or by the quart or gallon, if you please, what then? It is known what number of cubic feet are required for a given purpose, and that being known, it is easy to calculate the price per spindle, as every prudent man, before purchasing, would calculate; and I happen to have in my possession, documents from a source as perfectly responsible as your correspondent, or any other gentleman, stating that the calculation of $4 per spindle for water power at Lowell, is a correct one. As to what the Boott mill proprietors may have actually paid for *their* water power, I do not claim to know. It may have been given to them; but that alters not the case. What is the usual price of water power at Lowell, is the question. I am credibly informed that it averages $4 per spindle; and, on that price, I have founded my estimates. So much for "egregious error" No. 1.

I would here remark, that I consider water power at Lowell, at $4 per spindle, taking into view the numerous advantages of that city for manufacturing purposes, as cheap, or cheaper, than in any other part of this country with which I am acquainted.

Again, I stated that the amount of yarn spun in the Boott mills, was about 4 skeins per day per spindle; making 266,600 pounds (not 264,000 as stated by your correspondent) of cotton to be transported. But your correspondent states as a fact, that the amount of cotton spun, for the year ending July, 1840, was 388,000 pounds. But, in addition to the amount spun, as stated by me, I also added, in the account of transportation, 16 per cent. for waste, making 309,256 pounds, and as the Bartlett mills were spinning No. 40, I supposed the Boott mills to be doing the same, (although I well knew they were not) to make up the statement. I leave your correspondent to reconcile the difference between the amount given by him, and that given by me, as best he may. I have in my hands, a sample of the cloth from Boott mill No. 2. Knowing, as I do, the *exact number of skeins* to a yard, and that the number of yards manufactured there last year was but about 15,000,000, any school boy can readily tell how many skeins and pounds it contained. As your correspondent says they used 388,000 pounds of cotton, the number of the yarn must have been as much below the No. 40, as the proportion of 309,256 pounds, to 388,000. So much for the second error.

As to the cost of transportation, I am not fastidious, though I have reason to believe that my statement, as a general rule, was correct. Nevertheless, your correspondent may have it in his own way, and I will proceed, on his data, to compare statements; remarking, however, that your correspondent was applied to once for a statement of facts relative to the Boott mills, and *refused to give it*.

According to his estimate now, the expenses of the Boott mills are as follows:

Cost of Mill No. 2 including water wheels, shafting, belting and gearing, &c. ready to run machinery, $100,000

Interest on the above per annum,		$ 6,000
Cost of water power,	$ 15,000	
Interest on cost of water power, per annum,		900
Cost of fuel for all purposes,		1,600

As your correspondent has reduced the price of transportation from my estimate, and increased the quantity of cotton used, I set down

Cost of transportation,		512
Making a total amount of		$9,012

Thus, after making all the deductions your correspondent demands, the aggregate amount will give a fraction over six mills per yard for the cloth manufactured. The cost stated by me for the same objects, at the Bartlett mills, was $7\frac{1}{2}$ mills per yard. The gross amount given for the Bartlett mills, was $7,522 50. It must be borne in mind however, that the Boott mill cloths are less than a yard wide, while those manufactured at the Bartlett mills are from full yard, to a yard and a quarter. The latter are No. 40— the former somewhat coarser. On your correspondent's own showing therefore, the difference of a little more than a mill per yard in length in favor of the Boott mills, is fully compensated to the Bartlett mills, in width and fineness.

In the foregoing, it will be seen that I have not varied my former estimate as to the Bartlett steam mills, showing the cost of power and its requisites to be $7\frac{1}{2}$ mills per yard, which gave a balance in favor of steam power, of $3,300 per annum. Having now reduced the cost of water power, and requisites by the data of your correspondent, so as to make the cost per yard, a fraction over six mills, a trifle less than six per cent. on the market value, there is still left a balance per cent. on the market value in favor of the steam power, of the aggregate amount of $2,400 per annum.

When I spoke of "*incontrovertible facts,*" I did not mean to say "*facts*" which could neither be dodged by means of quibbles, nor evaded by means of subterfuge. I do not expect ever to find such facts in any case. I spoke with the common sense view of the subject, of facts not to be fairly disproved. Your correspondents have indeed contradicted some of my statements, but they have neither proved nor disproved any thing; and there is not a sentence in either of their communications, that any other person might not have penned, without an acquaintance with the subject. Much of both communications is entirely irrelevant to the subject, unfair, and destitute of facts to sustain it; to say nothing of the angry, and discourteous spirit too manifest on their face. But, I dismiss the subject, stating simply as I have before stated, that I aim at truth; and if in error, shall be happy to have my error pointed out. In the close, however, I would just intimate, that the power of corporations, exercised through their agents, to attack my veracity or character in any way, privately, or thro' the press, will not have the effect to silence me, or to stop my pen.

Letter of "B" printed in the *Boston Daily Advertiser and Patriot,* 27 April 1841.

STEAM AND WATER.—No. 2.

MR. EDITOR:—Since my communications published two or three weeks ago, I have been unable, from various circumstances, to pursue the subject, as I then proposed, until the present time. In my examination of the statements of "Justitia," I only endeavoured in that article to show that, in the two cases under consideration,—the Boott Mills driven by water and the Bartlett Mills by steam,—instead of a balance in favor of steam power on the tide waters, the balance was against it; even if we made no account of the extra cost of mule spinning on the one hand, and allowed for the extra cost of transportation from the tide waters on the other. There are however many situations where mills may be driven by water power in the immediate vicinity of navigation, such as those at Fall River, in Massachusetts, and Saco, in Maine, and many water-falls near the mouths of other rivers in that State, directly on the tide, where the expense of transportation will not exceed the estimate I have allowed the Bartlett mills, [of] one dollar per ton.

I proceed now to an examination of several of the statements of "Justitia," as promised in my last.—He observes, "from the first attempt to manufacture goods in the United States by means of steam power, most people have been quite skeptical as to the practicability of the profitable application of that power for such a purpose." "With people in general, on this, as on all other practical subjects, argument is of no avail." But he says further, "the great error in popular opinion relative to the employment of water or steam for manufacturing purposes, is the supposition that the power itself is the great desideratum in the list of expenditures, and which is to be looked at as a paramount consideration. This is a fallacy. The actual cost of steam power for the Bartlett Mills is but $7,181.67. The mere original cost of a moving power is, therefore, of secondary consideration." The above sum, which is the annual expense as estimated by "Justitia," is very nearly the interest of $120,000, which would not be considered a secondary consideration, by those who furnish the capital or pay the expense. We proceed to some examinations as to the correctness of his estimate of this item of expense.

He says the No. 1 Bartlett Mill contains 6,336 spindles, and has an engine of 75 horse power. No. 2 Mill 10,664 spindles, which in the same proportion would required 126 horse power, making 201,—therefore divide $7,181.67 by 201 gives $35.72 per year for each horse power. This divided by 310 working days, is about 11½ cents per day. Let us now enquire what has been the result of experience, both in this country and in Europe, as to the cost of steam power, and compare it with the above. Allen (Mechanics page 90) says Mr. Marshall of Leeds, consumed 14 pounds of coal per hour for each horse power.[42] The statement of others varied from 10 to 18 pounds. If we take the average of 14 pounds and multiply by 12 hours we have 168 pounds per day.—Baine(s) estimates the quantity of coal per day for each horse power at 180 pounds.[43] Dr. Cleland gives 298 pounds of inferior coal, or coal dross.[44] At six dollars and fifty cents per ton, the price assumed by "Justitia," 168 pounds would cost 48 cents per day for each horse power, for fuel only;—but after making any reasonable allowance, on account of the difference between the quality of the coal used in England and this country for this purpose, the fuel alone must cost three times as much as is allowed by "Justitia" for the whole expense. Allen, page 351, says, "In Manchester, where coals are as cheap as in most of the manufacturing districts of England, the total cost of steam power, including all charges, amounts to about 20 pounds per year for each horse power," which is equal to 31 cents per day, and says, "The actual expense necessary for operating a steam engine in England, all other things being equal, may be estimated at rather more than two fifths of what it is on the seaboard of the middle and Eastern States."[45] The average price of coals at Manchester and Glasgow may be taken at $2 to $2.50 per ton. At this rate 168 pounds would cost 17 cents. This sum deducted from the whole expense as above, 31 cents, would leave 14 cents for engineer, firemen, repairs, oil, tallow, &c. These charges would not be less in this country than in England. The lowest estimate as to fuel, is that of Boulton and Watt, who give ten pounds of good coal per hour, or 120 pounds per day, for each horse power. Now if we suppose the improvements in the use of steam have reduced the consumption of fuel to this quantity, this at $6.50 per ton would amount to 34 cents per day for fuel. To this add 14 cents for other expenses, and we have 48 cents per day for each horse power, and I have seen no fair calculation of the cost of steam power, including repairs and all expenses, which would give much less than 50 cents per day in this country where coal costs as much as $6 to $7 per ton.

We can now see something of the magnitude of the "error in popular opinion," which "Justitia" undertakes to correct. He has probably over-estimated the power of his engines about fifty per cent, that is, it would require about 134 instead of 201 horse power to drive his mills. If we allow him the benefit of this, it will bring his estimate of the cost of steam power to 17 cents per day, that is, about half the expense in England, notwithstanding the price of coal is more than double what it cost there. After all the credit given to the Yankees by "Justitia" for their improvements in the use of steam power, I was not prepared to suppose they had reduced the cost to one-third or one-quarter the cost in England, estimating fuel at the same price.

In making his estimate of the comparative expense of manufacturing in the Bartlett and Boott mills—"Justitia" says, "throstle spindles are supposed to require one-fourth more power than Mule spindles," but without stating his authority for this opinion I have made some examination to ascertain what is the usual estimate. I find that Drs. Ure, Brunton, Grier and Allen generally concur in stating that one horse power will drive 500 mule spindles, with carding and other machines for preparing the cotton for no. 40 yarn. Dr. Ure gives 180 throstle spindles for the same power, which it must be recollected are such as are generally used in England, and called *live* spindles, requiring much less power, and producing much less yarn than those used in this part of the United States. Allen gives 100 throstle spindles as the usual estimate in this country, and some experiments at Lowell and Saco, with what is called the *dead* spindle, such as were introduced at Waltham and generally used in this part of the country, give the number as low as 77 to 80 spindles. So that such spindles as are used in the Boott mills, which Justitia takes for his comparison, instead of requiring one-fourth more power, require at least five times as much power as those in the Bartlett mills; and I have not been able to find any authority or opinion to justify the estimate of a difference of only one-quarter between the power required for mule and throstle spindles.[46]

And here I would beg to say a word as to the authorities on which the statements of "Justitia" are founded. So far as my rec- ollection serves, for I have not the whole of the articles at hand, Justitia does not quote any authority for facts or estimates, except Mr. Montgomery's work, and it must allowed that he is rather singular in his manner of using this authority. His first three articles are written for the professed object of showing that this work is "deceptive" and "calculated to mislead" to such a degree as to be unworthy of confidence, and yet after all this, in pro- ceeding with his estimates, the only authority on which he relies for statements or calculations is the work, which he has been laboring so hard to discredit. If he had succeeded in convincing his readers of the truth of his charges against Mr. Montgomery, he has evidently failed to convince himself.

The great sin of Mr. Montgomery's book in the eyes of Justitia, is a statement that steam power costs in this country double what it does in Glasgow, and this he labors to disprove, and in his different articles more than once reverts to the subject. Mr. Allen, it may be seen above, considers the cost in England only two fifths as much as the cost here. Mr. Montgomery stated all the particulars of the cost of coal, wages, dimensions and construction of boiler and engine &c. making the expense of an engine of 40 horse power $12.20 per day for fuel, engineer and fireman, being 30½ cents per day for each horse power. This is exclusive of oil, tallow, repairs and other charges. Justitia neither denies the facts nor corrects the calculations; but remarks that there are other mills that cost less, and mentions the steam mill at Providence, with 10,500 spindles, driven at an expense of $14 per day. This mill has been often referred to as an example of the low cost of steam power in the United States. According to the preceding estimates, being mule spindles, this mill would require an engine of 45 horse power, but it is rated at 60.[47] It is said to require 46 cwt. of coal per day, which at $6.50 per ton, would amount to $14.95. The wages of engineer and fireman would be $3—making $17.95, or very nearly 30 cents per day per horse power,—differing less than one cent per day from the statement of Mr. Montgomery.

Most of the contradictory statements respecting the cost of steam and water power, may be accounted for by the want of attention to the difference between the mule and the throstle spindle. "Justitia" seems to be so little aware of this, by his cal- culations as to steam power, as well as those relating to the cost of manufacturing, as to lead one to suppose that he has had but little knowledge of the business. In all his comparisons between the Boott and Bartlett mills, he only states the respective number of spindles, and also between the Providence steam mill and the one mentioned by Mr. Montgomery, and takes no notice of any difference in produce or cost of spinning, or of power, except to allow, in the purchase of water power one-quarter more for throstle than for mule spindles.

The statement that steam power costs twice as much in this country as in Glasgow, one would suppose might be easily settled. The price of coal in Glasgow is 9 to 10 shillings sterling per ton—equal to $2.16 to $2.40. "Justitia" allows the average cost here to be $6.50. Is it strange, therefore, that steam power should cost twice as much? Yet "Justitia" remarks "there is not the great difference he (Mr. Montgomery) appears to imagine, between the cost of steam power in Glasgow and the U. States," and says, that at the moment he penned this statement he must have known it to be deceptive. This article has already extended to such a length, that I am unwilling to trespass farther in order to finish what I had to say, but will claim the privilege of resuming the subject if you or your readers consider it of sufficient importance.

B.

Letter of "B" printed in the *Boston Daily Advertiser and Patriot* 30 April 1841

STEAM AND WATER.—3.

MR. HALE—In my last communication, it was my endeavour to show that steam power must be much more expensive in New England, than in the manufacturing districts of Great Britain, on account of the great difference in the price of fuel, and that the statements of "Justitia" *prove a great deal too much*, when he attempts to show, that the cost of steam power here is less than half what the best writers on the subject, and the most experienced practical men, estimate the cost in England. In that country steam power was resorted to as a matter of necessity, because it was impracticable to procure sufficient water power, in any convenient location, for the rapid increase of the Cotton manufacture, and for other mechanical purposes; and up to the present time, water power continues to be used there in preference to steam, on account of its superior economy, even in situations where it is necessary to resort to the use of steam for part of the year, and where the cost of coal is less than half the price it bears in New England.

Allen (Mechanics, page 350) says,"Notwithstanding the abundance of coal found in England, and the very general use of the steam engine, water power is highly valued in all the manufacturing districts, and mills are erected on streams, which in many instances are sufficient to turn the water wheels, and operate machinery attached to them, during only a part of the year." And the returns to Parliament show a very considerable proportion of the Cotton mills in Great Britain, driven by water, and by water and steam together.[48] This gives us, most conclusively, the opinions of practical men in that country. Let us now proceed to a similar enquiry at home. "Justitia" quotes the price of water power at Lowell at $200 per horse power, and at Manayunk on the Schuylkill, at $1,016. A few words respecting this place will throw some light on the subject before us. It is situated about twelve miles from Philadelphia, with a ready communication, both by canal and railroad. The water power is sold, or leased, by the Schuylkill navigation company, and there are several Cotton mills in operation, in a pleasant and thriving village. A great part of the coal that comes to our market is brought down Schuylkill Canal, so near that it may be thrown from the boats into Manayunk mills, and certainly would not cost more than two thirds as much as the price in Boston, or Newburyport. Surely this would be the place to try the experiment of driving mills by steam. The transportation to Philadelphia, for goods or for Cotton, would cost less than between Newburyport and Boston. Why then should intelligent men give five times as much for water power as the price at Lowell, when steam would be even cheaper than in New England? Ask the managers of the mills at Waltham, Exeter, Dover, or any other place where they have had experience in the use of both water and steam. Perhaps it may be said that the experiment with steam power in the neighborhood of Philadelphia, has never been tried, and that they continue the use of water power, for the want of the light, which is just dawning upon us in the East. But the experiment has been very fully and fairly tried. The Philadelphia water works were driven by steam engines for some years, under the care of a very efficient superintendent, who, after a sufficient experiment, substituted water wheels for driving the works, which continue to operate to his entire satisfaction.[49] This was done on the score of economy, within a few rods of the landing place for the coal, brought down on the Schuylkill canal. If the opinion I have expressed should require any confirmation, the following extract from the report of the Lehigh Coal and Navigation Company, of January 13, 1840, seems to be directly in point. "It has been supposed by some persons that the power of steam, where coal is cheap, is preferable to water power, for the purpose of manufactures. The managers have, for some time, been examining this question, and present the following comparative views of the cost of using the two kinds of power, as the result of their inquiries.

At South Easton, the company charge an annual rent for water power $3 per inch, which for a sixty horse power, or 400 inches, under a three feet head and twenty feet fall, is	$1,200
Interest on cost of water wheel and allowance for wear,	200
Making the annual expense	$1,400
Being $23.33 per horse power per annum	
A sixty horse engine cost $7,000, on which interest per annum,	420
Repairs and perpetuating engine, 15 per cent	1,050
Engineers and firemen working day & night,	1,200
Eight pounds of coal, (the lowest estimate) per horse power per hour, is 1,825 tons per annum,[50] at $2 per ton, being the lowest price for coal at the coal landings,	3,650
	$6,320

Equal to $105.33 per horse power per annum. Showing an annual saving, by using water power instead of steam, of $82 on each horse power employed, or $4,920 a year on sixty horse power." It will be noticed, that in this calculation the water power pays a rent of $20 per horse power, equal to a valuation of $333—against $200, the price at Lowell; and that the price of coal is stated at only $2 per ton. It will be seen also, that the calculation is for the whole day of twenty four hours, for such works as require constant power.

After the question of the relative cost of steam and water power has been so well settled, by the practical good sense of the public, both in this country and Great Britain, it may seem unnecessary to go into any detailed calculation; but there is a love of paradox in some minds, which delights in the belief of impossibilities, and the pertinacity with which statements of the comparative cheapness of steam power have been brought before the public, has given them a sort of credit, because they were known by those best acquainted with the subject, to be too absurd for contradiction; therefore, as "Justitia" has made an array of figures to show that the Bartlett mills, with 17,000 spindles and 368 looms, operated by steam, have an advantage over the same number of spindles and looms in the Boott mills, operated by water, equal to $3,039.15 per annum, we may as well follow him through his calculations.

The difference, above stated, he leaves his readers to suppose, is an actual and necessary advantage in the use of steam power; though in making his estimate, a great part of this sum is made up of the cost of transportation, assuming that mills driven by water must be subject to an expense of two dollars per ton, and that those driven by steam do not incur any such expenditure whatever. This item of expense I shall leave out of view, as not being necessarily connected with the use of water power; and shall suppose that the mills, driven by water or by steam, use the same kind of spinning and other machinery.

To drive 17,000 throstle spindles, with machinery for carding and preparing the cotton, according to Mr. Allen's estimate, of 100 spindles per horse power, would require horse power ... 170

368 looms, with warping and dressing, at 8 looms per horse power, ... 46

For shafting, drums and geering, 15 per cent. ... 32

Making the power for two mills ... 248

Steam power equal to 248 horses, at 48 cents per day, multiplied by 310 working days, would amount per year to ... $36,902.40

Interest on cost of engines and gearing, $25,000 as stated for Bartlett mills, but altogether inadequate for the power for that number of throstle spindles, ... 1,500.00

Making the annual expense of the two mills driven by steam, ... $38,402.40

Water power for 17,000 spindles would cost, at $3 per spindle, ... $51,000

Wheels and gearing for two mills, as estimated by "Justitia," ... 42,500

Extra cost of foundation, for wheel pits, &c. ... 10,000

Being an outlay of capital equal to ... $103,500

The interest on this outlay of capital would be ... $6,210

Fuel for the two mills, 144 tons of coal at $7 per ton,   1,008
Making the annual expense of two mills, driven
   by water, $7,218

This sum, deducted from the expense of the two similar mills driven by steam, leaves a balance of $31,184.40 per annum in favor of water power. If we suppose half a million a sufficient capital for two such mills, the above balance will be more than six percent on the whole capital, or about equal to the average dividends on manufacturing stock.

As to the concluding article, on the comparative profit of the Boott and Bartlett mills, it may be sufficient to say, that the views and calculations of the writer are *about as correct*, as those relating to the cost of steam power. So far as relates to this branch of the subject, we may safely leave the matter to those most interested, and better acquainted with the operations of the Boott mills; remarking, however, that if Justitia is able to carry on the manufacturing business so much cheaper than others, as to be able to pay a dividend of profits, besides the extra expense of steam power, it must be on account of his superior talents and skill, and may go far towards overturning all former experience in conducting the business, but will not prove that steam is cheaper than water power.

P.S. Since the above was written, I have seen the article of Justitia in your paper of the 26th, in reply to my first communication, and am sorry that he is disposed to take offence at me, or others, who happen to differ from him, as to the comparative economy of water or steam power; particularly as this may place him in an attitude of hostility with a great portion of mankind, for he says himself, there is a "great error in popular opinion" on this subject. I intended to state my views with all proper courtesy to him, and did not even characterise his opinions as absurd, except by an allusion to a certain project for manufacturing by steam, which I was not aware proceeded from his pen, though he seems to assume the credit of it, and says, "the author of that project is ready to come under bombs, notwithstanding B's judgement on the subject, to carry it out in practice."

As it was only my intention to state my views respecting water power, and my reasons for them, which I supposed I might do with the same freedom used by others, I had no wish to enter into any controversy with Justitia; nor do I now find, after reading his article, anything to correct, or any argument to reply to; and will only ask that your readers will take my statements and the reasons for them, as brought forward by myself, rather than in the disconnected quotations of Justitia. I certainly did not undertake to convince him that he was wrong; and am perfectly willing that others should say that he is right.

I was truly sorry to learn, from the conclusion of his article, that there had been any conspiracy of "corporations" or others to "stop his pen," which no doubt will continue to be wielded as powerfully as it has been; but if such a conspiracy should succeed, I hope he will *still* be able to employ the pen of another.

Further Letters from Justitia
No. 2   *Boston Courier,* 11 May 1841

It was not my intention to again tax your patience or that of your readers; but circumstances compel me once more to take the field, and for the last time, in self-defence, against an opponent. In doing this, I am actuated by no selfish motive, but simply by a desire for the prevalence of truth. To promote that object, I propose now to notice two communications—one of which appeared in the Boston Daily Advertiser of April 27, and the other April 30, and both of them with the signature of "B."

In the commencement, by way of general comment, I have only to observe that the leading arguments of the writer of those articles are based on false premises, and are entirely fallacious; and if this shall be shown, as it is believed it will be, but a slight effort will be required to sweep away the entire fabric he has labored so hard to erect. To do this, permit me to refer to a statement of his. "He" (Justitia) "says the No. 1 Bartlett Mill contains 6,336 spindles and has an engine of 75 horse power. No. 2 mill, 10,664 spindles, which, in the same proportion, would require 126 horse power, making 201—therefore, divide $7,181.67 by 201, gives $35.72 per year for each horse power. This, divided by 310 working days, is about 11½ cents per day. Let us now, says he, inquire what has been the result of experience both in this country and in Europe, as to the cost of steam power, and compare it with the above. Allen, says he (Mechanics, page 90) says Mr. Marshall, of Leeds, consumed 14 pounds of coal per hour for each horse power. The statement of others, "B" says, varied from 10 to 18 pounds. If we take, says he, the average of fourteen pounds, and multiply by 12, we shall have 168 pounds per day. Baine(s), he says, estimates it at 180." "At $6.50 per ton, he continues, "the price assumed by

Justitia, 168 pounds would cost 48 cents per day for each horse power, for fuel only;—but after making any reasonable allowance on account of the difference between the quality of coal used in England and this country, for this purpose, the fuel alone must cost three times as much as is allowed by Justitia for the whole expense." This is a very plain case, and perfectly correct; and, had I been stating the cost of coals for Mr. Marshall's engine, as it was when Mr. Allen's notes were taken, sixteen years ago, I should have come to the same result. But "B" should recollect, it is the engines in the Bartlett mill, *now in operation*, that we speak of, and not one in Leeds, sixteen years ago.

His mode of argument is very much like the old mathematical problem—"If a jack-knife cost ninepence, what will a bushel of turnips cost?" "B" inquires what "*has been?*" What do we care? Our object is, to ascertain what *is* the "result of experience"?—not in Europe, but in this country. And, as that is the object, why travel across the Atlantic for the trial, when we may just as well make it at home? If one should inquire "what *has been* the result of experience" in regard to speed and the expense of fuel, in steam boat travelling, he may be told that the speed "*has been*" about eight miles per hour, and the consumption of wood, for boats of a certain burthern, about 45 to 50 cords per 200 miles. But, as to the present result, he might be told, with truth, that double that speed is obtained with less than two thirds the former amount of fuel. I do not contradict the statement of the amount of coal consumed by Marshall's engine, sixteen years ago. I offer, as specimens, those in the Bartlett mill; and if they will justify what I have said, it is sufficient. Let any gentlemen, competent to it, examine them, and test my statements by actual results. Is not that fair?

Agreeably to "B's" implied wish, he shall have some of my authorities. First, I have offered the engines themselves, as the best and most conclusive. Second, my own experience as a practical man, during fifteen years of laborious study and mechanical operations, one half in constructing and operating all kinds of cotton machinery by water, and the other half by steam; while he and others like him have speculated in water power, and other commodities, without ever having gone through the regular grades of practical mechanic science, and imagined results, to give energy to their remarks. In addition to these, he shall have testimony from one of the highest authorities in the scientific world, to show the astonishing difference in the practical results of steam power

JUSTITIA CONTROVERSY   329

now and formerly, in consequence of the great improvements that
have been made, both in construction and management. By this
mode, "B" may also learn, that what was true of the steam engine
sixteen years ago, requires much qualification to make it applicable
to that noble machine at the present day. He will also be able to
discover that then, even as now, some engines were operated at
an expense of fuel at least fifty per cent. less than that required
for others.

In the year 1774, the celebrated Smeaton introduced improvements, by which he succeeded in reducing the amount of fuel
required, by one third. In 1775, he designed the chase-water engine, with a cylinder seventy-two inches in diameter, and a stroke
piston of nine feet. The power of this engine was equal to that of
108 horses. Its consumption of fuel was 1,136 pounds of Newcastle
coal per hour. [See treatise on Manufactures and Machinery of Great
Britain, published in 1836, by Peter Barlow, Esq. F. R. S. and Charles
Babbage, Esq. F. R. S. which constitutes three vols. of the Encyclopedia
Metropolitana—p. 164, vol. 1.]Here, then, is the notice of an engine,
built sixty-six years ago, which consumed but about ten and half
pounds of coal per hour for each horse power. Subsequently, as
may be seen by reference to the work above quoted, page 189,
Messrs. Boulton and Watt effected still further improvements, for
which they obtained letters patent. So great were those improvements, that the patentee sold the right to use them for one third
of the amount saved in the expense of fuel, when compared with
the amount required with the common engines of that time. By
means of these improvements, the amount of fuel was reduced
to one half the quantity consumed even by Smeaton's engine; and
of course reduced to five and one fourth pounds of coals per hour
for each horse power. These engines, however, were used only
for raising water at the mines, and hence there was less machinery
attached than in the manufacturing engine; the latter having the
addition of the crank shaft, &c. to produce a rotary motion. Hence,
in the latter there is more friction to overcome, and a greater
consumption of fuel is necessary, all other things being equal, to
produce the same available power. If we allow one third for this,
the quantity of coals required would be about fourteen pounds
for Smeaton's engine and about 7 pounds for that of Boulton and
Watt.

In 1827, by experiments made by Mr. Davies Gilbert, a gentleman high in the scientific world, he found that four and one tenth
pounds of coals would generate steam equivalent to the power of

one horse, per hour, without any impediments to overcome. But, adding the power necessary to operate the air-pump and the piston, and to make up the loss by friction, imperfect vacuum, &c. the estimate reaches about five pounds.

The race of improvement in the steam-engine did not terminate with the career of Boulton and Watt, nor yet with the year 1825, when Mr. Allen's notes were taken. The work I have above referred to, published in 1836, says the improvement made in respect to the power of engines and economy in fuel, "is indeed very astonishing." A table of actual results of the engines at Cornwall is then presented, to illustrate the truth of this remark. The table embraces the period from the year 1813 to 1828, inclusive. In the year 1813, the number of engines stated is twenty-four. The average power, taking the best and the worst, is expressed by the number 19,456,000 for each. The average of the best is 26,460,000. Thus, even at that period, in only twenty-four engines, there was a difference between the average, between the whole number, *including the best*, and the best exclusively, of about 19 to 26; and between the best and the worst the difference must have been about one half. Up to the year 1828, the number of engines added was thirty, making the number fifty-four. These thirty probably embraced most or all recent improvements in construction and management, and the result is obvious. In 1828, the average of the best, when compared with the average of the whole, was as 76½ to 37; and the average of the best in 1828, compared with the average of the best in 1813, was as 76 to 26! Well might it be said that the improvements were indeed very astonishing. In the year 1815–16, over 5,000 bushels of coals were required to produce a given result; and in 1817, that consumption was reduced to 3,102 bushels—and that important change was effected entirely by a change in the *management!*

Another table in the same work contains valuable information as to fuel, &c. It ranges from one horse power to two hundred and twelve. By that table, a one horse power engine consumes twenty pounds of coal per hour; an engine of fifteen horse power, six and two-tenths pounds per hour for each horse power; and of one hundred horse power, five and a half pounds. And the same work has the results of experiments with five engines, by which it appears, from the amount of labor and fuel, the quantity of coal consumed varied from the former quantity, to less than three pounds.

These statements bring us up to the year 1836. They are published under the sanction of one of the first scientific societies in the world, and made up from official reports. I trust "B" will pardon me, if I prefer such authority as to steam power in Great Britain, to any thing he has advanced.

But suppose not one word of all this be true—and suppose the consumption of fuel in England were double what even "B" has stated—it would matter not in this discussion. My business is, and his business should be, with steam power in *this country*—not what "*has been*" in *England*, but what *now is* in *this country*. The latter he can test experimentally, if he pleases, at the Bartlett mills, and end all dispute. Will he do it? But "B" finding he could not escape the results of experience, as to the expense required to run the Bartlett mills, now faces about, and underrates the power of the engines, in order to increase the expense per horse power—about one half. But he must not escape thus. To meet him fairly on this ground, I present him below with a statement and estimate, made on scientific principles, to show him his mistake. Engine No. 1— Area of cylinder 173.166—average pressure on the boilers, 80 pounds—acting on the piston, 40 pounds—length of stroke, 4 feet − number of strokes, 80. Thus—173.166 × 40 × 80 ×

$$4 = \frac{2216524.800}{3300[0]} = 67\frac{17}{100} \text{ horses' power.}$$

Engine No. 2—Area of cylinder, 245.424—pressure on the boilers, 80 pounds—acting on the piston, 40 pounds—length of stroke, $4\frac{1}{2}$ feet—number of strokes, 76. Thus, 245.424 ×40 ×

$$76 \times 4\frac{1}{2} = \frac{3357400.320}{3300[0]} = 101\frac{73}{100} \text{ horses' power.}$$

The latter engine, when operating at a low pressure, as every one must consider the above rate to be for a high pressure engine, will not consume a quantity of coal per hour, for one horse power, to exceed $2\frac{1}{2}$ pounds; and which, at $6.50 per ton, and the expense of engineer and firemen, will amount to no more than 11 cents per day, per each horse power. These are the facts, and, in the face of "B's" estimates and authorities, I am still ready and willing to give bonds to furnish an engine that shall come up to this standard. Three pounds of first quality Anthracite coal is quite sufficient to furnish one horse power, with an engine rating 60 horses or upwards, and may be considered a practical rule, provided the engine is of proper construction, and properly managed.

Thus, it will be seen, "B's" estimate of 50 cents per horse power, which figures so largely in his calculation, is reduced to the point of insignificance.

After these expositions and offers, which are tendered to "B" in all candor and sincerity, it is hoped he will not close his eyes against them, and that, before he makes any further attempts at argument, he will take the trouble to satisfy himself by ocular demonstration. The Bartlett mills are open to inspection; and he can there arrive at conclusive mathematical results.

Determined to find fault, however, "B" next turns his attention to the Providence steam mill; and, with a dash of the pen, cuts down the power of its engine to 45 horses. As respects that engine, the facts are—Area of cylinder, 254.457—length of stroke, 6 feet—number of strokes, 52—pressure on boilers, 80 pounds to 100—pressure on piston, 40 pounds. Thus 254.457 × 40 ×

$$52 \times 6 = \frac{3175623.360}{3300[0]} = 96 \frac{23}{100} \text{ horses' power. Daily consump-}$$

tion of coal, $3\frac{19}{100}$ pounds per horse power.

And yet there are other engines in Rhode Island, which run at less expense than this. I would refer particularly to one at Olneyville, on the same principle as those in the Bartlett mills, and belonging to Messrs. John Waterman & Co.

Having thus given to my opponent a fair statement of the facts according to my knowledge, and also the challenge to test their correctness or incorrectness, by practical experiment, *on the spot,* I must take his speculations as being quite as worthless as brass farthings, till he shall have adopted this, the only certain rule. And this point I feel constrained to press upon him, with the firm conviction that, if he wishes any thing like fairness, he will conform to it.

I put the question to "B" as an honest man, if it is fair to test the cost of steam power in this country in 1840, by the amount of fuel required by a very few engines, or even all the engines in England, in 1825?—When it is attempted to be shown what can be done in the way of manufacturing by steam in this country, by the use of the best engine, under the best management, is it fair— is it honest—to bring up as examples, by way of opposition, old, and in many instance useless machines, as proofs? Would it not be equally fair to contend that the most improved water mills at this day could not manufacture at a profit, at present general prices, because the same could not be done by the worst mills formerly, when the weaving alone cost as much as the manufactured article is worth now? Away with such logic—it is not worthy the honest man. Let the question be decided then by the experiment, and I shall be content.

The estimates I have made, it will be perceived, waive all dispute about the difference between mule spindles and throstle spindles. The only true object is, the cost of power. After that power is obtained, "B" may apply it to mule spindles, throstle spindles, or grindstones—it is immaterial. This cost of a given amount of available power by steam, is now the question at issue. Decide that question by actual experiment, which "B" is again invited to do, and then he may talk and speculate about spindles. Till then, his speculations on that point have nothing to do with the subject, and assume not the least importance.

I might here come to a close. But I cannot refrain from saying a word on the report of the committee of the Lehigh Coal Navigation Company, which "B" has brought forward to aid his efforts. Whether that were a *packed committee,* and most deeply interested in the sale of coal or water, I undertake not to say. But certainly the report itself smacks strongly of ignorance, or of a deep interest in water power. The following is the report:—

"At South-Easton, the company charge an annual rent for water power of $3 per inch, which for a sixty horse power, or 400 inches, under a three feet head and twenty feet fall, is	$1,200
Interest on cost of water wheel and allowance for wear,	200
Making the annual expense	$1,400
Being $23.33 per horse power per annum.	
A sixty horse engine cost $7,000, on which interest per annum,	420
Repairs and perpetuating engine, 15 per cent,	1,050
Engineers and firemen working day and night,	1,200
Eight pounds of coal (the lowest estimate) per horse power per hour, is 1,825 tons per annum, at $2 per ton, being the lowest price for coal at the coal landings,	3,650
	$6,320

Equal to $105.33 per horse power per annum."

These estimates absolutely cap the climax of all the *humbuggery* I have yet seen in that line; and it is difficult to believe that any one acquainted with the business should not have detected its utter fallacy, on the slightest perusal. How, permit me to ask, was the

committee's mill to be erected, put into operation, and perpetu-
ated, without foundations, dam, flume, raceway, wheel pit, annual
repairs, &c.? And yet, not a solitary one of these items is included
in the estimate. Might not the water and the wheel have been
omitted with equal propriety?. Is not "B" ashamed, as a *manufacturer*, to sanction and adopt such an estimate? In the face of that
estimate, I submit another for the consideration of the practical
and well-informed manufacturer. Let any such one peruse both,
and decide which is nearest the truth:—

Rent of water power (60 horses) as above	$1,200
Interest of money for cost of wheel, wheel pit, race way, flume, &c. together with repairs on the same, and cost of perpetuating them,	1,500
Interest on additional cost of building and foundation on a river or mill stream, when compared with a steam mill on a selected spot,	300
Cost of heating the mill,	300
Total expense per annum,	$3,300

A engine of 60 horses' power, of the best construction, and in complete running order, to consume but three pounds of coal per hour per one horse power, can be had for $5,500, interest on that amount,	330
Repairing and perpetuating the same	220
Engineer and fireman, day and night,	1,000
Coal, three pounds per hour per horse power, 598 tons per annum, at $2 per ton,	1,196
Total expense per annum,	$2,746

Showing a balance in favor of steam power, of $554.

But now "B" wheels about, and assumes another position. He
says—"There are, however, many situations where mills may be
driven by water power in the immediate vicinity of navigation,
such as those of Fall River in Massachusetts, and Saco, in Maine,
and many waterfalls near the mouths of other rivers in that State,
directly on the tide, where the expense of transportation will not
exceed the estimate I have allowed the Bartlett mills, or one dollar
per ton."

And what has this to do with the question? To mills, the situation
of which on tide waters precludes the necessity of inland transportation, my remarks have in no single instance been applied by
me. On the contrary, those, and those only, which are located in
the interior, and to which inland transportation is absolutely nec-

essary, have been taken into the account. "B's" remarks are therefore superfluous, and useless. But to counterbalance his item, of transportation at one dollar per ton to *Saco*, which seems to be his favorite spot, he graciously allows the same rate for the steam mills at Newburyport. Suppose the proprietors of the steam mills might choose to transport all their commodities to and from New Orleans, by land, by the way of Nova Scotia, would that procedure affect my position at all? Every one knows that, in a commercial seaport, inland transportation is not necessary; but, if mill owners do not avail themselves of the advantage thus afforded, the location certainly is not in the fault. I have made the comparison as to the difference *necessarily* existing between the *interior location,* and the *location contiguous to the ocean.* My premises remain sound, and my conclusions just.

As to having quoted Mr. Montgomery's book after having questioned its authority, there was not the least impropriety in doing so. I only questioned its authority in certain parts, and conceded its authenticity in others. I had therefore as good right to quote parts I did not question, and which I believed to be correct, as tho' full and entire credit had been given to the whole. It is what would have been done by any person.

As most of the remaining portions of "B's" communication consists of declamation, mere assertion, and what he probably considers sly hits, and none of them having a direct bearing on the question, they will be permitted to pass.

After having come to the conclusion to submit the subject for decision, to those interested, and better qualified to judge than either of us, "B" thus terminates his communication, bearing the postscript:

"So far as relates to this branch of the subject, we may safely leave the matter to those most interested, and better acquainted with the operations of the Boott mills; remarking, however, that if Justitia is able to carry on the manufacturing business so much cheaper than others, as to be able to pay a dividend of profits, besides the extra expense of steam power, it must be on account of his superior talents and skill, and may go far towards overturning all former experience in conducting the business, but will not prove that steam is cheaper than water power."

"Justitia" will not affect to misunderstand the sneer thrown at him in the above quotation, nor will he feel much resentment at it; as such things generally owe their birth to *small causes.* Nor yet will he, as an offset, descend to boasting. He is not conscious of coveting what is not his own; and therefore is perfectly willing to be tried as a mechanic and manufacturer, by his works; and then, in point of "talents and skill," to abide the issue of the trial on comparison with his opponents. And as to "overturning all former experience, &c." "B" must recollect that all improvements of any considerable importance ever have produced, and ever will produce that affect; and however successful in the result, such improvements ever have been, and probably ever will be met at first with the same skeptical pertinacity, as now characterises the opposition to manufacturing by steam power.

It is not because "B" opposed my views, and called in question the correctness of my estimates, that he has been followed up with some degree of severity. Fair and manly discussion is at all times beneficial; but that species of controversy, in which a writer exercises all his ingenuity, by far-fetched arguments, irrelevant matter, interlarded with dark hints, contemptuous sneers, and ungenerous innuendoes, to prevent a certain result, is not exactly in keeping with the love of truth. It has been my object, in all I have written on steam and water power, to state facts according to my best knowledge and belief; and through that medium, if practicable, to arrive at the truth. I do not profess infallibility, and may have erred. But if so, the satirical, domineering, and sophistical manner in which my remarks have been met, has not been at all calculated to convince me or others of the fact. I profess to be a rational being, and a reasoning one; and to stand open to conviction. Give me arguments drawn from correct sources, to outweigh those advanced by me, and they will not be resisted. As yet, nothing worthy to be called argument has appeared against my premises and conclusions.

JUSTITIA.

Explanatory Notes

[1] In *CSMA* 1, 11–20, JM described the physical dimensions of this mill, but in *CSMA* 3, retitled *The Theory and Practice of Cotton Spinning*, 248–256, he presented estimates of operating costs over a fortnight and employed an accounting structure identical to the one published in *CM* 1, 114–125. The mill he described in *CSMA* 1 measured 145 x 37 feet within the walls, six stories high and intended for 23,000 spindles driven by a 40 or 50 horse power steam engine. In *CSMA* 3, he described a smaller mill, "of five stories, each about 10 feet high, and 90 feet by 38 within the walls," containing 2,400 mule and 2,100 throstle spindles and 128 powerlooms, all driven by a 24 horse power steam engine: clearly the mill used by JM in his *CM* 1 estimates.

[2] James made this point in every defence of steam-powered manufacturing that he published. See James, *Lecture*, 18, 24; idem, *Practical Hints*, 50; idem, *Letters*, 22, 29. He expatiated the point most fully in *Practical Hints:* "machinery can be driven by steam with a more equable motion, than by water. This imparts to the cloth a more equable and uniform texture. In the second place, the manufacture of cotton requires a certain degree of humidity in the atmosphere, as well as a proper degree of warmth. Having plenty of steam at hand, such a state of the atmosphere is always and readily attainable, and easily kept up. The consequence is, the cotton is smoothly and evenly wrought, and the goods will accordingly present a more beautiful finish."

[3] *Statistics of Lowell Manufactures* was an annual broadsheet summary of the capacity, employment, and output of the Lowell mill corporations as of 1 January of each year. Started in 1835, it ran throughout the nineteenth century.

[4] Justitia's arithmetic was faulty; the weight would be 7.83 tons a week or 407.2 tons per annum (52 weeks).

[5] Again incorrect: in fact 70 tons.

[6] In fact $3,382.8.

[7] This sounds like an overestimate; JM allowed $17,000 for the water wheels and gearing of his 5,000 spindle mill, so 6 x $17,000 with 25 percent deducted leaves $76,500.

[8] Justitia's figures add up to $204,244.

[9] Justitia's figures add up to $8,579.5.

[10] Again Justitia's addition was faulty; $21,344.14 is the correct figure ignoring the error noticed in footnote 8.

[11] Another error: this should read $106,372.

[12] The correct figure is $8,579, though a calculation free from Justitia's earlier errors gives $8,302.80.

[13] Untrue: half of 29,248 is 14,624, which is less than 17,000 by 2,376. This one must be a printer's error.

[14] Should be $1,285.9 ($1/7$th of $9,001).

[15] Wages of $4.4 per day amount to $1,364 for a 310 day working year.

[16] In fact 8,640 x 4 x 310 divided by 40 lb. = 267,840 lb.

[17] That is, 0.7 cents per yard.

[18] For a 310 day year the correct figure is $3,358.33.

[19] $2.2 a day for a 310 day year comes to $682.

[20] Justitia's arithmetic is seriously incorrect here. The market prices for the two mills' goods are 10.77 cents per yard for Boott Mill No. 2 goods (0.7 cents, the cost per yard to manufacture, being $6\frac{1}{2}$ percent of the market price) and 16.66 cents per yard for Bartlett Mill No. 2 goods (0.75 cents, the cost of manufacture, being $4\frac{1}{2}$ percent of the market price) according to Justitia. The Boott output of 1.5 million yards per annum will therefore yield $161,550 and the Bartlett output of one million yards per annum, $166,600, making a difference in favor of the Bartlett steam mill of $5,050, not $3,300.

[21] See explanatory footnotes to *CM* 2 for JM's use of White's *Memoir of Samuel Slater*.

[22] Montgomery should have known better than to add this remark: Baines, *Cotton Manufacture* (1835) 395, a work with which JM must have been familiar, quoted the Factory Inspectors' figures of 33,000 horse power from steam engines and 11,000 horse power from waterwheels in British textile mills in the mid 1830s.

[23] Should read $4,608.

[24] Surely on 6 January.

[25] William Grier, *The Mechanic's Calculator, or Workman's Memorial Book* (2nd ed., Glasgow: Blackie & Son., 1835).

[26] Robert Brunton, *A Compendium of Mechanics, or Text book for Engineers, Mill-wrights, Machine-makers, Founders, Smiths &c. containing Practical Rules and Tables* (5th ed., Glasgow: John Niven, 1831).

[27] Zachariah Allen, *The Science of Mechanics, as Applied to the Present Improvements in the Useful Arts in Europe, and in the United States of America* (Providence, Rhode Island: Hutchens & Cory, 1829).

[28] See *CM* 2, "Notices of the Various Machines," Note 18 for Girdwood.

[29] See ibid., Note 54, quoting Ure, *Cotton Manufacture* 1: 304–306.

[30] See *CM* 2, "Miscellanies, " Note 1 for Batchelder's dynamometer.

[31] Watt first rated the power of his steam engines by reference to nominal horse power, which he calculated from cylinder size. This sufficed until the efficiency of the steam engine was radically improved, as by lagging, high pressure, and so on. The only accurate rating for these improved engines was indicated horse power, measured by a pressure gauge and estimated from the indicator diagrams first drawn in 1796 by John Southern. See von Tunzelmann, *Steam Power*, 26–27; R. J. Hills and A.J. Pacey, "The Measurement of Power in Early Steam-driven Textile Mills," *Technology and Culture* 13 (1972) 25–43.

The formulae for calculating the power of a steam engine set out in the reference works quoted in the Justitia controversy utilised "effective pressure," which yielded a lower nominal horse power rating than if boiler pressure were used. See Grier, *Mechanic's Calculator* (7th ed., 1839) 280–284.

[32] Formulae for calculating horsepower were to be found in Grier and Brunton (works cited at notes 25 and 26). In this controversy Watt's measure of horse power (power required to raise 33,000 lb. a height of one foot in one minute) is recognised and has been assumed by this editor.

[33] Nathan Hale (1784–1863), editor of the *Boston Daily Advertiser* 1814–1854, for whom see the *DAB* entry.

[34] Incorrect: the previous four figures in this column add up to $221,333.

[35] This is 6 percent of $212,744, but 6 percent of the correct figure, $221,333, is $13,279.98.

[36] This should be $1,214.57 if the calculation is $\dfrac{8,996.82}{17,600}$ x 2,376.

[37] This should be 6,045.

[38] If this figure was based on 52 weeks, it should be $624; if based on a 44 week (310 days) year, it should be $528.

[39] This should read $7,781.67.

[40] This letter is a response to Justitia, *Boston Courier*, 16 March 1841.

[41] This is 266,600 lb. in ibid.

[42] Allen, *Science of Mechanics*.

[43] Actually Baines cited John Kennedy's estimate of 1817. See Baines, *Cotton Manufacture* (1835) 369.

[44] James Cleland, *An Historical Account of the Steam Engine, and Its Application in Propelling Vessels* (Glasgow, 1825), reported in the *Franklin Journal* I (1826) 106–111; on p. 111 Cleland's report of a 30 horse power engine working 10 hours a day and consuming 4 tons of coal (= 298.67 lb of coal per horse power per day) was repeated.

[45] Allen, *Science of Mechanics*.

[46] See above, notes 25–27, 29.

[47] If one horse power were needed for 500 mule spindles, then in theory the Providence mill with 10,500 mule spindles required a minimum of 21, but surely not as much as 45, horse power. If the mill in question was the Providence Steam Mill, started by Samuel Slater in 1828, it ran 70 horse power in steam by 1838; while its spindleage at that date is unknown, Slater had planned to have at least 12,000 mule spindles in the original mill. See *Woodbury Report*, 88; *McLane Report*, I, 951.

[48] See note 22 above.

[49] For the Philadelphia waterworks, built in 1799–1801, and their chief engineer, Benjamin Henry Latrobe (1764–1820), an English immigrant, see Pursell, *Early Stationary Steam Engines*, 31–33 Hunter, *Steam-Power*, *522–530. DAB.*

[50] This is on the basis of 355 days' operation per annum.

[51] Peter Barlow, "A Treatise on the Manufactures and Machinery of Great Britain," *Encyclopaedia Metropolitana* (29 vols., London: B. Fellowes et al., 1817–1845) VI, "Mixed and Applied Sciences" (1836).

Index

Allen, Zachariah, 247, 292, 319, 320, 323, 325, 327, 328, 337
Almy, Brown & Slater, 157
Almy & Brown, 154
American cotton goods, 6
American cotton machinery, 134
American Factories, 83
Amoskeag, NH, 53
Andover, MA, 8
Anglo-American power costs, 249
annual celebrations, 181
annual operating costs, 30
Anthony, Daniel, 153
anthracite, 54
Appleton Company, 20, 172, 174
appletree, 132
Appomattox, VA, 192
Arkwright, Sir Richard, 155, 158
Arkwright-type mills, 1
Arnold, Aza, 9, 234, 241
arrangement of machinery, 61
Ashley, Lord, 150
Axton, Mr., 102

B, 247, 251, 252, 255, 304, 310, 312, 313, 314, 318, 322, 327, 328, 329, 331, 332, 333, 335, 336
Babbage, Charles, 329
back rollers, 71
Baines, Eward, Jr., 210, 250, 327
bale, 65
Baltimore, 191
Barings, 14
Barlow, Peter, 250
Barnes, Charles, 241
Barnstaple County, MA, cotton manufacture in, 161
Barr, Alexander, 152
Barr, Robert, 152
Bartlett, Dr. Elisha, 177
Bartlett, William, 231

Bartlett Mills, Newburyport, MA, 240, 251, 276–283, 305, 306, 308, 310, 313, 322, 325, 326, 331, 332, 334
Batchelder, John M., 217
Batchelder, Samuel, 15, 17, 18, 19, 20, 21, 24, 25, 33, 233, 234, 242, 247, 293
Bates, Joshua, 14
Batesville Manufacturing Company, 28
batting, 84
Beaumont, James, 233
Bellows Speeder, 91
Belper, Derbyshire, England, 156
belt drives, 231
belt power transmission system, 58
belts, 56, 60, 61, 86, 211
Berkshire County, MA, cotton manufacture in, 161
Beverly, 152
Beverly Company, 153
Biddeford, 201
Bigelow, John P., 161
bing, 65
bituminous, 54
Blackstone, MA, 53
Blackstone River, 184
blankets, 84
Blantyre, Scotland, 1
Blantyre Cotton Works, 3
Blantyre Mills, 1
bobbin and fly frame, 103
Bodmer, Johann Georg, 9
Boot, Kirk, 166
Boot Cotton Mills Copany, 171, 174, 248, 251, 252, 272, 273, 276, 277, 280, 281, 282, 304–309, 311–313, 315, 320, 322, 326, 335
Boston, 13, 18, 20, 35, 130, 145, 146, 150, 166, 191, 269, 273, 281, 314, 323
Boston Manufacturing Company, 33, 242
Boulton & Watt, 240, 329, 330
breaker (carding engine), 68, 71, 84

339

Brewster, Dr., 110
Brewster, Gilbert, 110, 235
bricklaying, cost of, 260
bricks, 259
bricks, cost of, 259
Bridgewater, MA, 153
Briston County, MA, cotton manufacture
  in, 161
British cotton machinery, 134
British factories, 83
British steam costs. See also Steam power
  costs, 250
Brown, Joseph, 232
Brown, Moses, 154, 159
Brunswick, ME, 196
Brunton, Robert, 215, 292, 320, 337
Buffalo hides, 130
Butler, Wheaton & Company, 184

Cabotsville, CT, 53
calender rollers, 84
Calton, Scotland, 12
canals, 269
cans, 71
cap spindle, 97
capital, costs of, 148
card clothing; see filleting, sheet
card grinding, 76, 77, 78, 81
card stripping, 37, 71, 76, 81
carders, 72
carding, 61, 68–85, 111, 144
carding, bad, 75
Carding and Spinning Master's Assistant (by
  JM), 8, 9, 11, 32, 34
carding engine speeds, 72, 75
carding engines, 55, 70, 205, 206, 208,
  210, 211
carding engines, costs of, 204–211
carding engines, maintenance, 76
cards, 70
Carolina, 194
Chelmsford, MA, 168
Chelmsford Neck, MA, 166
Chemnitz, 43
Chesapeake, 194
Cheshire, England, 4
Chickopee, 185
Chickopee River, 185
Chili, 147
Cincinnati, 192
Clark's Mill, Virginia, 193
Claud Girdwood & Company, 233
cleaners (on carding engine), 70
cleaning (carding engine), 66, 75
Cleland, James, 247, 319, 338
clock; see watch clock, 55

cloth, width, 315
cloth beam, 127
coal, 212, 304, 307, 319, 331
coal consumption, 279
coal consumption for heating a cotton factory, 217, 219
coal costs, 273, 278, 281, 283, 319, 322,
  323
coals, 319
coarse goods, 65
coarse heavy goods, 65
Cocheco Manufacturing Company, 240
Colburn, Warren, 234
Colt, Peter, 188, 239
Columbus, 191
comforters, 84
commercial information, 19
common mule, 106
comparative advantages for British manufacturers, 295
Concord River, 166
Connecticut, 53, 97, 185, 186, 188
cops, card, 76
Cornwall, England, 330
cost calculations, 251
cost calculations, 252, 274, 275, 276, 277,
  278, 280, 288, 289, 305, 315, 317, 319,
  320, 326, 333
cost of buildings, 136
cost of living, 150
cost of manufacturing, 134, 147
cost of raw material, 147
cost of steam; see Steam power costs
cost of the buildings, machinery, 144
cost of water power: See Water power
  costs
costs, 11, 31, 35, 38, 39, 42
costs, comparative, 143
costs, overhead, 143
cotton, 144
cotton, costs, 145
cotton factories, 54
cotton factories, location, 146
cotton manufacture in Massachusetts, 161
Cotton Manufacture (CM1, by JM), 32, 34,
  38, 42, 43, 285
Cotton Spinner's Manual (by JM), 11, 21
cotton store, 135
cottons, superior, 65
Couillard, Samuel, 14
Coventry, RI, 53
cradle warper, 113
Craig, William, 233
Craig, William, & Company, 91
Crane, Mr., 239
Cranston, RI, 185

Springfield, MA, 185
squirrels, 72
Staffordshire, 18
Stalybridge, England, 6
static electricity, 86, 87
steam, 66
Steam Cotton Manufacturing Co., 245
steam engine, 54, 61, 136, 193, 211, 251,
    267, 268
steam engine, Cornish, 330
steam engine, output, 303, 329–332
steam heat, 66, 67
steam mills, capital productivity, 282
steam power, 61, 151, 269
steam power, profitability, 317, 326, 332
steam power capital costs, 279
steam power costs, 212, 250–252, 257,
    261, 266–268, 270, 275–278, 281, 297,
    301, 305, 306, 309, 314, 317–319, 321,
    324, 325, 328, 332–334
steam power costs, British, 319, 321, 322,
    327, 328, 329
steam power costs at Newburyport, MA,
    211–212, 254
steam power costs at Newport, RI, 212–
    213
steam v waterpower debate, 243–338
steampower costs at Newburyport, MA,
    211
steampower costs at Newport, RI, 213
steam-powered cotton mills, 185, 246,
    247, 269
Steep Falls, Saco River, 199
Stetson Avery & Company, 145
Stirling & Beckton of Manchester, 54
stocking yarn, 110
Stockport, England, 7, 102, 118, 211
Stone, 129
Stone, Amos, 127
stop motion, drawing frame, 89–90
stop motion, lap doubler, 85
stop motion, twisting frame, 117
stop motion, warping frame, 115–118
stop motions, 17, 19, 37, 90, 115
straikes, 78
stretcher, 93, 94, 205, 208
strickles, 78
strikes, 106
strippings, card, 77
Strutt, Jedidiah, 156, 157, 158
Suffolk County, MA, cotton manufacture
    in, 161
Suffolk Manufacturing Company, Lowell,
    20, 170, 172, 174
systems, carding, 76

take-up motion, 37
Tar River, 192
Taunton, MA, 53, 100, 103, 207
Taunton Manufacturing Company, 233
Taunton speeder, 91, 92, 93
technical specifications, 19
temperatures, 66, 67, 124, 259, 307
temples, (loom), 18
Tennessee, 191, 192, 194
textile technology, American, 37
Thames Company, 186
Theory and Practice of Cotton Spinning (by
    JM), 8, 11, 91
Thomas, Edward, 18
Thomson, George, 19
Thorp, John, 100
throstle frames, 55, 93, 94, 97
throstle frames, prices, 103
throstle spindle, power requirements, 101,
    250, 251, 273, 291–294, 302, 314, 320
throstle spindles, 96, 250, 277, 280, 291,
    293, 308, 321, 333
throughput, 39, 144, 282
throughput, spinning, 142
throughput, weaving, 142
time, factory, 55–56, 179–181
top stripper, 76, 77
tops (carding machine), 68
Toulmin, Hazard & Company, 145
transportation costs, 269, 270, 273, 277,
    281, 304–306, 308, 311, 314, 316, 335
treadles, 130
Treadwell, Daniel, 13–21, 24, 233, 247
Tremont Manufacturing Company, Low-
    ell, 20, 170, 172, 174
tube frames, 91

urchins, (carding machine), 72,
Ure, Andrew, 102, 133, 148, 150, 247,
    292, 320
Ure's *Dictionary*, 34
USA cotton manufacture (1831); capital
    investment, spindles and output, 163
USA cotton manufacture (1831); mills,
    looms, employees and wages, 164
USA cotton manufacture (1840); invest-
    ment, employment, output, 165

varnish for harnesses, 130, 131
Vermont, 53
vertical integration, 110
Virginia, 191, 193, 194

wages, 139, 140, 144, 149
wages, comparative (Anglo-American),
    149